Cengage Learning Series in
Renewable Energies

SMART GRID HOME

QUENTIN WELLS

Cengage Learning Series in
Renewable Energies

SMART GRID HOME

QUENTIN WELLS

CENGAGE
Learning®

Australia • Brazil • Japan • Korea • Mexico • Singapore • Spain • United Kingdom • United States

CENGAGE Learning

Smart Grid Home
Quentin Wells

Vice President, Careers & Computing: Dave Garza

Director of Learning Solutions: Sandy Clark

Acquisitions Editor: Stacy Masucci

Managing Editor: Larry Main

Senior Product Manager: John Fisher

Editorial Assistant: Kaitlin Murphy

Vice President, Marketing: Jennifer Baker

Marketing Director: Deborah Yarnell

Senior Marketing Manager: Erin Brennan

Associate Marketing Manager: Jillian Borden

Senior Director, Education Production: Wendy A. Troeger

Production Manager: Mark Bernard

Senior Content Project Manager: Kara A. DiCaterino

Senior Art Director: David Arsenault

Technology Project Manager: Joe Pliss

Library of Congress Control Number: 2011942789

ISBN-13: 978-1-111-31851-2

ISBN-10: 1-111-31851-4

Delmar
5 Maxwell Drive
Clifton Park, NY 12065-2919
USA

Cengage Learning is a leading provider of customized learning solutions with office locations around the globe, including Singapore, the United Kingdom, Australia, Mexico, Brazil, and Japan. Locate your local office at: **international.cengage.com/region**

Cengage Learning products are represented in Canada by Nelson Education, Ltd.

To learn more about Delmar, visit **www.cengage.com/delmar**

Purchase any of our products at your local college store or at our preferred online store **www.cengagebrain.com**

Notice to the Reader

Printed in the United States of America
1 2 3 4 5 6 7 16 15 14 13 12

TABLE OF CONTENTS

Preface ix
About the Author xi
Acknowledgments xiii

CHAPTER 1 1

Introduction to the Smart Grid

Introduction 2
Electricity—The Essential Power 2
Demand for Electricity Continuously Increases 3
Electric Power Generation 4
The Traditional Electric Grid 5
An Electromechanical System 6
Problems in the Power Grid 8
Electrical-Demand Pressures on the Traditional Grid 10
Environmental Factors That Affect the Electrical Grid 12
What Is the Smart Grid? 14
The Home Smart Grid 16
Home Energy Efficiency 17
Alternative Energy Power 18
Summary 20
Key Terms 21
Review Questions 23

CHAPTER 2 25

Smart Meters and In-Home Displays

Introduction 26
What Is a Smart Meter? 26
Analog Electric, Gas, and Water Meters 26
Digital Smart Meters 29
How Soon Will Your Home Have a Smart Meter? 30
How Utility Companies Use Smart Meters 30
Electric Smart Meters 31
Gas and Water Smart Meters 32
Advanced Metering Infrastructure (AMI) 34
How an IHD Lets You Use a Smart Meter 34
IHDs That Receive Smart Meter Signals 35
IHDs That Use Current Transformers 37
Other IHDs and Monitoring Devices 38
Aztech In-House Display 38
Black & Decker Power Monitor 38

DreamWatts 38
EcoDog 38
EcoFactor 38
em•power by Smart System Technologies 38
Energate 39
Energy Engage 39
EnergyHub 39
Energy Joule 39
Energy Owl Electricity Monitoring System 39
EnviroScape Energy Management by BreezePlay 39
GreenQuest 39
GroundedPower 39
Home Automation, Inc. (HAI) Smart Grid Solutions 40
Kill A Watt 40
PowerCost Monitor by Blue Line Innovations 40
PowerHouse Dynamics—eMonitor 40
PowerNab 40
Rainforest Automation Inc. 40
Shaspa—Social Energy Meter 40
Silver Spring Networks—Smart Energy Platform 40
Tendril 41
The Meter Reader by Energy Monitoring Technologies 41
WattsUp? Meters 41
Installing an In-Home Display 41
Installing an IHD to Work with a Smart Meter 41
Installing an IHD without a Smart Meter 42
Summary 49
Key Terms 50
Review Questions 52

CHAPTER 3 53

Developing a Home Energy-Efficiency Plan

Introduction 54
Home Energy Use 54
Installing and Setting Up an IHD 57
Reading and Analyzing Data on an IHD 64
Utility Partnerships with Google PowerMeter 64
Power Meter Partnerships with Google PowerMeter 64
Base-Load Power 66
Continuously Operating Appliances 67
Regularly Operating Appliances 68
Variable-Load Power 68
Increasing Home Energy Use Efficiency—A Two-Part Approach 69
Reduce the Total Amount of Energy Used 69
Turn Off When Not in Use 70
Replace Inefficient Appliances 70
Manage the Timing of Energy Use 71

Summary 72
Key Terms 73
Review Questions 74

CHAPTER 4 75

Conducting a Home Energy Audit

Introduction 76
Conduct an Energy Use Audit 76
Measure the Electrical Energy Used in the Home 77
Create an Electrical Appliance Chart 77
Make a Home Lighting Chart 77
Measure the Power Consumption of Home Appliances 78
Determine the Gas Fuel Energy Used in the Home 84
Determine the Oil, Coal, or Other Fuel Energy Used in the Home 85
Conduct an Energy Loss Audit 85
Determine Air Leaks in the Home 86
Determine Heat Transfer into Building from Exterior 88
Determine Heat Loss through Walls and Ceilings 90
A Professional Energy Audit Report 94
Summary 98
Key Terms 100
Review Questions 102

CHAPTER 5 103

Upgrade Home Energy Efficiency by Reducing Heat Transfer

Introduction 104
Reduce Heat Transfer in and out of the Home 104
Seal leaks 104
Seal Outside Leaks 105
Seal Inside Leaks 108
Insulate 109
Energy-Efficient Windows, Doors, and Skylights 114
Energy-Efficient Windows 114
Energy-Efficient Doors 117
Energy-Efficient Skylights 118
Energy Efficiency Outside the Home 119
Awnings and Screens 119
Landscaping for Your Location 120
Shade 120
Windbreaks 121
Summary 123
Key Terms 124
Review Questions 125

CHAPTER 6 127

Taming the Energy Hogs in Your Home

Introduction 128
Design and Operation of HVAC Systems 128
Central Heating Systems 129
Furnace Fuels 131
Central Air Conditioning (A/C) 131
Single- and Multiple-Unit HVAC Systems 133
Zoned and Non-zoned HVAC Systems 133
Improving the Energy Efficiency of a Home HVAC System 136
Seal and Insulate Ducts 136
Maintain the HVAC System 138
Automate the HVAC System 139
Adding Zones 139
Install Programmable Thermostats 139
Space Heaters 140
Room Air Conditioners 142
Water Heaters 143
Tempering Tanks 145
Lowering Water Temperature 146
Tankless Water Heaters 146
Clothes Washers, Dryers, Dishwashers, Stoves, and Pumps 149
Clothes Washers 149
Clothes Dryers 149
Dishwashers 150
Stoves 151
Pumps 152
Sump Pumps 152
Well Pumps 154
Summary 155
Key Terms 157
Review Questions 159

CHAPTER 7 161

Energy-Efficient Lighting and Entertainment Systems

Introduction 162
Incandescent Lights 162
Incandescent Inefficiency—Light versus Heat 163
Power Consumption 163
Energy-Efficient Lighting 165
Compact Fluorescent Lights (CFLs) 165
How CFLs Work 166
Cost and Energy Efficiency 166
Advantages and Disadvantages 168

Light-Emitting Diodes (LEDs) 171
How LEDs Work 171
Cost and Energy Efficiency 173
Advantages and Disadvantages 176
Energy-Efficient Electronics 177
TVs, Stereos, and Home Theaters 178
Game Consoles 179
Computer Systems 179
Other Power-Hungry Appliances 180
Refrigerators and Freezers 180
Exercise and Fitness Equipment 181
Workshop Equipment 182
Summary 182
Key Terms 184
Review Questions 186

CHAPTER 8 187

Home Automation and Lighting Control

Introduction 188
Home Automation and Lighting Control Design 189
Load Requirements and Grounding 190
Home Automation Plan 190
Lighting Zones and Scenes 192
Home Run-Connected Lights and Daisy Chains 193
Types of Automation Systems 195
Power-Line Control Systems 196
X-10 Lighting Control Systems 196
UPB Lighting Control Systems 198
Other Power-Line Systems 199
Wireless Control Systems 199
Wire Runs and Wireless Zones 200
Automation Control Components 201
Command Modules and Controllers 202
Remote-Access Controllers 203
Controlled Outlets 203
Light Switches and Dimmers 204
Fixtures (Luminaires) 204
Automated Window Treatments 204
Sensors 205
Automation System Installation and Setup 205
Safety Cautions for Wiring and Connecting Automation Components 205
Installing Control Modules 206
Installing Exterior Lighting Zones 207
Installing Security Lighting Zones 207
Programming the Automation System 208
Troubleshooting an Automation System 209
Summary 211
Key Terms 212
Review Questions 213

CHAPTER 9 215

Solar Power for Your Home

Introduction 216
Solar Radiation 216
  What Is Solar Power? 218
  Using the Sun's Free Energy 220
Passive Solar Energy Use 220
Solar Water Heating 222
  Active and Passive Heating Systems 223
  Installing a Solar Water-Heating System 225
Solar Electricity Generation 227
  Types of Photovoltaic (PV) Modules 227
  Installing a Grid-Connected Solar Power System 228
    Power Conditioning Equipment 230
    Safety Equipment 231
    Meters and Monitoring Equipment 232
    Connecting to the Grid 233
    Legal Considerations 234
  Installing a Stand-Alone Solar Power System 235
Summary 238
Key Terms 239
Review Questions 241

CHAPTER 10 243

Wind Turbine Power for Your Home

Introduction 244
Wind Power from Solar Radiation 244
  How Wind Turbines Work 245
  How Much Wind Is Enough? 246
  How Large a Wind Turbine? 248
  Vertical-Axis Wind Turbines 250
Wind Turbine Towers 253
Balance-of-System Equipment 254
  Safety Equipment 256
  Meters and Monitoring Equipment 257
  Connecting to the Grid 257
  Legal Considerations 258
Hybrid Solar Panel and Wind Turbine Generating Systems 259
Summary 260
Key Terms 263
Review Questions 264

Glossary 265
Index 277

PREFACE

The terms "smart grid" and "energy efficiency" have been part of our technology lexicon for several years, and these terms both have specific meanings for professionals and academics who work with electricity and other forms of energy. But for most people not involved directly in an energy industry, the smart grid appears to have little relevance and the concept of energy efficiency is confined to the miles per gallon of the cars they drive. This book is designed to fundamentally change the reader's perception of the smart grid and his or her understanding of the broad applications of energy efficiency in the home.

Every person in the United States is now or soon will be affected by the implementation of smart grid technology in the production and distribution of electrical energy. As national and worldwide demand for electric power rises and the cost of producing it increases, the smart grid has become a critical necessity in containing costs and providing adequate power when needed and where needed. This essential function of the smart grid will expand as digital technology takes over more of the electromechanical grid of the twentieth century in the future.

But the smart grid does not end at the electrical service connection of each consumer's residence. The same technology principles that are making the production and distribution of electrical energy more efficient across the national grid can be applied within every home to reduce overall power use and make better use of every watt consumed. The chapters that follow will teach how to do this in three basic ways.

Before you can improve the energy efficiency of your home or office, you must first determine how much energy you are using and what you are receiving for the expenditure of this energy. You must also determine the cost of the energy being consumed and whether or not this cost is affected by the time frame in which the energy is used.

Once you have begun to monitor electrical and other forms of energy use over time, you can begin to improve the energy efficiency of your home in two ways: by turning off electrical appliances and other energy users when not needed and by shifting the use of appliances that are needed to the times when using them will be most cost effective.

Finally, you can reduce the amount of energy consumed in your home by reducing or blocking completely the transfer of heat from one part of the home to another. Heat transfer accounts for up to 90% of the energy consumed in a home. Minimizing or preventing heat transfer always saves energy, whether it is done by improving the home's insulation in the walls and roof, reducing the amount of hot water kept in storage, or cooking with microwaves in place of an oven.

THE INTENDED AUDIENCE

A home can be made into an energy-efficient unit and connected to the smart grid by professional technicians who contract to do the necessary work for the homeowner, or the upgrades can be performed by the do-it-yourself homeowner. This book is intended to assist both groups in the implementation of basic smart grid and energy-efficient technology. Professionals and homeowners alike will find it a practical guide to improving any home's energy-efficiency performance and obtaining the best return on energy-related upgrades for the home.

The vast majority of American homes need energy-efficiency upgrading, and many job openings are now available for technicians who can perform insulation and structural upgrades; install more energy-efficient appliances; upgrade HVAC equipment and entertainment systems; install renewable energy electrical generating units; and integrate digital automation, monitoring, and control systems. These professional workers will perform most of the smart grid and energy-efficiency improvements in residential communities, but the dedicated homeowner can also acquire the needed skills to competently upgrade his or her own residence. Whether you are a do-it-one-time homeowner or in training to become a professional smart grid technician, this book will serve as a useful guide and reference.

Each chapter begins with Learning Objectives to be achieved by studying the text. These are followed by an Outline that defines the main headings and subheadings in the text. Following the main chapter text, a Summary highlights the major points covered and a list of Key Terms defines important terminology introduced in the chapter. Careful review of these sections will assure that the reader has mastered the material thoroughly. In addition, 20 Review Questions at the end of each chapter help students recall key concepts and procedures and commit them to memory.

SUPPLEMENTS

An Instructor Companion Website containing supplementary material is available. This site contains an Instructor Guide, test banks, an image gallery of text figures, and chapter presentations done in PowerPoint. Contact Delmar Cengage Learning or your local sales representative to obtain an instructor account.

Accessing the Instructor Companion Website from SSO Front Door

1. GO TO http://login.cengage.com and log in using the instructor email address and password.
2. ENTER author, title, or ISBN in the **Add a title to your bookshelf** search box, then CLICK on the **Search** button.
3. CLICK **Add to My Bookshelf** to add Instructor Resources.
4. At the Product page, click on the **Instructor Companion site** link.

NEW USERS

If you're new to Cengage.com and do not have a password, contact your sales representative.

ABOUT THE AUTHOR

Quentin Thomas Wells is a 1965 graduate of the University of Utah with a magna cum laude degree in International Politics. He began his professional career as a clandestine Intelligence Officer for the Central Intelligence Agency (CIA) working as a case officer in the Soviet Bloc Operations Division and as an analyst in the Directorate of Intelligence.

Upon leaving the CIA, he became the director of a full-service advertising agency and later a motion picture writer and producer. He worked as production manager for Stockdale Corporation, then as president of his own firm, Executive Services, Inc., which specialized in technical writing, documentary production, and educational films. He later joined Career Research Corporation, a manufacturer of technology education systems, and became general manager of the firm. In 1989 he left Career Research to become president of the Sandy Area Chamber of Commerce.

In 1991 Wells joined the staff of Salt Lake Community College and became director of the Program Innovation Department. He later moved to the Communications Department as director of the college's Student Media Center. He taught courses in video and audio production and supervised the student media outlets (newspaper, television channel, radio station).

Wells' writing credits include over 30 documentary video programs and films, more than 40 technical instruction manuals, and dozens of television commercials. He wrote, produced, and directed 10 multi-part video courses for television and the Internet and over 40 CBE courses for college classroom use, as well as numerous corporate training programs. He co-authored the 2006 *Residential Integrator's Certification* book with Todd B. Adams and is the author of the *Guide to Digital Home Technology Integration*, *HTI+ Home Technology Integration in Depth, HTI+ Guide to Home Networking*, and the 2000 novel *Final Payment.*

ACKNOWLEDGMENTS

Many generous individuals and companies have assisted me with their time and professional expertise in the writing of this book. Among those who provided research material, product specifications, and graphics are Google.com (Google PowerMeter), The Energy Detective (TED), Electronic Educational Devices, Inc. (WattsUp?), Echo Green Energy Products, Rocky Mountain Power Co., Questar Gas, General Electric, Lutron Lighting, X-10.com, Hummer Wind Power, Guangzhou HYenergy Co., Ltd., ITS Insulation, and Manwill Heating and Air Conditioning, Inc.

The first draft of *Smart Grid Home* was reviewed by Christine Hertzog, Menlo Park, CA; Sikandar Khatri, Santa Clara University, Santa Clara, CA; Peter Kiser, Henry Ford Community College, Dearborn, MI; and Isaac Slaven, Attica, IN, each of whom provided valuable additional information, insights, and corrections.

I have been guided through the project by the Cengage Learning Staff, particularly Acquisitions Editor Stacy Masucci, Editorial Assistant Andrea Timpano, Copy Editor Lucas McIntire, Project Manager Mahendran Mani, Marketing Specialist Maria Klimek, and above all my Senior Product Manager John Fisher. Without their capable assistance, the *Smart Grid Home* would have never appeared in print and certainly would not have had the professional presentation you will see in the following pages.

CHAPTER

1

INTRODUCTION TO THE SMART GRID

OBJECTIVES

After studying this chapter, you should be able to:

- Understand the reasons why the use of electricity is increasing throughout the world.
- Describe how electricity is generated from other forms of energy.
- Identify the main components of the traditional electric grid and how they operate.
- Describe how new smart grid technology enhances the capacity and efficiency of the traditional grid.
- Describe how smart grid technology can be adapted to home use to increase energy efficiency and reduce the consumption of electric power.

OUTLINE

I. Introduction

II. Electricity—The Essential Power
- A. Demand for Electricity Continuously Increases
- B. Electric Power Generation

III. The Traditional Electric Grid
- A. An Electromechanical System
- B. Problems in the Power Grid
- C. Electrical-Demand Pressures on the Traditional Grid
- D. Environmental Factors That Affect the Electrical Grid

IV. What Is the Smart Grid?

V. The Home Smart Grid
- A. Home Energy Efficiency
- B. Alternative Energy Power

VI. Summary

INTRODUCTION

In this chapter you will learn how electricity is generated from other forms of energy and how it is moved from the point of generation to the place where it is used. You will also learn about the traditional electric grid that has transmitted power across the country for a century and why the grid needs to be updated and expanded with new technology. You will be able to identify the elements that change the traditional grid into a smart grid and how these components increase the grid's capacity and efficiency. Finally, you will learn how the smart grid benefits electricity users and how your home can become a part of the smart grid by using its features to reduce energy consumption, maximize energy efficiency, and perhaps produce some or all of your own electrical power.

ELECTRICITY—THE ESSENTIAL POWER

ELECTRICITY IS CREATED by converting other forms of energy. Most electric power today is produced in generating plants that use fuels such as coal, natural gas, or nuclear rods. Heat from these sources produces high-pressure steam to drive generators that deliver power to high-voltage transmission lines. A minor, but increasing source of heat for power generation is solar energy, which is converted directly into electricity without using steam power.

Coal-burning power plants are the predominant source of electricity generation in the United States. They produce about half (50%) of all U.S. electric power. Natural gas plants and nuclear-powered plants each produce about one-fifth (20%) of the nation's power. The remaining 10% comes from renewable energy sources, mainly hydroelectric plants, followed by wind-generated power and solar power units.

World power generation is skewed more toward hydropowered plants, with somewhat less power generated from coal-fired and nuclear plants than in the United States. But other nations also use more oil as fuel for electrical generation, and this means that hydrocarbon fuels still constitute about 67% of the total energy used for world electric power generation. Figure 1-1 shows the major sources of energy for producing electricity in the United States and in the world.

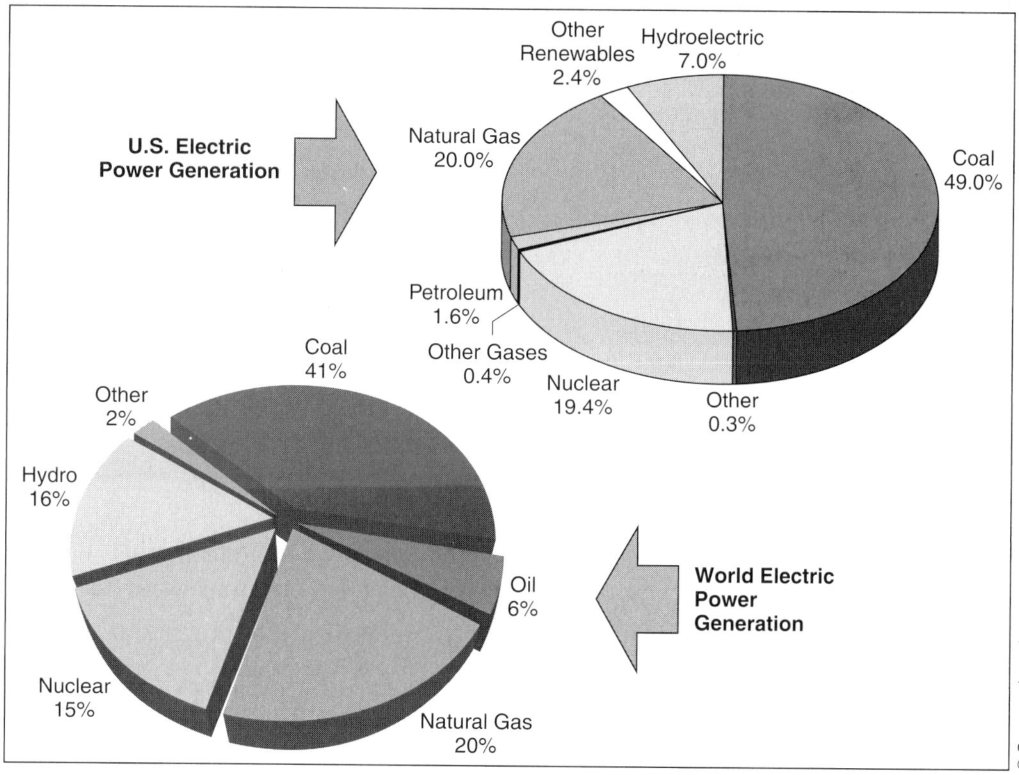

FIGURE 1-1

U.S. and world electric power generation energy sources.

Demand for Electricity Continuously Increases

From the time the commercially practical electric light and power generation system was invented by Thomas Edison, the use of electric power has continuously increased throughout the world. Advances in technology over the past century have resulted in the development of hundreds of electrical devices for industrial and consumer use.

The largest demand for power comes from electric motors that do much of the heavy work in residential and industrial applications. They drive air conditioners and furnaces (Figure 1-2), woodworking and metal-fabricating machines, elevators and escalators, winches and ski lifts, plus thousands of smaller tools, appliances, and convenience items—everything from your alarm clock to your garage door opener.

Electric lighting, which now illuminates cities and hamlets around the world, also uses enormous amounts of electric power. Incandescent bulbs, light-emitting diodes (LEDs), fluorescent tubes, mercury or sodium vapor lamps, and neon displays all require a continuous electric power source to operate.

Electronic equipment, computers, stereo systems, video displays, printers, scanners, and hundreds of other devices require not only continuous but clean electric power, free from voltage changes or other disturbances.

FIGURE 1-2

The growth in the use of residential air-conditioning systems like this one has greatly increased the demand for electric power in the United States.

New demands for electric power are constantly being developed. Chief among them in recent years are electric cars, which operate on power from batteries that must be frequently recharged from a continuous power source.

Electric Power Generation

At present there are about 10,000 central power plants in the United States that generate most of the country's electricity. Another 5600 smaller plants, called distributed generation facilities, provide lesser amounts of power. These are usually backup plants for large industrial or municipal users or combined heating and electric-power-generating plants.

Coal-burning power plants, gas-fired plants, and oil-burning plants all produce electricity by converting the heat energy of burning fuel into steam that drives a turbine engine. The turbine is connected directly to a generator that produces electricity as it rotates. Figure 1-3 shows the process of generating electricity in a coal-fired system. Gas- and oil-fired plants use an almost identical process but with an alternate heat source.

Some power is also produced by generators driven by renewable energy sources. These include moving water impounded in dams or flowing in streams or tides, and moving air that turns wind turbines. Figure 1-4 shows how a hydrogenerating plant uses water flowing rapidly under high pressure to power a rotating turbine that drives a generator to produce electric power.

The ever-increasing use of electrically powered equipment has correspondingly increased the demand for reliable electricity in every community. A complex distribution grid has been built up over the years to distribute electricity to end users from the point of generation.

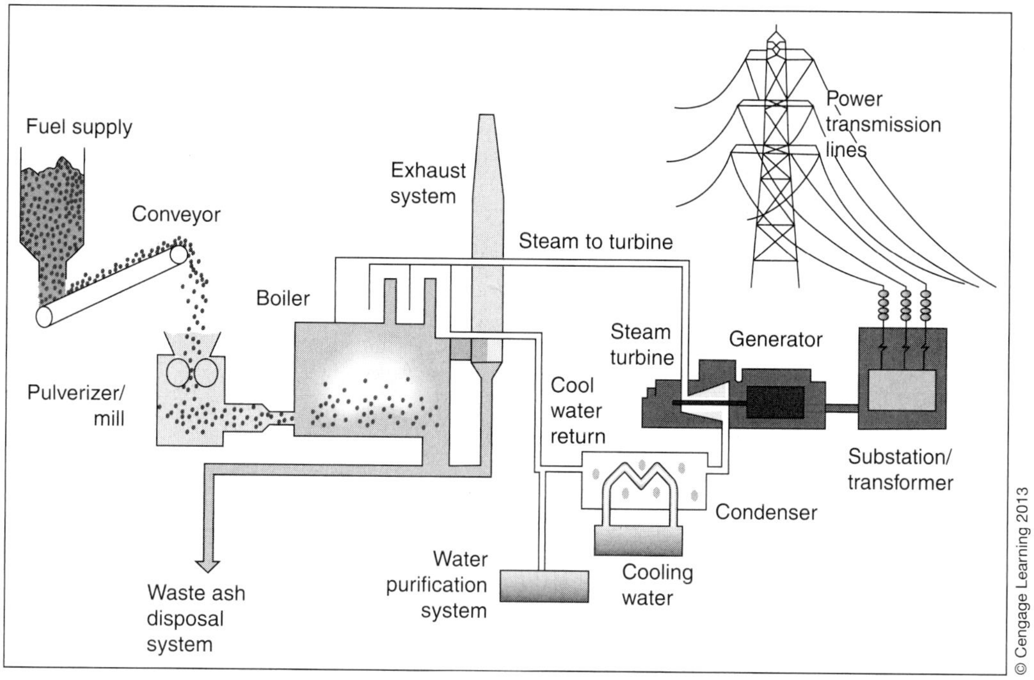

FIGURE 1-3

A coal-fired generating system uses heat to produce steam that drives a turbine.

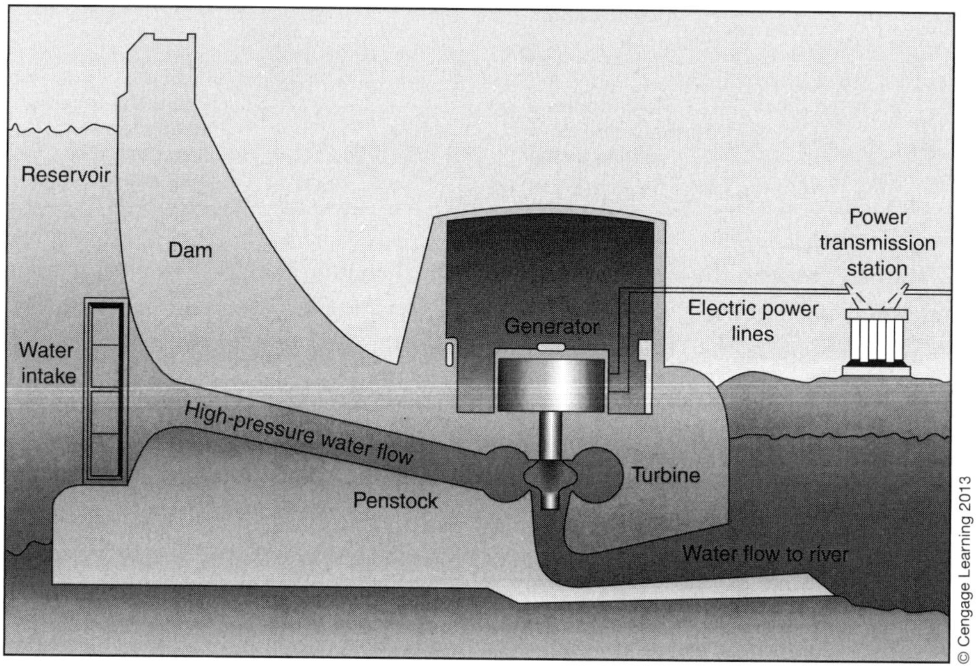

FIGURE 1-4

Hydrogeneration uses high-pressure water to drive a turbine that is connected to a generator.

THE TRADITIONAL ELECTRIC GRID

Alternating current, the dominant type of electricity used in the United States and throughout the world, cannot be stored. It has no "shelf life." Once generated, it must be instantly delivered to the end user through transmission and distribution lines.

No matter how it is produced, electric power moves from the generation point to the user through the wires of the electric grid. In the United States, about 200,000 miles of transmission lines transport high-voltage (bulk) power from generating stations to substations throughout the country. The substations reduce the voltage and distribute the power to over 100 million end users through millions of miles of distribution lines. Figure 1-5 shows a graphic representation of the traditional electric grid.

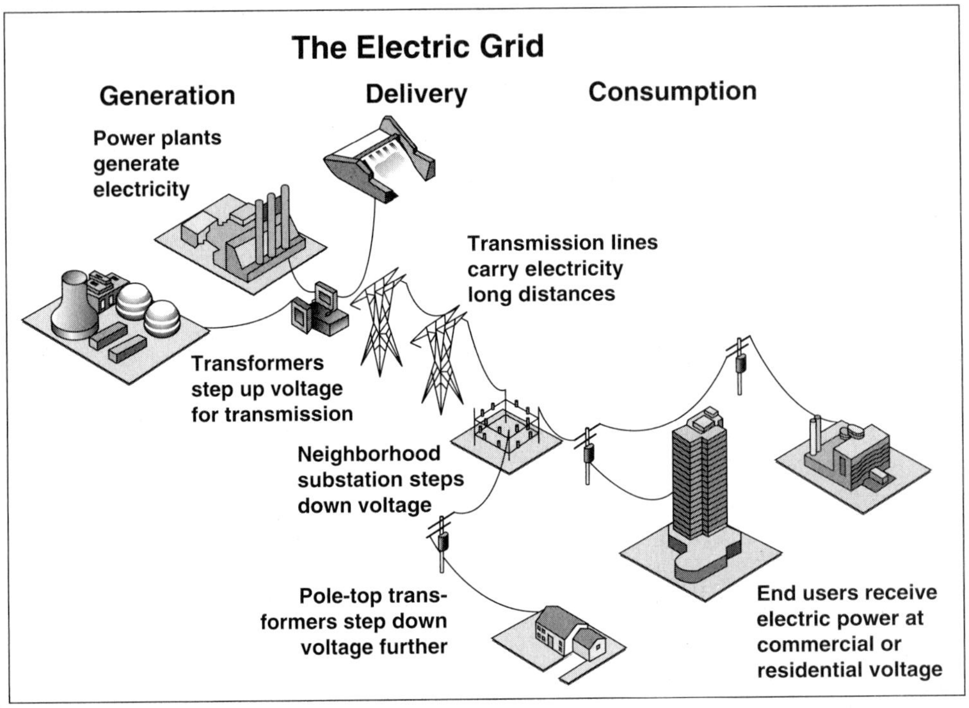

FIGURE 1-5

The electric grid delivers power from generating facilities to end users through a complex system of transmission lines, transformers, distribution lines, and substations.

The end points of the electric grid are electric meters. A meter is placed at the point where generated electricity (electric service) passes from the power company's distribution lines into the internal lines of every end user. These meters measure the amount of electricity used by each consumer, and charges for electric service are determined by meter readings.

The traditional electric grid, composed of generating facilities, transmission and distribution lines, meters, and the related administrative, construction, and maintenance units, is one of the largest and most capital-intensive industries in the United States. The total asset value of U.S. electrical facilities is more than $800 billion and the industry's annual sales are more than $300 billion, according to the U.S. Department of Energy.

The electric power industry is a proportionally large segment of the economies of other developed nations throughout the world and is becoming an ever-larger segment of developing economies as their use of electrically powered technology increases. Even the least-industrialized countries are rapidly increasing their power use as demand for electric and electronic technology increases among their populations.

An Electromechanical System

The traditional grid is an electromechanical system. It is made up of hardware: generators, wire transmission lines, transformers, switches, meters, and other components.

It was designed in the 1950s, incorporating many components and facilities that were already more than a half-century old (Figure 1-6), and constructed to meet the objective of delivering adequate power to meet the demands of all end users. With only minor exceptions, it is a one-way system. Power moves from large generation facilities down the grid to end users. There is no means of communication or any interactivity between end users and the grid; there is no way that end users can receive information about power usage from the grid or feed any information back along the grid. There is also no means whereby end users can feed power back into the grid.

Since the traditional electric grid was constructed, the telecommunications network has been completely transformed from an electromechanical to a digital system. During the same period, many other digital networks have been created, including cellular phone networks, cable networks, satellite TV networks, the global-positioning system (GPS), and the Internet. All of these digital systems have bidirectional communication and can be managed, diagnosed, and even repaired remotely.

FIGURE 1-6

This substation is an early part of the traditional power grid and has been in operation continuously for more than a century. The inset shows the year it went into service.

The traditional grid has not kept pace with the digital revolution. It has barely maintained its basic objective of keeping the lights on as demand for electricity has skyrocketed due to the huge increase in electric appliances and the proliferation of electronic devices. At times of peak demand the grid must function at its maximum generating and transmission capacity and, on a few occasions, it has been unable to do so, resulting in large-scale power failures in parts of the country. Of the U.S. gross domestic product, 60% now depends directly on electric power (compared to 20% in 1950), but the traditional grid supplying that power is still an electromechanical system with no digital communication capability from producer to end user or in the other direction.

Problems in the Power Grid

Several basic deficiencies in the traditional electric grid increase its vulnerability to failure as power use continues to rise. These include its high rate of power loss from long-distance transport, its limited ability to respond to rising peak-load requirements, and its lack of backup capacity.

Because electricity cannot be stored, whatever amount of power is needed at a given moment must be generated and transported to the end users at that moment. This means that the traditional electric grid must always be capable of generating at peak-load capacity even though that peak load may only be required for a few hours a day on perhaps 60 days per year. In fact, the grid must have some excess generating capacity above any anticipated peak because if demand rises at any time above the system's ability to respond with adequate power, the result will be some type of power failure in the system. A failure may include voltage and amperage fluctuations, brownouts, and possibly a major blackout as the grid shuts down to avoid damage from excess demand that it cannot meet. Major blackouts occurred in California, Oregon, and Washington in 1996 and in the northeastern United States and parts of Canada in 1965, 1977, and 2003.

Peak demand in one part of the grid may be met by transporting power into the high-demand area from another part of the grid where demand is lower, but this is an inefficient and potentially dangerous method of meeting peak-demand needs. The further electric power is transported over high-voltage transmission lines, the more power is lost due to resistance in the wires. This transmission loss means that more generating capacity is required to meet peak demand by transporting power from a distant point than would be required if the power were generated locally. Moving large amounts of power around the grid thus increases the peak-load generating capacity required for the whole system, not because more power is required by consumers, but because large amounts of power are lost in transmission.

Meeting peak-demand needs with imported power also assumes that peak demand will not occur in many areas simultaneously. Although this is usually true, exceptions have occurred during heat waves when air-conditioning demands soared above predicted levels in large parts of the country. Any part of the grid required to meet its own peak demand has little power available to send elsewhere. If a large enough area hits peak demand, even for a few minutes, the traditional grid may not be able to generate enough power to keep the system from going down.

Although the Great Recession of 2008–2009 temporarily reduced the rate of increase of electric power consumption in the United States and other industrialized nations, generating capacity in the North American grid is presently very little above peak-demand requirements. Projections by the North American Electric Reliability Corp. (NERC) show that demand for electricity is outpacing the installation of new generating capacity. Nationally, there was about 13% more generating capacity than the peak summer demand for 2010. This margin in generating capacity, however, will fall to 6% by 2016, which will increase the risk of localized power shortages and power quality problems. Figure 1-7 shows the potential excess demand versus generating capacity in various areas of the country.

One analyst has theorized that the increased use of energy-efficient devices, particularly fluorescent and LED lighting, will have little effect on the continuing rise in

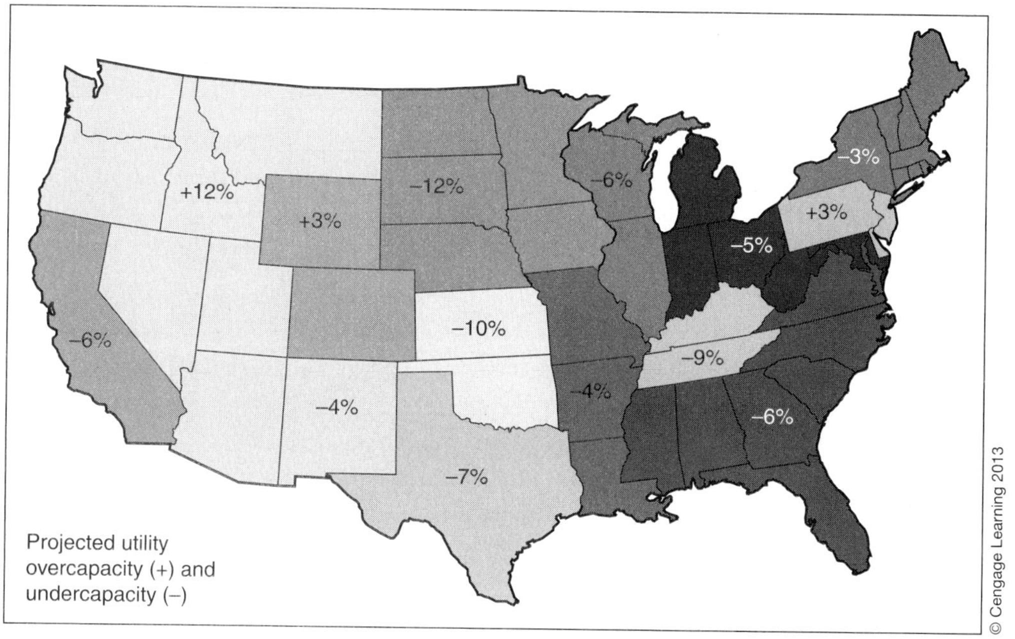

FIGURE 1-7

The overcapacity (positive percentages) of generating capacity in the Northwest and Mountain areas of the country may not be enough to provide for the undercapacity (negative percentages) in the rest of the country.

electric power demand and could even result in greater power consumption than would occur if the less-efficient devices remained in service. Low-energy lighting, he reasons, costs less than incandescent lighting and will result in the increased use of light fixtures in both residential and commercial applications.

A light that requires 2 kilowatt hours (kWh) of electricity per day to operate is likely to be turned off whenever possible or perhaps not installed at all because of the expense. An energy-efficient light, requiring only a quarter or less of the older light's power, will be used more and more such lights will be installed in areas once deemed too costly to illuminate. Rather than decrease the demand for electricity, energy-efficient lighting will therefore simply increase the demand for light—higher light levels in areas previously lit dimly, and new lights in areas not formerly illuminated at all because of the cost. Increased use of lighting will occur in developed nations, and installation of lighting will accelerate in developing countries where the high cost of power has previously limited the widespread use of electric lighting.

As noted, moving electric power over long distances to meet peak demand is also inefficient due to the power loss that occurs in transmission. The amount of loss varies due to a number of factors, including the transmission voltage, the distance traversed, the size and composition of the transmission line, and weather conditions along the transmission route. Higher voltages, longer distances, and inclement weather all increase the amount of loss, which in a worst-case scenario can amount to 10% or more of the total power being transported.

International transmission of electricity on the grid can introduce still more complexities into the system. Canada, in addition to transferring electric power among its various provinces as needed, is also an exporter of electric power, primarily to the United States. This transfer of power across the international border benefits both countries, but key factors will limit the increase of future power imports from Canada to the United States.

More than half of Canadian electric power is produced from hydrogeneration at present. This percentage will fall to slightly less than half by 2030, but Canada's large hydropower capacity will remain the country's largest power source for at least the next 20 years. Figure 1-8 shows the actual electric generation in Canada and the projected generation for the next 20 years.

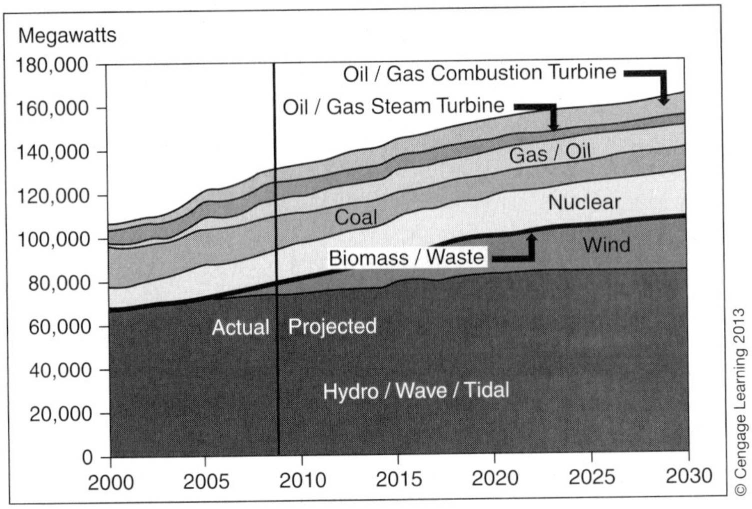

FIGURE 1-8

Sources of actual and projected Canadian electric generation.

Hydrogeneration plants are expensive to build and must be located where rivers can be dammed to provide the water power that drives their turbines. This means that the power produced is often far removed from high-demand areas, requiring the construction of new transmission lines to carry the electricity to users.

Because hydrogeneration plants and the transmission facilities that connect them to consumers are costly and time consuming to build, there is little incentive to construct them when alternate sources of power are available at lower initial cost. For this reason, it is unlikely that Canada will build new hydrogenerating plants to produce power for export. The same is true for other nations with great hydrogeneration potential. Other types of generating plants are cheaper and faster to build and can be located nearer to where the power is needed.

Finally, the traditional grid makes the building of plants for increased peak-generating capacity a poor investment. Power companies are reluctant to build new power generation plants solely to meet high peak demand because they are expensive to construct and unlikely to be used often enough to be profitable. Company managers know that peak demand in most areas will occur for only a minor fraction of the days in each year and will persist for only a few hours during each of those days. The rest of the time a plant built to supply peak-capacity power will stand idle, generating no electricity and earning no revenue because its power is not needed at the moment and cannot be stored for future use.

Electrical-Demand Pressures on the Traditional Grid

As the traditional grid approached and occasionally exceeded its capacity in the late twentieth century, several elements in the global economy mandated the need for change. The three most pressing factors driving the need for change in the traditional electric grid are:

- Expanding regional (Figure 1-9), national, and world population, which will increase the need for some level of electric service in all areas

FIGURE 1-9

Hoover Dam and its reservoir Lake Mead supply water and electricity to nearby Las Vegas, but rapidly growing population in recent years has increased demand for water and electricity and the level of the lake has fallen so low as to imperil the city's water and power supply.

- Increased expectations of people in developing areas for more and better electric service
- Development of whole new generations of electrical and electronic devices that nearly all people want to own or use

More people need electric power service now than ever before. The world population passed 6 billion persons in 2000 and, as of December 2008, stood at 6.7 billion, according to the U.S. Census Bureau. In July of 2011, United Nations estimates showed that it reached 7 billion. Population growth is greatest in those areas with the least-developed generating facilities, which means that demand for additional power is growing even faster than the overall population.

Less-developed countries and agricultural economies are rapidly industrializing, and demand for consumer electric products is expanding exponentially among their people. The per-person use of electricity in these areas is increasing even faster than their population growth, thus further stressing the traditional grid to provide the required power resources.

Finally, the demand for electric power is growing because everything in the industrial and consumer segments of the economy is being electrified. Electricity is the preferred form of energy because it is clean, easily controlled, and equally well suited for micro and macro applications. Electricity now functions in every facet of our lives. The average home contains 86 electric motors, over 20 data processing and control devices, and more than a dozen video and audio system components. New ones are in constant development, as are the advertising campaigns to create demand for them.

All of these devices and applications require more power from the traditional grid than it can deliver without significant upgrade. The U.S. Energy Information Administration (EIA) predicts that worldwide electric power generation will nearly double between 2004 and 2030. Figure 1-10 shows this increasing demand projected to 2035 and the types of fuel that will be used to meet it if present trends continue.

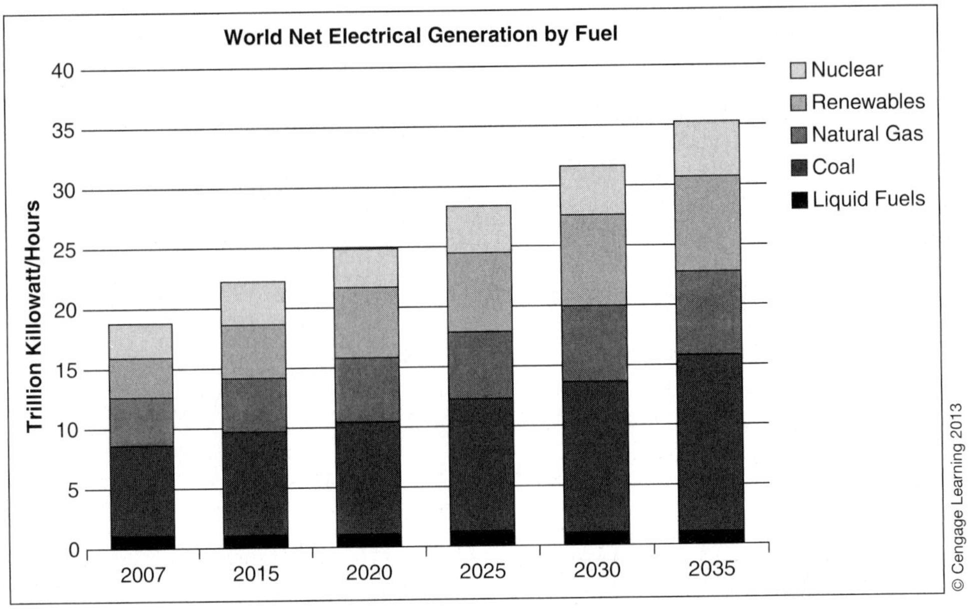

FIGURE 1-10 World electrical energy demand and the energy sources that will be used to generate it.

Environmental Factors That Affect the Electrical Grid

Two environmental factors also increase the need for change in the traditional grid:

- The rapid development of alternative electrical energy sources
- The need to conserve water use and prevent water pollution

As Figure 1-1 shows, coal is the fuel most used for electric power generation in the United States, accounting for about half of all generating capacity. Worldwide, coal supplies over 40% of the energy for electrical generation, and this percentage is growing, especially in China, where electric power generation has tripled since 2000 and coal may account for 70% of that nation's electric energy fuel by 2030.

Coal is the dominant energy source for electric power because it is relatively inexpensive compared to other fuels, it is plentiful, and it is widely dispersed around the world. Most of the industrialized nations have coal resources within their borders, as do many developing nations. This means that the availability of coal in quantities large enough to meet demand is fairly stable over time. Economic conditions, political events, and natural disasters are much less likely to affect or interrupt the supply of coal than is the case with oil or nuclear fuels.

Coal is also the most polluting energy source. It produces high levels of carbon dioxide (CO_2, the primary greenhouse-effect gas responsible for global warming) and particulate emissions that cause respiratory problems in humans. Particulates can be cleaned from the waste gases released by burning coal, and most governments, including the U.S. government, require that new coal-burning power plants meet stringent limits on particulate emissions. Upgrading of older, dirtier plants to meet the new standards has been somewhat slower, but is now being accomplished at a steady rate. Figure 1-11 shows a coal fired plant that has been upgraded to meet new federal pollution-control standards.

Carbon dioxide, however, cannot be removed from the waste gases because it is a by-product of the oxidation of carbon that is the basic chemical reaction of burning coal and other carbon fuels. Carbon dioxide can be captured and stored in a variety of ways, but all of them significantly raise the cost of using coal as a fuel and so have not been widely adopted.

FIGURE 1-11

The Hunter Power Plant, located near the town of Castle Dale, Utah, has three generators driven by coal-fired steam turbines with a combined capacity of 1472 megawatts.

Some coal also contains sulfur, which is released during combustion of the fuel as sulfur dioxide gas. This gas, when combined with water in the atmosphere, produces a sulfuric acid solution that falls to earth as acid rain, polluting freshwater streams and lakes as well as forests and agricultural land.

Natural gas (methane) also produces carbon dioxide when it burns and can contain small amounts of sulfur. It does not produce as high a concentration of pollutants as coal, but both of these fuels are dirtier than other energy sources such as hydroelectric or wind power.

Nuclear-fueled power plants (Figure 1-12) are also cleaner than those fired with carbon-based fuels. Their only pollution is spent nuclear fuel rods and radiation-tainted work materials that must be disposed of in carefully controlled waste-management facilities.

The development of alternative nonpolluting methods of generating electricity, such as solar cells, wind turbines, geothermal generators, and tidal motion turbines, has increased pressure to reduce or eliminate pollution-producing generation plants. But alternative energy sources like these usually produce electric power in smaller amounts than a large fuel-fired plant. They also must be located in areas where wind or water power is available or where tides or geothermal activity is sufficient for power generation. These source areas are often remote from end users of electricity as well as from large fueled power plants, the only points at which power can be fed into the traditional grid.

Because the traditional grid makes no provision for accepting power feeds from anywhere except central generating facilities, the expanding development of alternative energy sources has required a basic change in the grid's structure. As solar and wind-generated power sources multiplied in widely dispersed locations, a way to integrate these power

FIGURE 1-12

The Enrico Fermi Nuclear Generating Station is a nuclear power plant on the shore of Lake Erie near Monroe in Frenchtown Charter Township, Michigan. It has a generating capacity of 1098 megawatts.

sources into the grid had to be found. Renewable-source power offers a means of augmenting centralized power generation, especially during peak-demand times, that cannot be overlooked as expanding power demand threatens to exceed generating capacity.

Another environmental pressure for change in the traditional grid comes from the increasing need to preserve scarce water resources. Hydroelectric power plants use enormous quantities of moving water to turn turbines and their connected generators. They do not consume the water, but as it moves from storage behind a dam through the hydrogeneration process, it passes downstream and to a much lower level in the river system, where it is often difficult to use again for agricultural or municipal purposes without expensive pumping.

All steam-driven power plants, whether fueled by coal, gas, or nuclear energy, consume large quantities of water both in the generating process and in their cooling facilities. Some water is recycled, but much is returned to the environment in altered condition, either by the addition of large amounts of heat energy, which can adversely affect life forms in or near the water, or by being discharged as steam into the atmosphere, where it is lost to the surface environment and may affect weather patterns. Figure 1-13 shows how water is used in a nuclear generating facility.

Solar and wind-generated power sources, on the other hand, are almost water neutral. Neither requires much water for generation or for cooling. Increased use of these types of power generation offers a means of reducing water use and preserving water resources for other functions.

WHAT IS THE SMART GRID?

The concept of a smart grid emerged about the same time that digital communication and control technology was changing other major segments of the economy, from transportation and publishing to entertainment and medicine. In place of a top-down,

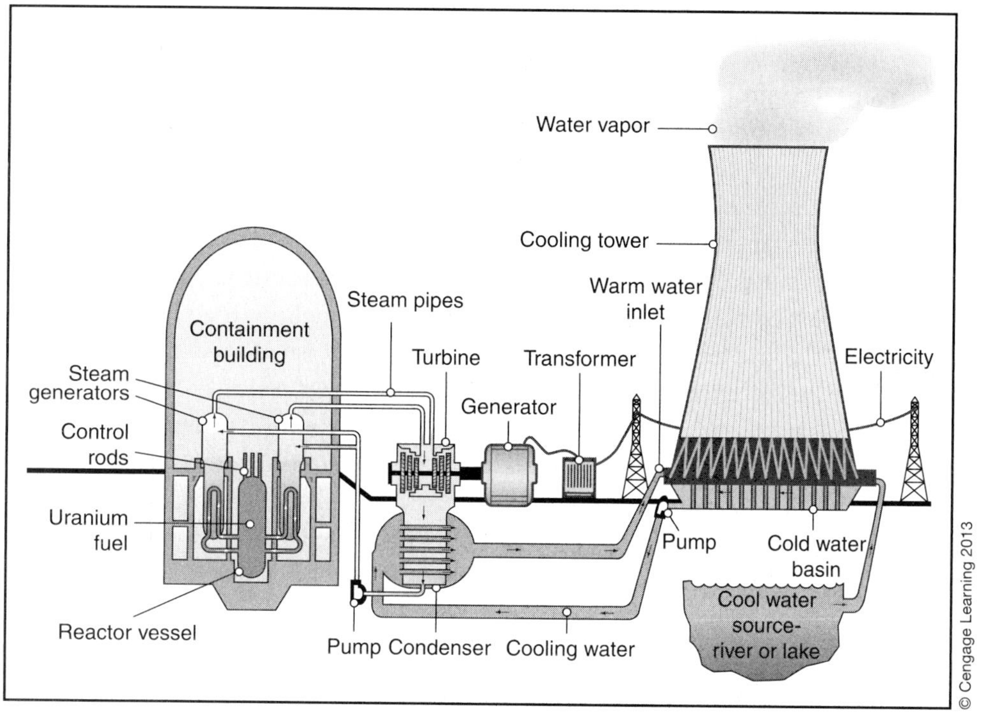

FIGURE 1-13

A nuclear generating plant uses water for steam turbine generation and to cool its nuclear reactor.

one-way system of electric power generation from a central plant to end users, the smart grid is being built as an integrated and interactive power network. Its generating capacity is decentralized and spread more evenly around all segments of the grid. Generating units are brought closer to the consumers who use the power and who, in many instances, may also be generators of power, both for their own use and for feeding back into the grid for use by others. Figure 1-14 shows a typical segment of a smart grid.

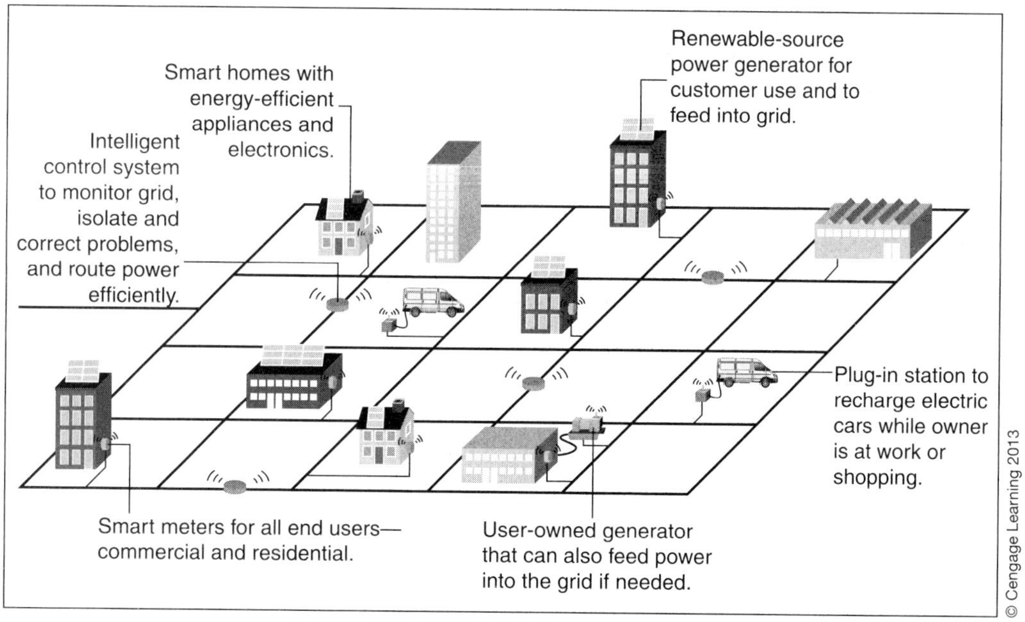

FIGURE 1-14

The smart grid permits electricity to be fed into it from many sources and provides digital management to maximize efficiency.

In this model, some large central power plants remain, but they are augmented by many smaller generating facilities distributed throughout the grid and by renewable end-user energy sources that can feed power back into the grid. Even more important, the grid gets "smart" by digitally communicating information about power supply, demand, usage, and capacity to and from every part of the grid.

In addition to the traditional grid's generating facilities and transmission network, the smart grid consists of three new components:

- Smart control and measuring devices
- Digital communications systems
- Computer software programs

Smart devices include computer-controlled generators and other power sources, as well as meters, monitors, and intelligent electronic devices that gather information about the demand for power, its availability from various sources, the delivery capacity of each part of the grid, and the flow of power throughout the system. Computer programs for the grid help it allocate electricity efficiently, isolate and control problems, and monitor the electrical functions of generators and consumers.

The U.S. smart grid is developing primarily using radio frequency communication to transmit information. Many parts of the grid communication system use Wi-Fi technology similar to that in local area computer networks (LANs). Other parts of the system are hybrids, incorporating Wi-Fi with cellular, satellite, fiber-optic, or other data transmission methods. Home segments of the smart grid often use power-line technology that sends communications over the same lines that distribute AC power.

The control devices in the smart grid enable it to respond to data received from all components in the system. Power production can be controlled and drawn from the closest available sources to minimize loss through long-distance transmission. Peak demand can be reduced through limiting controls placed on end-user devices and average total use can be minimized by timing noncritical power consumption to occur during low-demand periods. Figure 1-15 shows the control room of a digitally-controlled generating and power transmission facility.

Digital communication and control is what makes the grid smart. Sensors that monitor and report conditions on the grid enable switches and other controls to respond instantly. Real-time information permits system operators to predict, diagnose, and mitigate issues that might previously have caused interruptions or serious power disturbances. With smart grid technology, end users have more control over their energy consumption and costs.

Like the traditional grid, the smart grid is privately owned and operated. Local, state, and national government agencies play a large part in it, principally through the development of standards and regulations that govern the production, distribution, and use of electric power, but the smart grid is a cooperative venture. Over 3100 public or privately owned utilities and cooperatives distribute power in the United States. Another 2000 non-utility companies also produce, market, or transmit power. All of these share responsibility for maintaining the interconnectivity and operating efficiency of their "piece" of the smart grid. They are being joined by an increasing army of end users who contribute to the smart grid by controlling their individual power usage and by feeding back available renewable-source power to the grid.

THE HOME SMART GRID

Although digital communication and control technologies have made the local, national, and even global smart grid possible, they have also opened new possibilities for homeowners to make their individual homes more energy efficient by reducing total

FIGURE 1-15

One of the three computer-equipped control rooms at the Hunter Power Plant that connect its generators with the smart grid and enable its operators to digitally control them.

energy consumption and by lowering peak demand. Accomplishing these two objectives turns the average residence into a smart grid home.

A further contribution to the smart grid can be made by homeowners who install renewable-energy-generating components to supplement the power they draw from the smart grid during peak-demand times and to feed power back into the grid when their production exceeds home requirements.

With this type of equipment, the home becomes an integral part of the smart grid and its occupants will be far better prepared to respond to future increased power requirements, such as recharging electric cars, maintaining essential home power during interruptions on the grid, and providing economical electric power for increasing home demand.

Home Energy Efficiency

Three devices are required to make a smart grid home more energy efficient. The first is a smart meter that measures power use through the home's electrical service connection in real time. The smart meter communicates this information by wireless transmission to the utility company for billing purposes, but can also send it to a receiver in the home.

The home receiver is the second device needed for the smart grid home. It allows the homeowner to monitor power use over time and confirm the effectiveness of energy-efficiency measures as well as efforts to reduce peak-demand power usage.

The third smart grid home device that will assist the homeowner in improving energy efficiency is a power-usage monitor (watt meter) that accurately measures power consumption by individual devices. The usage monitor reports the actual power consumption of a device when in use and when turned off but still connected (plugged in) to the AC power service.

Readouts from the smart meterdisplay and the usage monitor, such as the display shown in Figure 1-16, allow the homeowner to reduce power consumption by replacing high-wattage devices with more efficient low-power versions. They also show which devices should be disconnected from the AC power grid when not in use instead of just being turned off.

FIGURE 1-16

A simple home display unit shows power usage in real time and provides warning lights to alert the homeowner when peak load levels are reached.

The smart meter display also allows the consumer to time the use of appliances that require large amounts of power (washers, dryers, compressors, etc.) for non-peak-demand hours. This results in significant cost savings and at the same time reduces peak demand on the main power grid.

Energy efficiency in a smart grid home can be further improved, often dramatically, by reducing heat transfer out of the home during cold weather and into the home during warm weather. Improved sealing and insulation reduces the demand for heating and air conditioning, with corresponding reductions in power consumption and cost. These savings can be realized regardless of whether the HVAC system is all electric or fueled by oil or natural gas.

Alternative Energy Power

Although the cost of alternative-energy-generating equipment is substantial, the savings obtained over time will generally recover the investment and realize a cost savings. Solar panels are less expensive than wind generators and are suitable for installation in most residences. Wind generators can produce more power than the average-size home solar generating unit, but they cost more and are cost effective only in areas where wind velocity is sufficient and its duration long enough to operate the generator for a substantial portion of the average day. Figure 1-17 shows a residential-scale wind-generation system in use.

FIGURE 1-17

This wind turbine suitable for residential installation can produce about 1000 watts of electric power in a 12-mile-per-hour wind.

Solar generating equipment for home installation is now available from several manufacturers and can be installed by a professional service or by the homeowner. Depending on the state where the home is located and the local electrical code requirements, all electrical connections for alternative energy sources must either be performed by a licensed electrician or inspected by one prior to being put online. Figure 1-18 shows a typical home solar panel installation.

FIGURE 1-18

This solar generating unit includes nine panels, each of which can produce about 2000 watts of DC electricity. The DC current must then be converted by an inverter to AC power for home use.

The following chapters in this book provide detailed information on how to increase the energy efficiency of your home by making the home itself more energy efficient and by using only energy-efficient appliances and devices in it. Other chapters describe how to time your energy usage to avoid heavy power consumption during peak-demand (high-cost) hours of the day. Finally, later chapters describe how to select and install alternative energy sources so as to generate all or part of your residential power and perhaps develop sufficient capacity to feed back power into the larger grid.

SUMMARY

AS THE USES OF ELECTRICITY for lighting, motor-powered equipment, and electronic devices have expanded, the demand for electric power has increased. Steam-powered turbines, fueled by coal, gas, oil, or nuclear energy, drive generators to produce electricity. Generators can also be powered by water-driven turbines or by wind. Electricity can also be created from solar energy. Electric power is delivered to users through the electric grid, a system of high-voltage transmission lines, transformers, switches, lower-voltage distribution lines, and meters. The traditional grid is an electromechanical system that operates in only one direction: from generator to end user. It does not allow users to communicate with producers, nor does it permit electricity generated by users to be uploaded into the grid. The traditional grid is now being upgraded into a smart grid by the addition of digital communication systems that permit all segments of the grid to communicate with one another and monitor the operation of the grid. The smart grid can also accept power input from many sources, which facilitates the integration of renewable energy power produced in many locations into the grid. As demand for electrical energy continues to grow, the smart grid will increase the capacity of the grid and at the same time increase its efficiency. Smart grid homes will contribute to the effectiveness of the smart grid by using more-efficient electrical devices that reduce energy consumption and by generating some or all of their own electricity using renewable energy sources.

KEY TERMS

Alternating current Electric current that reverses direction several times each second; the most common type of electric current for home and commercial use.

Amperage The amount of electric current flowing in a circuit; measured in amps.

Blackout An electric power failure over a large geographic area; usually caused by the demand for power exceeding the immediately available supply.

Brownout A drop in voltage that can damage electric equipment; usually caused by the demand for power exceeding the available supply.

Coal-fired power Electricity generated by steam turbines fueled by coal.

Digital communication network A real-time communication system for providing data monitoring and control in an electrical grid.

Direct current Electric current that flows in one direction only; batteries and photovoltaic cells supply direct current.

Distribution line An electric line carrying current from 480 down to 120 volts that connects a higher-voltage transformer to an end user.

Electrical capacity The maximum amount of electricity a generator is capable of producing or a transmission line is capable of carrying without damage or failure.

Electric meter A device for measuring the amount of electric current flowing in a circuit.

Electromechanical A system composed of mechanical parts (switches, relays, etc.) that are operated electrically.

Energy conservation The process of reducing the amount of electricity required to accomplish a given function.

Energy efficiency The ability of a building or device to perform its intended function using the minimum amount of energy; the less energy consumed over time, the more energy efficient the building or device is.

Fluctuation A variation from the normal flow of electric voltage or amperage.

Gas-fired power Electricity generated by steam turbines fueled by natural gas.

Generation The process of producing electricity by converting other forms of energy.

Generator A mechanical device similar to a motor that produces an electrical current when its shaft is rotated.

High voltage (user) Electric power at 120 to 240 volts; the voltage at which electric power is delivered to homes and commercial users.

High-voltage transmission The movement of electricity at voltages above 480 volts; transmission lines can carry as much as 700,000 volts.

Home receiver An electronic device for receiving output data from the smart meter that measures a consumer's electric power usage.

Hydrocarbon fuel Any fuel composed of hydrogen and carbon compounds; coal, oil, and natural gas are examples.

Hydroelectric power Electricity produced from generators driven by water-powered turbines.

LAN Local area network.

Low voltage (user) Electric current below 50 volts, usually in the range of 5 to 24 volts; battery-powered devices, electronic equipment (internal circuits), telephones, and low-voltage lighting are examples of equipment that runs on low voltage current.

Low-voltage transmission Movement of electricity in distribution lines at 480 volts or less; high-voltage power is stepped down by transformers to low-voltage levels for delivery to end users.

Nuclear power Electricity produced by turbine generators driven by high-pressure steam from a nuclear reactor.

One-way electric system An electrical grid that allows power and communication to flow in only one direction, from the producer to the end user; now being replaced by the smart grid that permits electric flow and communication in multiple directions.

Overcapacity The amount of generating capacity that exceeds the projected peak demand of an electric grid; some overcapacity must be maintained to prevent failure of the system if demand exceeds supply.

Peak-load capacity The maximum amount of electricity a system is capable of generating to meet demand.

Power-line technology (PLT) A communication and control system in which data signals are transmitted on AC wiring using a different frequency and voltage than the regular AC current flowing in the circuits.

Power-usage monitor A device for measuring the amount of electricity a circuit is using.

Real-time information Data about the condition and operation of a system that are sent while the system is working so that responses can be made quickly.

Renewable energy source A source of energy for producing electricity that replenishes itself within a reasonable period of time; solar energy, wind energy, and tidal flows are examples.

Smart electric grid A multidirectional electric grid that can accept multiple power inputs, has digital communication throughout, and is monitored and controlled by computer software.

Smart meter A meter that transmits a continuous digital data signal on a radio frequency that communicates how much current is flowing through it.

Solar power Energy from the sun transformed into electricity using photovoltaic technology or concentrated into a heat energy source to generate electricity and/or provide other heating functions.

Switch A device for controlling current flow in an electric circuit; a switch can interrupt current or change its flow to another part of the circuit.

Traditional electric grid The electromechanical grid constructed mainly in the 1950s that provides electric power across North America; it is currently being upgraded into a digitally controlled smart grid.

Transformer A device that steps electric voltage up to a higher level or down to a lower level.

Transmission The process of moving electricity from the point where it is generated to the points where it is used.

Transmission line A high-voltage wire used to move electricity at voltages above 480 volts over long distances; distribution lines then carry the electricity from step-down transformers to end users.

Transmission loss The loss of power that occurs when electricity is moved through a wire; the amount of loss varies with the voltage, amperage, temperature, wire material, and weather.

Turbine An engine with a bladed, rotating central shaft that is driven by high-pressure steam or moving water and connected to a generator to produce electricity.

Undercapacity The amount of potential peak demand in an electric grid that exceeds its peak generating capacity.

Voltage The electric pressure in a circuit measured in volts; the higher the voltage, the more current will flow through a given resistance.

Wi-Fi A wireless radio frequency technology that permits high-speed data and voice communication throughout a smart grid.

Wind-generated power Electricity produced from a generator driven by wind power turning a multibladed rotor.

REVIEW QUESTIONS

1. What three factors are mainly responsible for the continuing increase in demand for electricity?
2. What are three fuels used to produce steam for turbines that generate electricity?
3. Name two nonburning energy sources that are used to produce electricity.
4. What is the source of heat used to produce steam for the turbines of a nuclear power plant?
5. How long can alternating current (AC) be stored? What is its "shelf life"?
6. Where are electric meters typically placed?
7. What does the term *electromechanical system* mean as applied to the traditional electric grid?
8. What percentage of the U.S. gross national product depends on electric power?
9. What is peak-load capacity in the electric grid?
10. What is peak demand in the grid?
11. What is the usual cause of a large blackout?
12. What are some causes of power transmission loss?
13. Do hydrogenerating plants consume water?
14. Even though it is presently more expensive to generate, why is wind-generated electricity preferred over coal- or gas-fired electrical power?
15. Why is carbon dioxide that is produced by burning coal, oil, or gas referred to as a "greenhouse gas"?
16. What waste products do nuclear power plants produce?
17. How do coal, gas, oil, and nuclear generating plants use water?
18. What characteristics differentiate the smart grid from the traditional grid?
19. What are two main ways that a home can become part of smart grid technology?
20. What is a smart meter and what benefit can it provide to electric power users?

CHAPTER

2

SMART METERS AND IN-HOME DISPLAYS

OBJECTIVES

After studying this chapter, you should be able to:

- Understand what smart meters are and what they do.
- Describe how smart meters are used to measure electricity, natural gas, and water use.
- Describe how utility companies use data from smart meters.
- Describe how a homeowner can use information from a smart meter displayed on an in-home display (IHD).
- Describe different types of IHDs and specific application parameters of each.
- Describe how to install an IHD that uses current transformers (CTs) to measure electricity use.

OUTLINE

I. Introduction
II. What Is a Smart Meter?
  A. Analog Electric, Gas, and Water Meters
  B. Digital Smart Meters
  C. How Soon Will Your Home Have a Smart Meter?
III. How Utility Companies Use Smart Meters
  A. Electric Smart Meters
  B. Gas and Water Smart Meters
  C. Advanced Metering Infrastructure (AMI)
IV. How an IHD Lets You Use a Smart Meter
  A. IHDs That Receive Smart Meter Signals
  B. IHDs That Use Current Transformers
  C. Other IHDs and Monitoring Devices
V. Installing an In-Home Display
  A. Installing an IHD to Work with a Smart Meter
  B. Installing an IHD without a Smart Meter
VI. Summary

INTRODUCTION

In this chapter you will learn how smart meters differ from older analog meters in the way they measure electricity, natural gas, and water. You will learn how utility companies use smart meters to charge customers for the amount of energy or water they use and also for when they use it. You will understand how an in-home display enables consumers to reduce the amount of energy they use and adjust the time schedule on which they use it. You will be able to describe several types of in-home displays and identify the specific situations in which each should be used. Finally, you will also learn how to install an in-home display and the current transformers that supply its data.

WHAT IS A SMART METER?

TO UNDERSTAND THE advances that smart meters make in the measurement of electricity, natural gas, and water usage, you must first know how ordinary meters measure these utilities and what their limitations are.

Analog Electric, Gas, and Water Meters

Analog meters measure the flow of something over a given period of time. An electric meter, for example, measures how much electricity passes through it in a set amount of time, usually a month. The meter has no means of communicating its measurement information; it only records how much electricity has passed through it into the home or business where it is installed. Once a month, the electric utility sends a meter reader to each home or business to read and record the information on the meters. Charges for electricity are based on electric power usage for the month. Figure 2-1 shows a residential electric meter with analog dials showing cumulative electricity use.

FIGURE 2-1

A typical residential electric meter. The analog dials rotate in alternating directions, which means that the current meter reading is 94,995.5 kilowatt hours.

Electric utility companies charge for the amount of electricity used in a monthly period. The use is measured in total kilowatt hours consumed for the month. It does not matter which day of the month or which hour of each day the electricity is used; the meter cannot measure when the power is used, only the total amount used. Because of this limitation, the charge per kilowatt hour must be a set amount regardless of when the electricity was used, even though the cost of producing the electricity may be significantly higher during peak-load times than during off-peak hours. The limited information available from the meter prevents the utility company from charging for electricity according to the cost of producing it.

The same is true for ordinary meters that measure natural gas or water usage. These meters only measure the total amount of gas or water used in a month. They also do not communicate their information; they must be read by a meter reader each month. Users are charged the same amount for each gallon of water used or each cubic foot of gas consumed, regardless of whether it was consumed at the time of lowest demand or the time of highest demand. With ordinary meters, the utility cannot determine time of use and so can only charge per unit used.

The user of electricity, natural gas, or water measured by an ordinary meter also has no incentive to minimize use during high-demand periods because the charge is the same regardless of when the use occurs. For electric use, this means that it is no more costly to run the dishwasher and the washing machine in the afternoon while the air conditioner is also running at full capacity than to wait and run these appliances after dark when the air-conditioning demand for power is less. For gas and water use there is also little incentive to conserve during peak-demand periods because the charge is the same per unit regardless of when each unit is used.

With ordinary meters the only way to reduce the cost of electricity, gas, or water is to reduce overall use per month. And reducing total use of any utility is difficult when the user has no means of measuring how much is being used at any given time. We leave our lights burning, our computers running, and our sprinklers running at midday because we have no way of knowing what we would actually save if we turned them off or switched their use to a different time. Figure 2-2 shows a typical electric bill for a residence and the amount of power used daily during each month over the preceding year. The charge per kilowatt hour was a constant $0.075 for the first 400 hours used and $0.092 for the hours above 400 over the entire period, regardless of peak-demand months or times of day.

Your Balance With Us

Previous Account Balance	129.68
Payments/Credits	-130.00
New Charges	+62.94
Current Account Balance	**$ 62.62**

Payments Received

DATE	DESCRIPTION	AMOUNT
Sep 13, 2010	Payment Received - Thank you	130.00
Total Payments		**$ 130.00**

Detailed Account Activity

ITEM 1 - ELECTRIC SERVICE

Utah Residential Service Schedule 1

METER NUMBER	SERVICE PERIOD From	To	ELAPSED DAYS	METER READINGS Previous	Current	METER MULTIPLIER	AMOUNT USED THIS MONTH
50962265	Aug 31, 2010	Sep 30, 2010	30	30145	30801	1.0	656 kwh

Next scheduled read date: 10-29. Date may vary due to scheduling or weather.

NEW CHARGES - 10/10	UNITS	COST PER UNIT	CHARGE
Basic Charge - 1P			3.75
Energy Charge Summer 1st Block	400 kwh	0.0752920	30.12
Energy Charge Summer 2nd Block	256 kwh	0.0927490	23.74
Customer Efficiency Services		0.0482000	2.60
Home Electric Lifeline Program			0.23
Utah Sales Tax		0.0415000	2.50
Total New Charges			**62.94**

Historical Data - *ITEM 1*

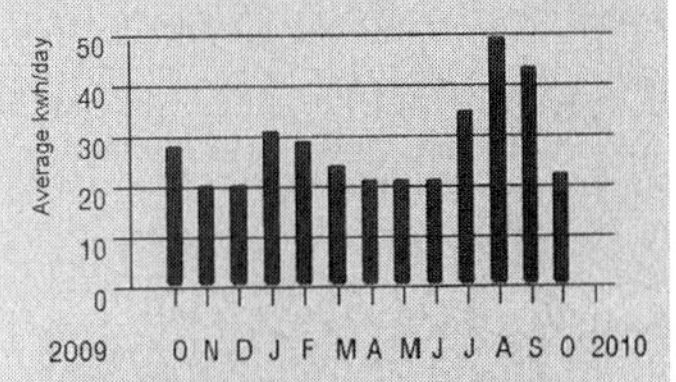

Your Average Daily kwh Usage by Month

PERIOD ENDING	OCT 2010	OCT 2000
Avg. Daily Temp.	68	70
Total kwh	656	835
Avg. kwh per Day	22	28
Cost per Day	$2.10	$2.62

Just in case of a power outage, keep our new Outage Hotline number handy - 1-877-508-5088. You also can use a wireless connection to access the Internet for outage information and updates by visiting rockymountainpower.net/outage.

FIGURE 2-2

A typical fixed-rate utility bill shows a constant charge for the amount of power used, regardless of peak-load times.

Digital Smart Meters

Smart meters change the dynamics of electrical and other utility use dramatically. Instead of reading the cumulative amount of electricity used in a month, a smart meter reads electric use every hour, or even more frequently. The solid-state digital meter records hourly meter readings and periodically transmits the data back to the electric utility company. These meters also monitor the condition of the electricity passing through them and communicate any disturbances (voltage sags or spikes, waveform distortions, etc.) or power interruptions back to the utility company.

Utility companies can also communicate with individual smart meters to install new software upgrades, remotely connect or disconnect power service, or query them for real-time data about the circuit they are monitoring.

Some smart meters transmit their data via a dedicated radio frequency (RF) network owned by the utility. Each meter in this technology is equipped with a radio that transmits its data to one of the utility's network access points. The meters and the access points are both components in an RF mesh technology system. This system allows each component to send and receive data from other nearby components. This permits meters and other sensing devices to securely route data to and from the utility via a "mesh" of RF network coverage. The system provides two-way communication between the meter and the utility for transmission of electric use information and for monitoring of the circuit in which the meter is installed.

Another method of transmitting smart meter data is over Internet Protocol (IP)–based public wireless networks. Using smart meters that function on this technology enables utilities to rapidly deploy secure and scalable smart grid solutions with no capital expenditures required for constructing a private communication network. Because they are built on open standards and leverage existing programs and tools, IP networks and solutions may be remotely upgraded to interface with new technologies.

A third type of smart meter is called a net meter. Net meters have all the functions of other types of smart meters, but include an additional capability. They allow homeowners to produce part or all of their electricity requirements on site and sell any excess back to the utility company that serves their area. With a net meter, the user is charged for the amount of energy that is purchased from the electric utility minus the cost of the energy that the user provides to the utility. The home uses the electricity generated on-site first, reducing the amount that must be purchased from the utility. If the electricity generated by the home's solar or wind facilities is greater than its power requirements, the excess is passed back into the electric grid after being measured by the net meter so that the homeowner can be paid for it.

Many incentives are available to encourage homeowners to install renewable-energy-generating facilities at residential sites. These incentives vary by state and by the electrical utility serving the area where the home is located. They will be discussed in more detail in Chapter 9 on solar power generation and in Chapter 10 on wind power generation.

Smart meters are programmable devices that can be upgraded remotely, providing the ability to implement future innovations easily and securely. Figure 2-3 shows an installed residential smart meter with a digital reading displayed on its screen.

FIGURE 2-3

A residential smart meter with a digital display showing cumulative kilowatt hours used. This type of meter records current use at least four times each hour and transmits its data to the utility company.

How Soon Will Your Home Have a Smart Meter?

As of April 2010 about 76 million smart meters were installed worldwide, mostly in North America, where 15% to 20% of homes now have them, and Europe. About 12 million smart meters are now installed in U.S. homes and businesses, with at least 15,000 new ones being added to the total daily.

The Swedish research firm Berg Insight says that the worldwide installed base of smart meters is growing at a compound annual rate of 31.1%. The firm projects that the total smart meter installed base will reach 302.5 million at the end of 2015. That figure will be almost 50% of households in North America and Europe and about 25% of homes in the Asia-Pacific area. Berg Insight projects that growth of smart meter installation will continue in the second half of this decade, with many markets approaching 100% penetration by 2020.

Most smart meters now in service in the United States are located in the western half of the country, with about 7 million installed in California alone. Texas will have over 3 million by the end of 2012 and other western utility companies are installing smart meters as rapidly as their service resources permit. Eastern utilities have been slower to adopt smart meter technology, but are now beginning to install them in business and some residential locations.

HOW UTILITY COMPANIES USE SMART METERS

For the electric, gas, or water utility company, the smart meter provides a means of billing its customers not only on the basis of how much energy or water they use, but when they use it. As they install smart meters, electric companies are changing their billing practices to charge more for power that is used during periods of peak demand. The reasoning behind this timed pricing for electricity is simple: peak-load power costs more to produce than off-peak power, and so the customer should pay more for it. Water companies are also charging more for water used during peak periods of demand, and gas companies may soon follow their example.

Electric Smart Meters

To understand why peak-load power is more expensive, imagine a generating facility with four turbines turning four generators. Each generator requires the same amount of input energy to produce electricity, but all four generators do not operate continuously. When demand is low during the hours from midnight to six in the morning, only one generator may be needed to produce electricity and the other three sit idle. As demand rises during the day, the second and third generators are put online and begin producing power to meet the increased need. Finally, in the evening all four generators may be required to meet the peak load of air-conditioning, lighting, and other electrical demands. Figure 2-4 illustrates this example and shows in the graph the number of hours each generator may be required to operate even though all generators must be kept online and ready to generate power when required.

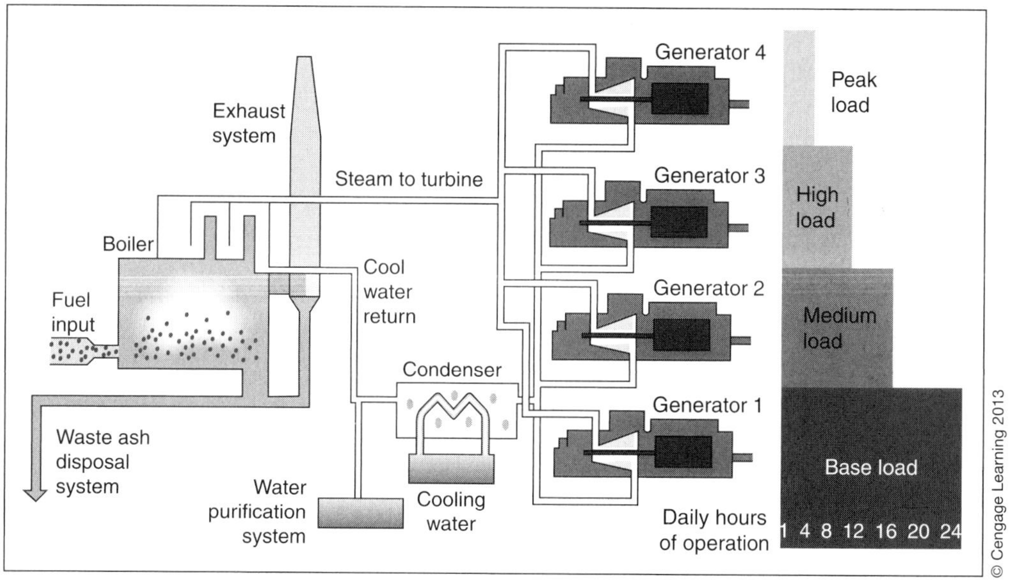

FIGURE 2-4

Peak-load power is more expensive to generate than base-load power because the generators that produce it operate only a few hours daily but must be kept online continuously.

Even though the fourth generator operates for only 4 hours a day, it costs as much to build, install, and operate as the first generator that runs continuously. The second and third generators also cost the same to build and operate as the others even though they run for only a part of each day. The three generators that operate for less than 24 hours each day will consume less fuel than the generator supplying base-load power, but nearly all of the other costs associated with power generation will be about equal for all four generators. This is the reason why peak-load power, produced for only a few hours a day, is more expensive to produce than base-load power that is produced continuously. The peak-load generator must be online and ready to produce at any time when demand requires it, but its cost of operation must be paid for by the few hours of power that it actually generates. Each kilowatt hour of peak-load power produced by the fourth generator in this example costs significantly more than each kilowatt hour of base-load power produced by the first generator.

Electric power utilities know that large peak loads on their power grids are expensive to produce because they require the inefficient use of equipment and labor, but they must maintain peak-load generating capacity or risk blackouts and damage to their grids if peak demand ever rises above peak capacity. By charging more for peak-load power, the utilities are better able to recover their investment in equipment. The higher cost of peak-load power also serves as an incentive to consumers to change their electricity use habits so as to use less power during peak-demand times. Later chapters in this book discuss how electrical usage can be reduced during peak-demand times.

Smart meters also reduce costs for electric utilities in other ways. Because they transmit their data by radio, they do not have to be visited periodically by a meter reader. This reduces the company's labor costs and also eliminates errors in billing caused by inaccurately read meters.

Electricity can also be turned on or off at precise times by a smart meter without the need for any service personnel to be present. This feature further reduces labor costs and speeds up the connection process for users. It also helps prevent users from being charged for electricity after they have moved from a home. Residential power can be turned off instantly with a phoned request.

Gas and Water Smart Meters

Natural gas suppliers and water utilities are also installing smart meters on gas and water users' service connections. These types of smart meters provide many of the same advantages as electric smart meters, but they are not currently being used in quite the same way.

A gas smart meter measures the amount of natural gas passing through it in cubic feet and sends this data to the utility company. Like electric smart meters, gas meters record a reading every few seconds, and this information can be provided to customers to help them reduce gas consumption by adopting energy-efficient practices.

But unlike electricity, gas consumption rates cannot be easily altered by the customer. Slight improvements can be achieved by using less gas-heated water for washing dishes and clothes, or by reducing the temperature to which the water is heated, but few people are willing to tolerate cold showers for the sake of saving a few cents worth of gas. Similarly, the temperature inside gas-heated homes can be tuned down a few degrees to save fuel, but the resulting discomfort for the occupants will sharply limit how low the thermostat is set.

Utility companies do not price gas higher during period of peak load because the cost of supplying the gas does not increase substantially during those times.

Consequently, the savings that customers can achieve from reduced gas usage are also less than those with electricity.

Gas smart meters are valuable to the utility company because they help it monitor gas lines for leaks and breaks, spot trouble points, and instantly record any increase in demand so that additional supplies can be brought online. They also eliminate the need for meter readers to visit homes and businesses to read the meters for billing. With smart meters, billing can be computerized using the data supplied continuously by radio.

Smart water meters are also being installed by many water utilities, and these devices make it possible for companies to charge for water use on an upward sliding scale. Water is a finite resource in most areas, and water utility companies want to limit how much is used during high-demand periods. Smart meters measure water passing through the supply line continuously and radio this information back to the utility. Many suppliers, especially in areas where water is scarce, charge customers a set rate for the first "block" of water they use in a given time period and a progressively higher rate for each "block" of additional water used in the same period thereafter.

Figure 2-5 shows a residential water bill for a summer month with water use billed in "blocks" of 100 cubic feet at four different rates. The first level, 1-block water, the lowest billing rate, is $1.26 per block. The next level, 2-block water, is billed at $1.93 or 153% of the rate for 1-block water. The 3-block rate is $2.67 or 212% of the 1-block rate.

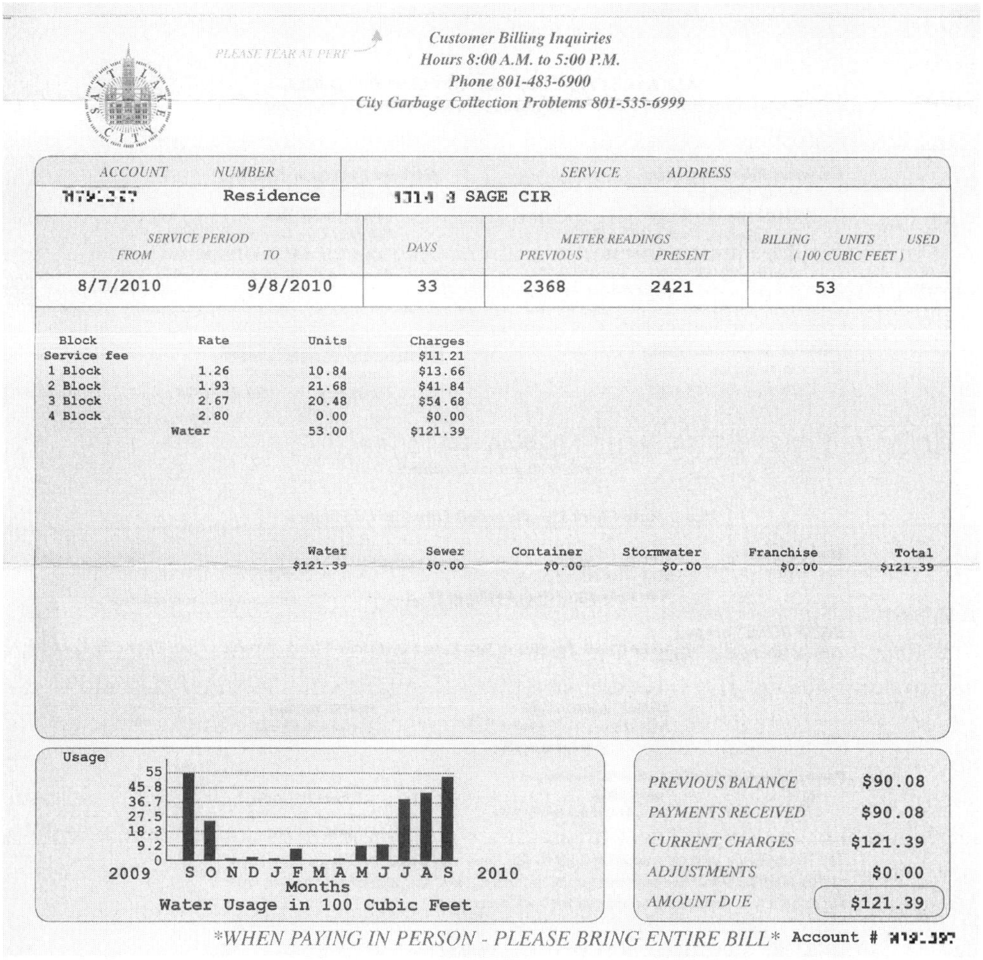

PLEASE TEAR AT PERF

Customer Billing Inquiries
Hours 8:00 A.M. to 5:00 P.M.
Phone 801-483-6900
City Garbage Collection Problems 801-535-6999

ACCOUNT NUMBER		SERVICE ADDRESS
[illegible]	Residence	[illegible] SAGE CIR

SERVICE PERIOD FROM	TO	DAYS	METER READINGS PREVIOUS	PRESENT	BILLING UNITS USED (100 CUBIC FEET)
8/7/2010	9/8/2010	33	2368	2421	53

Block	Rate	Units	Charges
Service fee			$11.21
1 Block	1.26	10.84	$13.66
2 Block	1.93	21.68	$41.84
3 Block	2.67	20.48	$54.68
4 Block	2.80	0.00	$0.00
	Water	53.00	$121.39

Water	Sewer	Container	Stormwater	Franchise	Total
$121.39	$0.00	$0.00	$0.00	$0.00	$121.39

PREVIOUS BALANCE	$90.08
PAYMENTS RECEIVED	$90.08
CURRENT CHARGES	$121.39
ADJUSTMENTS	$0.00
AMOUNT DUE	$121.39

WHEN PAYING IN PERSON - PLEASE BRING ENTIRE BILL Account # [illegible]

FIGURE 2-5

A residential water bill shows the four-tier rate that this utility company uses to charge for water. The highest monthly charge for the year is $121.00; the lowest is $12.00.

The 4-block rate, which this customer did not use enough water to reach, is $2.80 per block.

Note also the 13-month water usage chart at the bottom of the bill. It starts and ends with the month of September, which for this user is the month of highest water usage. During the winter and spring months of November through April, water use for this residence is less than 900 cubic feet per month and would all be billed at the lowest rate. In the summer and fall months of May through September, the use rate rises dramatically to a peak of 5300 cubic feet in September. Nearly all of this additional water, which accounts for more than 80% of total water use during the year, is used for outside yard and garden irrigation. Water utilities charge more for this "outside" water than for essential "inside" water used for drinking, washing, and sanitation. The higher charges encourage users to conserve water by reducing the amount used for "outside" purposes. Chapter 5 discusses ways in which water use can be curtailed and its cost reduced.

Advanced Metering Infrastructure (AMI)

The overall technology system that utilizes smart meters, communications links that connect them with utilities and consumers, and other hardware components and software programs that analyze and act on the data provided is known as Advanced Metering Infrastructure (AMI). AMI systems measure and analyze energy usage and interact with automated control devices through various communication media to implement either on request or by predefined energy-use schedules. This infrastructure includes hardware, software, communications, consumer energy displays, and energy-use controllers. Network communication between the measurement devices, analytic programs, and controllers supplies collected information to customers, utility companies, and service providers. This enables providers and consumers of electricity, gas, and water to participate in demand-response solutions, products, and services. The AMI helps bring about a change in energy and resource usage from normal consumption patterns, either in response to price incentives designed to encourage lower energy use at peak-demand periods or simply as a means of conserving scarce resources and minimizing each user's impact on the local and global environment.

HOW AN IHD LETS YOU USE A SMART METER

The data accumulated and transmitted to the supplier utilities by electric, gas, and water smart meters is also available to the homeowner or business owner through the use of an in-home display (IHD). This information is available to the consumer, either directly from the smart electrical meter or through some form of communication link with the providing utility. The IHD displays the information in a manner that allows the consumer to analyze it and adjust energy and resource use to maximize efficiency and minimize cost.

An IHD allows electric utility customers to track their energy usage in kilowatt hours in real time using a chart or graph format. The display provides most types of energy awareness information that a customer may require: total energy consumption over a given time period, real-time by-the-minute usage in kilowatt hours, matching real-time pricing, and comparison with typical rates of energy usage.

An IHD is essentially a display that communicates with utility meters (electric, gas, water), processes the data it receives, and displays it in a format that the consumer can understand and use to increase energy efficiency. An IHD can also receive custom real-time messages from the utility companies regarding billing, energy prices, and other service bulletins. Most IHDs currently available are portable devices for placement in existing dwellings, but they are now also being installed as wall-mounted permanent fixtures in many new homes.

In addition to its display functions, an IHD may be equipped to communicate with automated appliances in the home to monitor and control their operation for maximum energy efficiency. Typical wireless communications standards are ZigBee (802.15.4 2.4 GHz), which is used in the United States), Wireless MBus (sub 1 GHz), which is more common in Europe, Wi-Fi, or power-line technology.

The most user-friendly IHDs are very simple to use and are designed for use by consumers who are not technology proficient or willing to spend a lot of effort on home energy efficiency. They provide simple tools for cutting energy consumption (gas and electricity) by 5% to 10%. The touch-screen display of a simple IHD is built around a clock rather than a graph that shows real-time electricity consumption. Using the clock, a user can program a thermostat and participate in utility demand-response programs where electricity consumption is curbed during peak times. Once installed and programmed, this type of IHD requires no input or monitoring from the user.

A simple IHD connected only to the home's heating, ventilating, and air-conditioning (HVAC) system can cut electricity use by about 10%. This projection assumes that utility companies offer time-of-day pricing, which is not yet the case in most of the United States. A smaller savings can be realized even in areas that do not have time-of-day pricing. In both cases, the IHD responds to the utility's demand-response program by communicating with a smart meter, if the home has one installed, or directly with the utility through a dedicated communications link. The IHD can direct the home's HVAC system to reduce energy use during peak-demand periods by shutting down for short intervals during the heating or cooling cycle.

IHDs That Receive Smart Meter Signals

One currently popular IHD in use in the United States is the PowerPortal Home display, which provides customers with data on current energy usage, energy price information, and utility messaging. The PowerPortal Home is battery-operated, weighs less then 5 ounces, has a battery life of up to 2 months, and is one of the most portable IHD devices available. Figure 2-6 shows the PowerPortal unit with its display operating.

The PowerPortal Home is an AMI-ready display. It connects wirelessly to most AMI systems that measure, collect, and analyze energy usage data, and can be programmed to interact with electric meters and electrical appliances either on request (on-demand) or on predefined schedules. It is designed for the ZigBee smart energy wireless profile, which allows for broad compatibility with many AMI systems. PowerPortal Home has a magnetic backing that will attach to any flat steel surface and a rechargeable battery. It also has an integrated base so it can be placed on a table, countertop, or even on the front of a refrigerator.

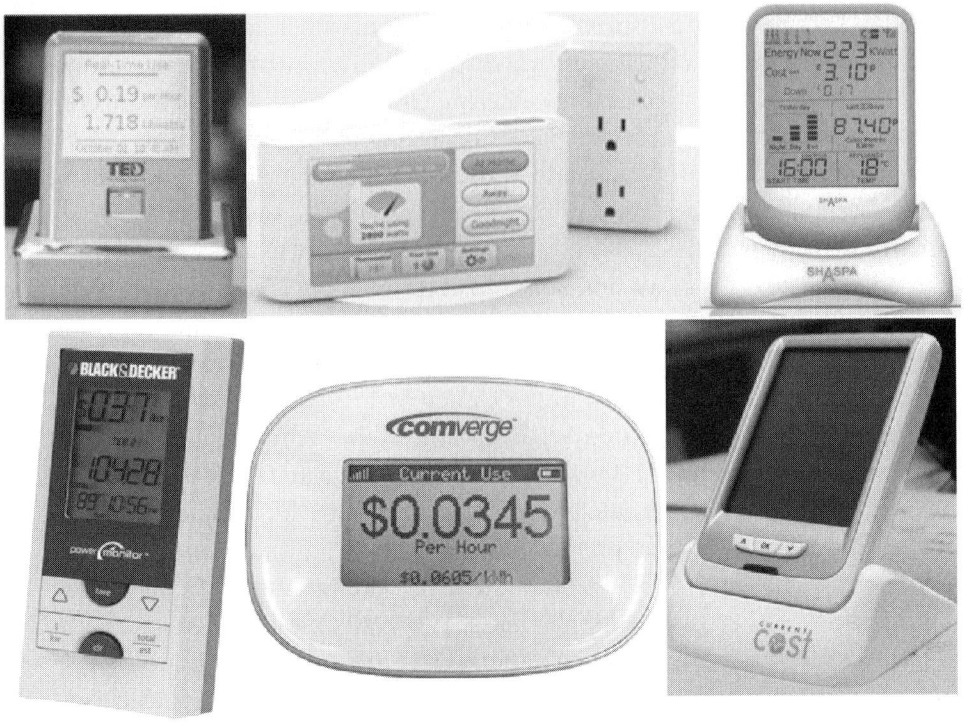

FIGURE 2-6

The PowerPortal Home display unit, seen here in the lower center of the picture surrounded by other similar IHDs, allows electrical customers to read data from their home's smart meter and other measuring devices.

ZigBee is the trade name of the global wireless language that is used to connect different control and monitoring devices to work together. It is based on the IEEE 802.15.4 physical radio standard that operates in unlicensed bands worldwide at 2.4 GHz (global), 915 Mhz (Americas), and 868 Mhz (Europe). The ZigBee standard allows raw data throughput rates of up to 250 Kbs and can transmit over a distance of 10 to 100 meters, depending on power output and environmental characteristics.

ZigBee technology is supported and promoted by the ZigBee Alliance, an association of several hundred companies that have agreed to develop and market products that are compatible with one another because they all use the ZigBee communication standard. ZigBee is now used in a wide range of products and applications across consumer, commercial, industrial, and government markets worldwide. It has become the leading wirelessly networked sensing and control standard for use in energy, home, commercial, and industrial areas. For more information about the ZigBee Alliance and technical information about the ZigBee standard, visit http://www.ZigBee.org.

Many other brands of IHDs are now available that can receive and interpret smart meter data signals, but not all IHDs can receive signals from all smart meters. A list of IHD brands and the companies that manufacture them appears later in this chapter. The brand of smart meter installed on the home's power service determines which IHDs can be considered for in-home use. Some smart meters use proprietary communication protocols that can only be received and read by IHDs designed and built to that standard. Others are ZigBee compliant or operate on a Wi-Fi standard that many IHDs can accept. In all cases, it is important to know what brand and model of smart meter is installed at the home before purchasing an IHD, because the IHD must conform to the smart meter's communication protocol, never the reverse.

IHDs That Use Current Transformers

Another type of IHD is based on a direct hardware connection to the inside service panel of the home's electric service. This type of energy monitoring system includes a hardware module that measures electricity usage in real time and transmits its data to the home's IHD either through a wired connection or by radio.

An example of this type of IHD is the TED 5000, one of several IHDs produced by The Energy Detective (TED). The TED 5000 uses a hardware-based gateway to receive signals from voltage and current sensors that are installed at the home service panel. The data from the hardware module is sent to a built-in web interface through a standard 10BaseT RJ-45 Ethernet connector. The TED 5000 gateway can also establish a link to Google's PowerMeter portal website to archive its data.

The TED 5000 supports up to four current/voltage sensor modules, but most homes will require only one or two, one for each service panel through which power is received from the utility. Each sensor module comes with two clamp-on current transformers (CTs) that connect to each leg of the split-phase, 120/240-volt home electric service. These modules communicate with the gateway hardware using power-line communication through the AC wiring. The modules send their data signals using the voltage-sensing connections at the panel. The gateway can be plugged into any wall outlet in the house and receives the transmitted information there for relay via a network connection to the home's IHD. Figure 2-7 shows a TED 5000 PC-based display.

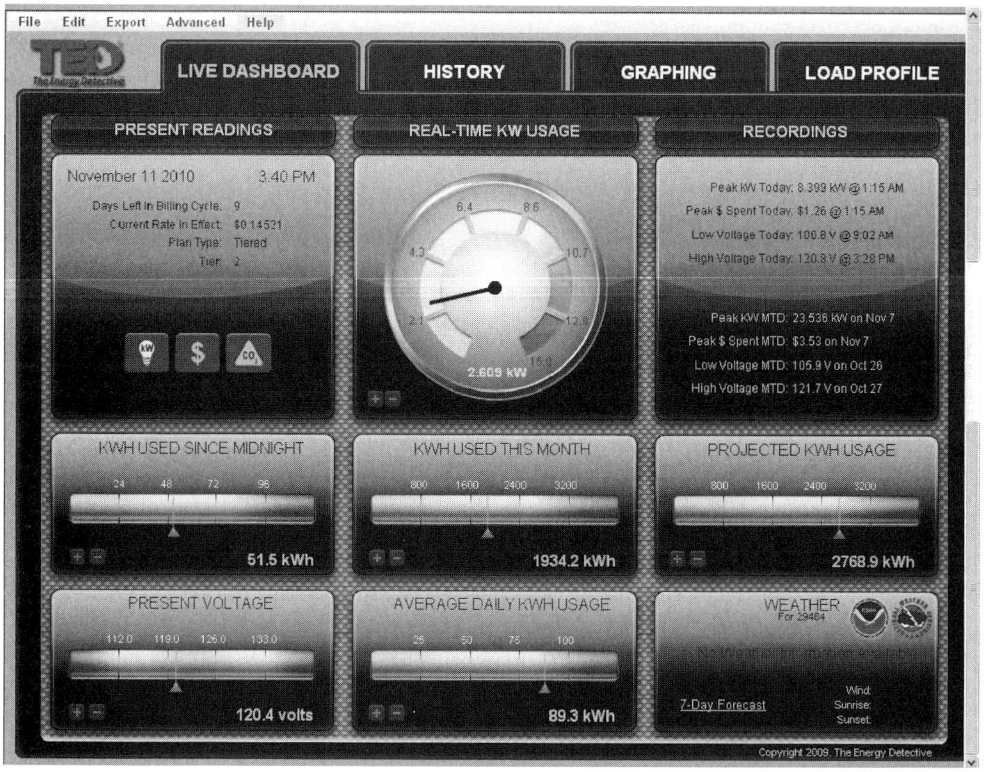

Photo Courtesy of TED, Inc.

FIGURE 2-7

The in-home display on a PC for the TED 5000 shows real-time power use, cumulative use, average daily use, cost figures, and other data.

In addition to sending data to a local IHD, the TED 5000 can also send data to the Google PowerMeter portal for display on a user's account there. Google PowerMeter is a free energy-monitoring tool that enables a consumer to view home energy consumption from anywhere online using energy information provided by monitoring systems such as the TED 5000, utility smart meters, or other energy-monitoring devices.

Microsoft Hohm is another web-based energy-monitoring service. Users who sign up for Hohm are provided with a home energy report and energy-saving recommendations tailored to their individual homes.

Other IHDs and Monitoring Devices

Several other IHDs and energy-monitoring displays are available in the United States and other countries. Some of those available for North American electrical service are described next.

Aztech In-House Display

Collects data from the smart meters and allows homeowners to conveniently monitor their real-time energy consumption patterns and estimated costs in home.

Black & Decker Power Monitor

The Power Monitor consists of a wireless transmitter that easily attaches to the electric meter, and a wireless handheld display that uses data from the home's electric meter to show information about energy consumption.

DreamWatts

The DreamWatts Solution is a ZigBee radio-enabled thermostat communicating with an Internet gateway connected to a home or commercial-site Internet router. It communicates with PowerMand's central Network Operating Center servers. Users access information through any device with web access (PC/Mac, laptop, smart phone, etc.) and can use their DreamWatts account to modify temperature settings or create and maintain custom energy management profiles.

EcoDog

EcoDog's FIDO Home Energy Monitoring System oversees energy use throughout the home, on a room-by-room basis. Using patent-pending algorithms, the software displays recommendations for consumers to reduce electric costs.

EcoFactor

EcoFactor calculates each home's individual thermal characteristics, or "dynamic signature," using existing hardware such as a two-way communicating thermostat and/or a gateway device to a broadband Internet connection. With each home's dynamic signature, EcoFactor learns how much energy is required to heat and cool the home.

EcoFactor operates with off-the-shelf hardware and is hosted on-demand, so there is no software to install or maintain.

em•power by Smart System Technologies

The em•power system helps control and coordinate HVAC systems and home appliances so the homeowner doesn't have to. The em•power system features multiple interfaces to control electrical appliances in the home.

Energate

Energate's home energy management solutions let consumers and utilities manage energy use and reduce peak demand without sacrificing comfort or convenience. The system is a home energy management partner that is platform-agnostic, ZigBee compliant, future-proof, and feature-rich.

Energy Engage

Energy Engage™ is a state-of-the-art, Web 2.0 consumer engagement solution that allows utilities to provide individual customers access to relevant, up-to-date information regarding their electricity usage and costs.

EnergyHub

EnergyHub is the consumer face of the smart grid that helps homeowners reduce energy consumption and save money. EnergyHub produces smart, simple, and cost-effective energy management tools that connect consumers and utilities and help solve energy problems.

Energy Joule

Demand for energy and the price of energy change throughout each day. The Joule shows customers the crucial information needed to change power-use behavior. It provides information on when to use fewer lights and appliances.

Energy Owl Electricity Monitoring System

The Energy Owl E.M.S. simply "watches" the home's electrical meter using an optical sensor. The scanner sends the data to the desktop hardware that then either transmits the data to a personal computer or stores it for later retrieval. When the Energy Owl software is open, it provides you with real-time data, updated once a minute. The Energy Owl E.M.S. is available only for digital electrical meters, specifically the Itron Centron, the most popular digital meter for residential applications.

EnviroScape Energy Management by BreezePlay

The EnviroScape Energy Management and Control Panel is the first residential consumer product on the market that is monitored 24/7 to provide actual usage and rate data, and that can provide controls to automatically adjust systems to affect consumption and automate savings.

GreenQuest

GreenQuest is a free-to-use online energy dashboard that enables households and commercial building owners to track their utility bills. Local and regional sponsors provide GreenQuest to their customers or community through a privately branded GreenQuest website, placing a link on the organization's homepage, and promoting the GreenQuest link.

GroundedPower

GroundedPower combines real-time monitoring with a highly intuitive Interactive Customer Engagement System that increases energy efficiency, creates persistent behavior change, and develops sustained savings.

Home Automation, Inc. (HAI) Smart Grid Solutions

HAI has a full suite of products to cover any type of HAN application, including programmable communicating thermostats (PCTs), in-home displays (IHDs), and load control modules (LCMs).

Kill A Watt

Connect any appliance to the Kill A Watt™, and it will assess how efficient the device really is. Users can calculate their electrical expenses by the day, week, month, or even an entire year. With the amazing Kill A Watt™, you'll know "Watts" killing you.

PowerCost Monitor by Blue Line Innovations

With this technology, an easy-to-install sensor unit is attached to the outside of the home's electric utility meter. The sensor reads the meter and transmits the information (range 100 ft) wirelessly to the Power Monitor. The Power Monitor shows real-time electricity usage, highest amount spent in a 24-hour period, time, temperature at meter location, signal strength, and the battery strength of both the display and sensor units.

PowerHouse Dynamics—eMonitor

PowerHouse eMonitor enables a homeowner to measure each individual circuit. This allows the homeowner to know, in real time, how much energy every appliance in the house is using, how much electricity is being produced by the solar panels on the roof, and, for example, how much power it took to charge the electric car last night.

PowerNab

PowerNab is a software platform that integrates commercial and home electronic and electrical devices, allowing the entire building's electronic and electric circuitry to communicate. Its ability to remotely control these devices can result in receiving precise information to achieve targeted power-saving goals.

Rainforest Automation Inc.

Rainforest Automation Inc. enables two-way communication between utility companies, homeowners, and their power meters. These products allow utilities to communicate power information in real time to their customers, collect power meter information for the utilities, and provide an energy management platform that benefits both.

Shaspa—Social Energy Meter

The Shaspa Smart Home Kit is an open platform that utilizes a network of intelligent sensors that serves to provide information about the state of the home. Through the Shaspa Bridge, this information is processed and made available through a number of access methods, including touch-screens, mobile phones, and browsers.

Silver Spring Networks—Smart Energy Platform

Silver Spring enables utilities to rapidly deploy advanced metering or distribution automation and focus on demand-response or electric vehicle projects. Silver Spring delivers the unified platform needed for these missions to succeed.

Tendril

The Tendril Residential Energy Ecosystem (TREE) gives the homeowner insight into and control over the household energy footprint. The in-home wireless network allows appliances and electrical outlets to talk instantly and directly to a home energy monitor or web portal, enabling the user to understand and manage energy consumption. The user can monitor consumption in real time and conserve energy by adjusting usage according to personal energy needs and goals.

The Meter Reader by Energy Monitoring Technologies

The Meter Reader EM-2500 (SO) includes a monitor and two current clips. It measures total house load or individual 110-volt / 220-volt circuits, such as stove, dryer, air conditioner, pool motors, and so forth.

WattsUp? Meters

WattsUp? meters are "plug load" meters that measure the amount of electricity used by whatever is plugged into them. Fast, intuitive, and easy to use, WattsUp? meters quickly and accurately measure any 100- to 250-VAC device. The meters incorporate sophisticated digital electronics that enable precise and accurate measurements in an easy-to-use format.

INSTALLING AN IN-HOME DISPLAY

The first step in installing an IHD is to determine whether your home is equipped with a smart meter on the incoming electric service. The easiest way to do this is to contact your local electric supplier and ask whether a smart meter is installed at your address. If a smart meter is in place, find out what type it is and whether an IHD designed specifically for that type of meter is available from the power company or from a third-party supplier that the power company has contracted with.

Installing an IHD to Work with a Smart Meter

Using the IHD supplied or recommended by the power company is nearly always the best option. The IHD will likely be less expensive than if you have to buy a single unit at retail on your own and you can also be certain that the IHD will work correctly with the meter used by the power company. A few brands of smart meters have a proprietary output code that can be received and read only by a single type of IHD. If your meter is one in this category, you have no choice but to use the specified brand of IHD, but you don't necessarily have to buy it from the power company or its designated supplier. You might save some money by looking for an alternate supplier that will offer a better price in order to break into an otherwise closed market.

If your electric power company doesn't supply or recommend an IHD for home use, ask which brands work best with the smart meter you have installed. You should also ask for configuration information so you'll know how to set up the IHD to receive and display data from your smart meter. If the power company or its partner firm supplies the IHD, it will probably be configured correctly when you get it, but if you are buying from a third-party supplier, you will need exact programming instructions for the IHD, including:

- The identifying number of the smart meter on your home that the IHD is intended to receive from.
- The precise frequency or protocol to which the IHD should be set to receive data from the meter.
- The range of the meter (not the IHD). The strength of the smart meter's transmission signal determines how far away from the meter the IHD can be located and still receive its signals.

Some meters may require a transmission device attached directly to the meter to transmit data from it or to boost or translate the outgoing signal into a format that an IHD can receive and interpret. Most smart meters have built-in transmission capability and can communicate directly with an IHD within a range of 100 feet or so.

Once you know the specifics of the IHD you need, pick the brand that best meets all of the requirements at the most reasonable cost and buy it. If the power company is not supplying or recommending a specific IHD to work with the smart meters it installs, be sure that the one you choose has some local installations that are already up and running. Ask the supplier for two or three references that you can call to find out how the IHD works and any difficulties encountered in getting it set up properly.

Most homes at present do not have smart meters, although the number of homes with smart meters is growing at the rate of thousands daily. If your home is one of the majority without a smart meter, ask the power company in your area if it will install one. Some companies are willing to install an individual smart meter even though they have not adopted them as a company-wide initiative yet. Other companies may be in the process of installing smart meters over the entire service areas and just have not yet reached your neighborhood. If that is the case, you will have to decide whether to wait until your turn comes around in the regular installation schedule or ask to be moved to the top of the list so you can install an IHD immediately.

In either case, actually in every case, do not expect the meter or a power-company-supplied IHD to be free. The customer always pays for both devices, although you may be able to spread the cost out over a few months or even a year's billing cycles. And used properly, the smart meter and IHD will more than pay for themselves in saved electric costs. Once you have the smart meter installed, follow the preceding recommendations to choose an IHD for your power service.

Most IHDs are handheld and completely portable devices. Those intended for wall mounting or placement on a countertop are also wireless and battery powered so they can be placed almost anywhere in the home. If the IHD will feed data into a computer for screen display, it needs to be placed in a manner convenient for an Internet or LAN connection.

Installing an IHD without a Smart Meter

If you do not have a smart meter and are unlikely to get one soon, you can still use an IHD to monitor energy use. The IHD will receive its data input from current transformers placed on the incoming power lines at the service panel rather than from the meter outside.

Caution: Most state and local electrical codes require that any electrical work that requires opening the service panel of a residence must be performed by a licensed electrician.

Before attempting to install current transformers (CTs) yourself, check your local electric code to see if this requirement is in force in your area. If it is, do not attempt to install the CTs or do any other work inside the electric service panel yourself. Even if you feel qualified to do this work, performing it in violation of your local electrical code is illegal and could result in a fine or other penalty if discovered later. More important, performing electrical work in violation of the code may void the homeowner's insurance policy that protects you from loss in the event of a fire, whether or not the fire was caused by the out-of-code electrical work.

If the local electrical code does not require all service panel electrical work to be performed by an electrician, you can install CTs on the incoming electric cables by following these steps:

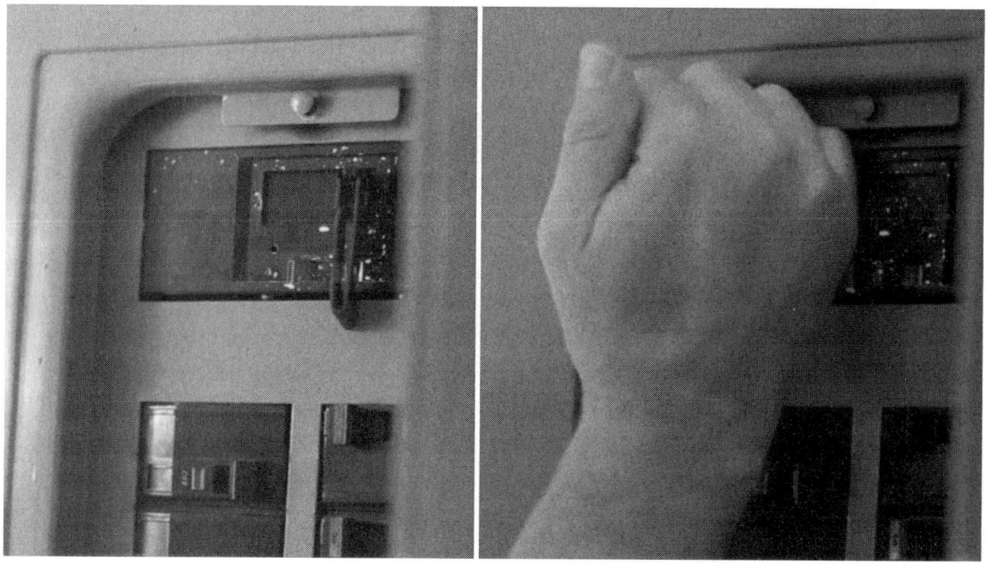

1. Turn off the main breaker in the service panel. Do this before any other step.
2. Remove the service panel cover to expose the internal wiring in the service box.

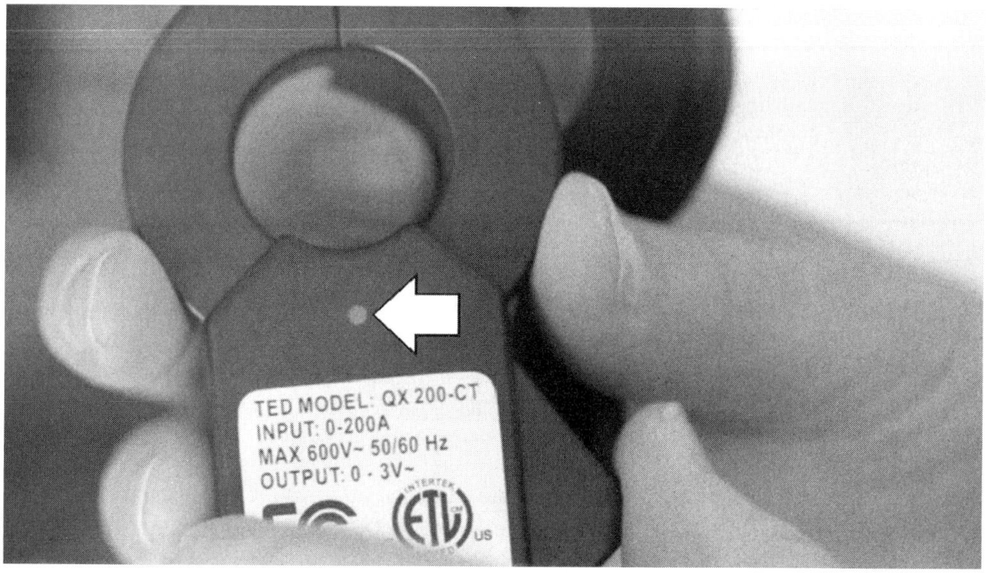

3. Identify the polarity marker on the CT. This marker must be on the side of the CT that is toward the main breaker when the CT is installed.

4. Clip one CT completely over one of the incoming lines of the electric service. The CT must completely encircle the line, but it does not need to grip the line snugly. It can be loose on the line as long as it is installed around it and with the polarity mark toward the main breaker. Do not touch the hot lug to which the incoming line is connected.

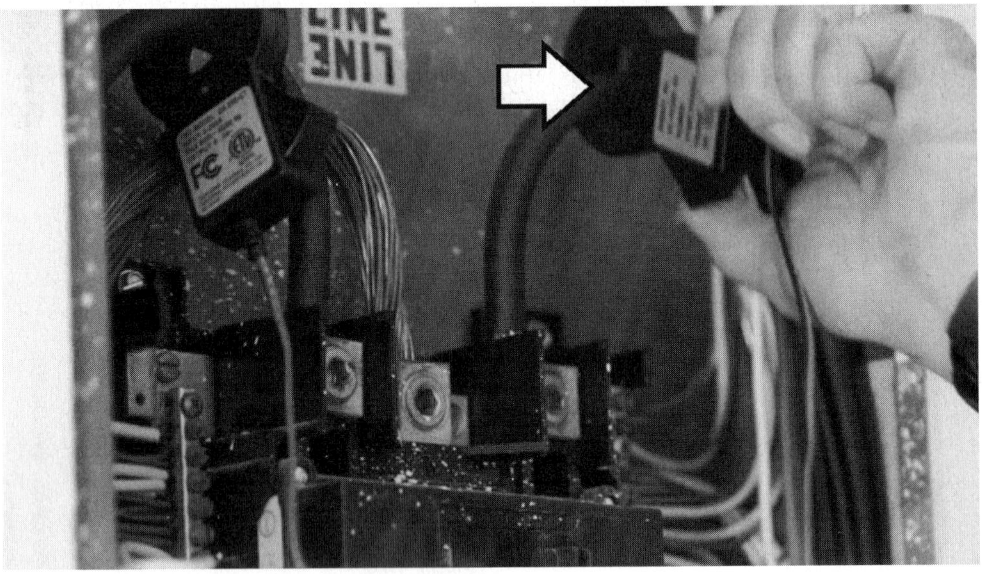

5. Connect the second CT in the same manner, again with the polarity marker facing toward the main breaker in the direction of current flow.

6. Connect the white lead of the measuring transmitting unit (MTU) to any open position on the neutral bus inside the service panel. Again be careful not to touch any hot lug on the incoming lines.

7. Select adjacent "A" phase and "B" phase circuits to which you will connect the red/black MTU leads; do not connect the black and red to the same circuit or circuits on the same phase. They must be on different phases, but you can use a circuit with a 15-, 20-, or 30-amp breaker for each lead.
8. The main breaker should be turned off already, but turn the breakers off on the circuits you selected as a precaution.
9. Connect the red and black leads from the MTU power cable to the selected "A" and "B" phase circuit breakers.

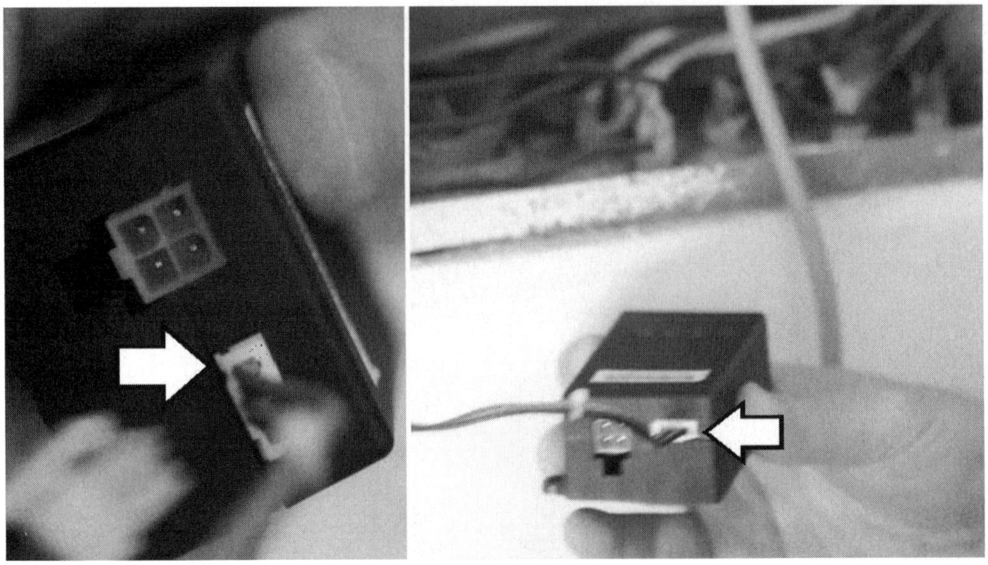

10. Connect the MTU cable (with red, black, and white leads) to the MTU. This is a snap-in connection and the plug will only fit one way. Don't force it. If it doesn't slip in easily, turn it over and try the other way.

11. Connect the power cable base to the MTU. There is only one way it can be connected. Line up the connector and snap it in to the MTU.

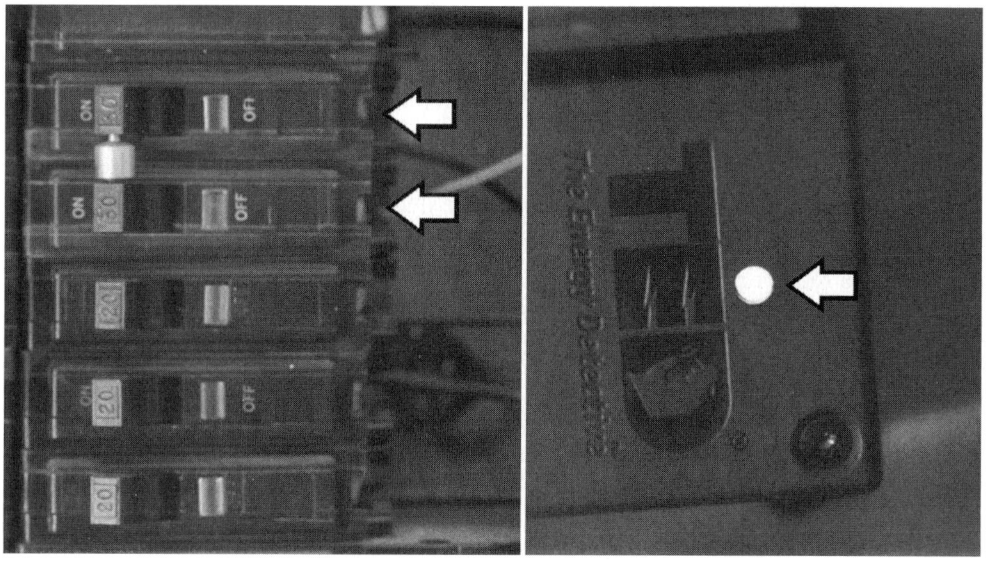

12. After all the connections are made and the MTU is positioned inside the service panel, turn on the main breaker and the branch circuit breakers. The green LED on the MTU should light up to show that it is working correctly.

13. After the CTs and MTU are installed, the next step is to plug the gateway that receives the MTU's data into a wall outlet near the computer or IHD that will display the data. The data is then received by the gateway from the AC wiring it is plugged into and relayed to the computer or IHD using a regular network (RJ45) connection or wireless technology.

The CT installation described above is for the type of MTU that communicates with a gateway or an IHD using power-line technology. The connections to the two circuits (one “A” phase and one “B” phase) are to permit data transmission over the home’s AC wiring. If the MTU uses wireless communication technology (Wi-Fi or ZigBee), no connections to the AC wiring are needed and the MTU transmits directly to the IHD or gateway.

For wireless transmission of data, it may be necessary to install the MTU outside the electric service box in order to avoid interference or signal-strength loss. If outside installation is required, the setup will be similar to that shown in Figure 2-8. The CT connections to the MTU are run through an approved bushing in the side of the box. The MTU is wall mounted near the service box.

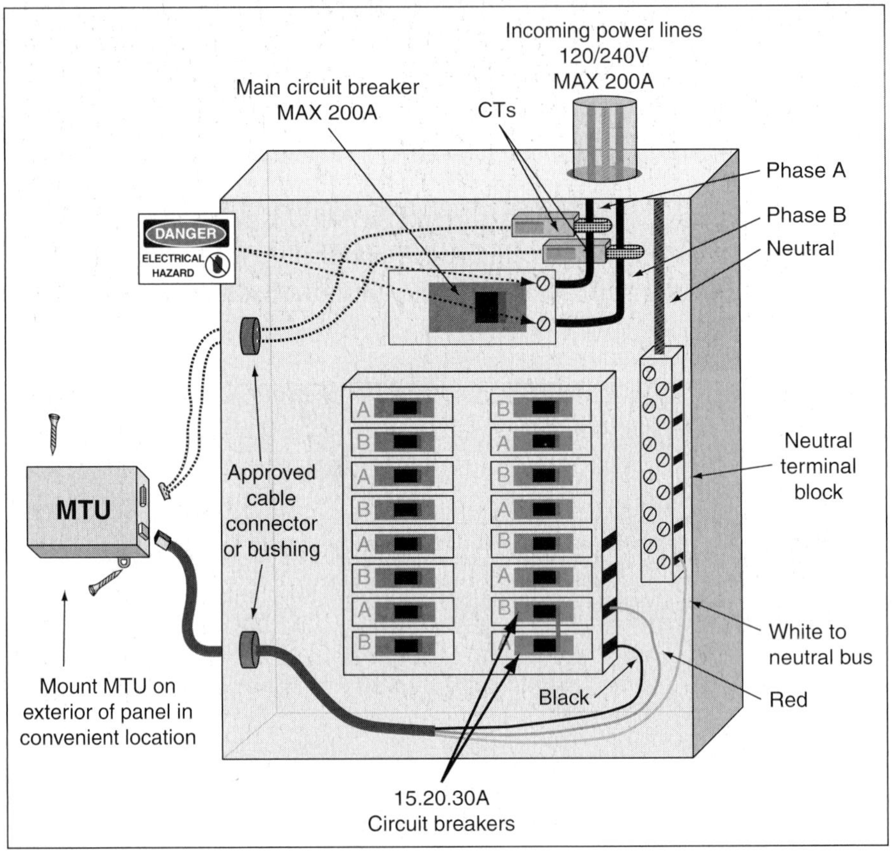

FIGURE 2-8

Diagram of a CT installation with the MTU located outside the electric service panel.

Once the CTs are installed and the MTU is working properly, the electric service panel faceplate should be replaced in position. Before doing this, it is a good idea to test the gateway and IHD to be sure they are receiving and displaying data correctly.

SUMMARY

Analog electric meters measure the flow of current over a given period of time. Analog gas and water meters do the same for those commodities. These measurements are cumulative and utility companies can only use them to charge for the total quantity of energy or water used. Smart meters measure electric gas or water use continuously and send their measuring data in real time to the utility companies. This enables them to charge for the amount used and for when it was used, in peak- or low-demand times. An in-home display (IHD) enables a consumer to read and analyze smart meter data to reduce electric, gas, or water consumption and make a home more energy efficient. Some IHDs are made to work with specific smart meters that use proprietary communication systems. Others can be programmed to work with a variety of smart meters. Still others do not require a smart meter at all, but can receive their measurement data from current transformers (CTs) placed inside a home or business electric service box and connected to a measurement transmitting unit (MTU). Installing CTs for an in-home display can be done by a homeowner in areas where local codes permit, but otherwise a licensed electrician should be used. Some MTUs can also connect to Internet sites that will display and analyze their data.

KEY TERMS

10BaseT RJ-45 Ethernet connection A rectangular connector on either end of an eight-wire cable commonly used to connect components (nodes) of a local area network (LAN) to one another.

Advanced Metering Infrastructure (AMI) The wireless communication network that allows two-way communication between smart meters, utility companies, and consumers.

Analog meter A meter that measures cumulative electricity, gas, or water use mechanically or electromechanically.

Base load The electric load in a home or business that is always in use by devices that are never turned off.

Circuit breaker An electrical device that allows only a set amount of current to pass through it. If the limit is exceeded, the breaker cuts off the entire flow.

Constant rate charge Charging for energy or water use at the same rate no matter how much is used or when it is used.

Controllers Devices for managing the operation of another device remotely, either through a wired connection or wirelessly; devices that send commands to control modules in an automation system.

Cubic foot (Ft^3) A unit of measurement used for natural gas and water. Equal to a block of water or gas 1 foot square and 1 foot high at a pressure of 30 inches of mercury (approximately 14.7 psi) and a temperature of 60 degrees Fahrenheit.

Current transformer (CT) A device that measures electrical use without being a part of the circuit it is measuring; a small measurement current is induced in the CT by the passage of the larger current being measured.

Dedicated radio frequency (RF) network A wireless network for connecting smart meters to a utility company. Can be a private network or part of a public network.

Demand-rate charge Charging for energy or water use according to the volume used and the time during which it is used; also called demand charge.

Digital smart meter A meter that measures electricity, gas, or water usage electronically and sends a constant digital signal of its measurement to a utility company.

Disturbance (electrical) Any variation in the voltage, amperage, or current characteristics in an electrical circuit.

Electric meter A device for measuring the amount of electric current flowing in a circuit.

Electric service box The metal box that contains the electric service panel and into which all AC wiring in a home or business is connected.

Electric service panel The panel inside an electric service box that contains the main and branch circuit breakers and the neutral bus.

Gas meter A device for measuring the flow of natural gas through a pipe in cubic feet.

Gateway An electronic device that receives data from a measuring device and relays the data to a display or recording device.

Google PowerMeter portal/website A free website where consumers can register to have their electric smart meter data or their monitored electric power use sent for display and analysis using a graphic display on a home computer.

Heating, ventilating, and air-conditioning (HVAC) system The integrated electromechanical system that supplies heating and cooling to a home or business. HVAC systems often use gas or oil in addition to electric power.

electric power.Infrastructure The integrated components of the delivery system in a utility network. In a power grid it includes transmission lines, distribution lines, transformers, meters, and so forth.

In-home display (IHD) An electronic device for displaying electric power use in real time and over designated time periods for analysis.

Interruption (electrical) The complete loss of voltage and current flow in a circuit for any amount of time.

IP-based public wireless network An RF network that is available for public use, such as Wi-Fi, mobile phone networks, LANs, and so forth.

Kilowatt hour (Kwh) A measure of electricity used; equal to 1000 watts of electricity used for 1 hour. It is the usual unit on which electric charges are based (for example, $0.08 per kWh).

Light-emitting diode (LED) A light source consisting of a solid-state diode made of a material that emits light when a current is applied to one side of its circuit; more efficient than an incandescent light.

Main breaker The circuit breaker that controls all incoming electricity to a home or business; if the breaker is tripped, all power is cut off.

Measuring transmitting unit (MTU) A device that receives measurement data from a meter and transmits it to a display or recording device.

Mesh network A wireless network composed of many nodes spread over a geographic area, each of which can send and receive data from those nodes near it.

Meter reader A person who physically visits utility meters to record their readings for billing by the utility. Not needed in areas where digital meters are installed.

Net meter An electric meter that measures current fed back into the electric grid as well as current being used by the consumer; the utility then charges only for the "net" electricity used.

Neutral bus The neutral (neither positive nor negative voltage) line on an electric service panel to which the neutral (white) wires of each circuit are connected.

Peak load The maximum amount of electricity a home or business can require; also the maximum amount of natural gas or water a consumer can require.

Polarity The direction of current flow in an electrical circuit; current flows from the positive pole of a circuit toward the negative pole.

Power-line technology (PLT) A communication and control system in which data signals are transmitted on AC wiring using a different frequency and voltage than the regular AC current flowing in the circuits.

PowerPortal Home display An in-home display made by Comverges Corporation and used in conjunction with smart meters also manufactured by Comverges.

Real-time data transmission The transmission of data at or near the same time as the data are created or measured.

Smart meter A meter that transmits a continuous digital data signal on a radio frequency that communicates how much current is flowing through it.

Two-phase current 240-volt AC current that is distributed to consumers in the form of two 120-volt "legs," each delivered through a separate wire and useable individually (120-volt circuit) or as a combined pair (240-volt circuit).

Water meter A device for measuring the flow of water in cubic feet through a pipe.

Wi-Fi A wireless radio frequency technology that permits high-speed data and voice communication throughout a smart grid.

Wireless MBus (sub 1-GHz) A wireless radio frequency technology that permits high-speed data and voice communication throughout a smart grid.

ZigBee (802.15.4 2.4 GHz) A wireless language that is used to connect different control and monitoring devices to work together; based on the IEEE 802.15.4 physical radio standard that operates at 2.4 GHz (global), 915 Mhz (Americas), and 868 Mhz (Europe).

REVIEW QUESTIONS

1. How does an analog electric meter work?
2. What measurement does an analog gas or water meter provide?
3. How are the data from analog meters (electric, gas, and water) transmitted to the utility companies?
4. What is constant-charge billing?
5. What are the major differences between smart meters and analog meters?
6. How does radio frequency (RF) mesh technology facilitate communication to and from smart meters?
7. What is the difference between a net meter and a standard smart meter?
8. What is peak-load capacity as applied to an electric generating facility?
9. What is peak-demand billing for electricity?
10. In addition to measuring electricity use continuously over time, what other advantages do electric smart meters have over the analog type?
11. What is AMI, and how does it improve the electric grid?
12. What is an IHD, and what is its main function?
13. What is ZigBee, and how is it used?
14. What is a CT, and how is it used with an IHD?
15. How does a power-line IHD communicate with the measuring device in the electric service box?
16. What is a PowerPortal Home Display, and how does it differ from a TED 5000 IHD?
17. What is the function of the main breaker in a home, and where is it located?
18. What is a neutral bus, and where is it located?
19. What is polarity, and how do you determine the polarity of a circuit?
20. What is an MTU, and what is its function?

CHAPTER

3 DEVELOPING A HOME ENERGY-EFFICIENCY PLAN

OBJECTIVES

After studying this chapter, you should be able to:

- Describe the major segments of home electric energy use and variations that occur from climate and other factors.
- Read and analyze electricity consumption data from an IHD.
- Describe base-load power and variable-load power and what devices contribute to each.
- Develop a basic plan to improve home energy efficiency and reduce total electricity consumption.

OUTLINE

I. Introduction
II. Home Energy Use
III. Installing and Setting Up an IHD
IV. Reading and Analyzing Data on an IHD
  A. Utility Partnerships with Google PowerMeter
  B. Power Meter Partnerships with Google PowerMeter
V. Base-Load Power
  A. Continuously Operating Appliances
  B. Regularly Operating Appliances
VI. Variable-Load Power
VII. Increasing Home Energy Use Efficiency—A Two-Part Approach
  A. Reduce the Total Amount of Energy Used
  B. Manage the Timing of Energy Use
VIII. Summary

INTRODUCTION

In this chapter you will learn about the factors that cause variations in the amount of power used by homes in various parts of the country. You will learn how to read and interpret an in-home display (IHD) screen to analyze the amount of power used in your home and develop methods of increasing home energy efficiency while decreasing the overall demand for power. You will be able to distinguish between base-power load and variable-power load and the electrical devices that contribute to each. And you will evaluate how best to minimize cost for electric power while still benefiting from its convenience and work-saving ability.

HOME ENERGY USE

ENERGY CONSUMPTION IN the average home is made up of several major segments. The largest of these is either heating or cooling of the home. Which one of these is larger depends on the climate where the home is located. In areas that have cold winters and hot summers, the cost may be about evenly divided between heating energy and air-conditioning energy. Warmer climates will have lower heating bills and higher costs for air conditioning. Colder climates will be the opposite.

An average home on the U.S. West Coast (California, Oregon, or Washington) consumes 35% less energy for heating and cooling than a home in Arizona, New Mexico, or most other areas of the country. The Pacific coastal areas have generally mild temperatures both winter and summer. This reduces the need for both home heating and air conditioning. On the other hand, the warm, humid climates of the South Central and South Atlantic regions require more electricity for air conditioning, whereas the cold winters in the Northeast and North Central regions result in much higher consumption of natural gas and heating oil for heating.

Nationally, the average residential use of energy of all types (electricity, heating oil, natural gas, coal), as measured by the cost of the energy, is divided into the following major categories by approximate percentage used:

- 34% for space heating the interior of the home
- 15% for heating water used in the home
- 14% for lighting inside and outside
- 13% for air conditioning the interior of the home
- 9% for refrigerator and freezer operation
- 6% for electronics (computers systems, entertainment centers, game sets, etc.)
- 6% for washing and drying (clothes and dishes, mostly for clothes dryers)
- 3% for other (pumps, workshop equipment, pool equipment, exterior heating, etc.)

In Canada the heating percentage is somewhat higher and air conditioning lower due to the more northern climate, but otherwise the percentages of energy use are similar.

These national averages are expressed as percentages and the amount of energy used per household varies widely in different parts of the country. Figure 3-1 shows the average energy use for various parts of the United States during the year 2000. Energy use has increased since then, but the different consumption rates per household in each area have remained constant.

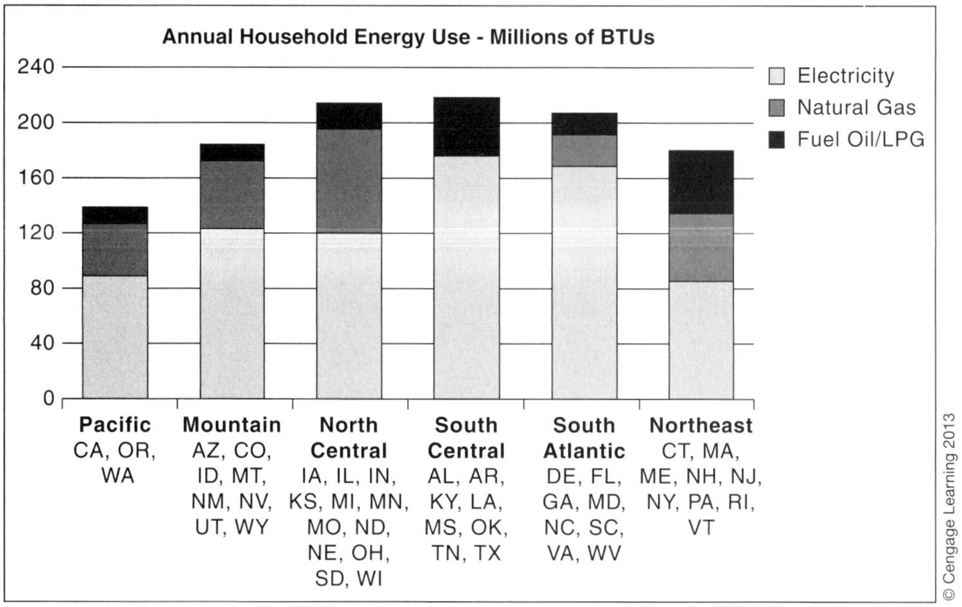

FIGURE 3-1

Household energy consumption varies considerably across regions of the country.

Major U.S. cities also show variations in per capita electrical consumption that are far in excess of the differences in total energy use found in various regions of the country. Climate variations alone cannot account for the fact that Dallas electric utility customers each use 16,116 kWh of power per year, more than three times the 4696 kWh used by New York City residents. Similarly San Francisco residents' 6753 kWh per capita electrical consumption is less than half that of Houston or Phoenix residents.

FIGURE 3-2

Electrical power consumption per residential customer in six major U.S. cities shows over 300% differences between high- and low-use figures.

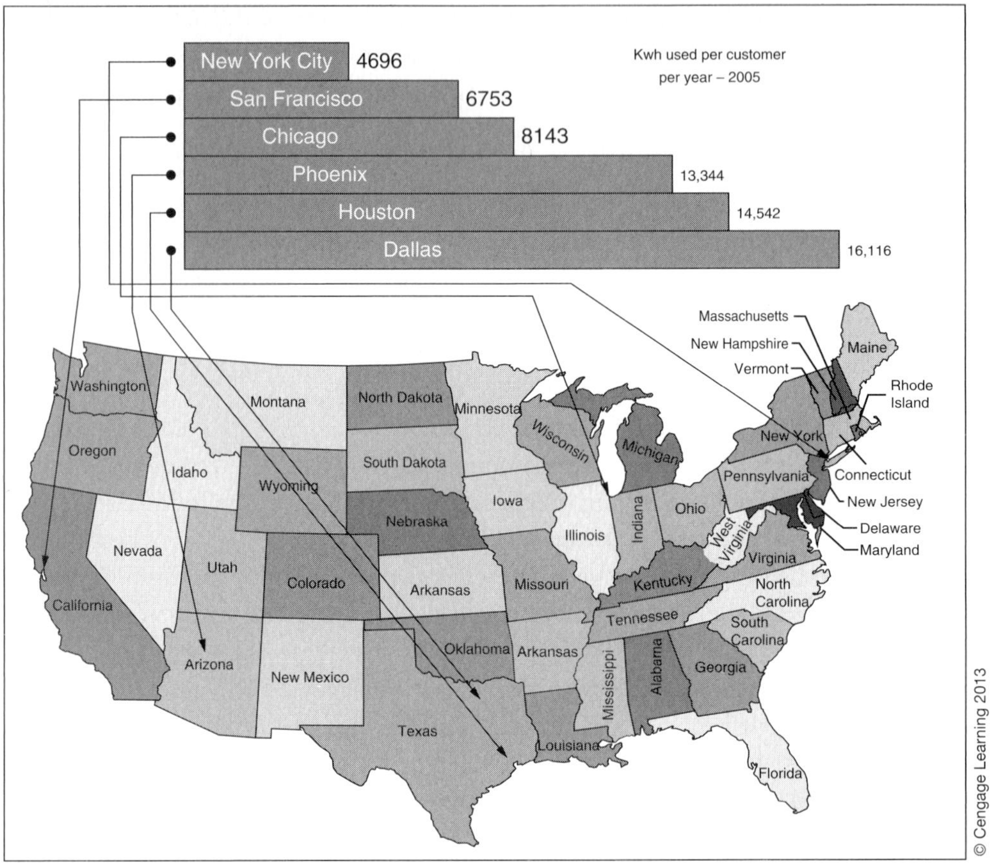

Figure 3-2 shows the magnitude of these differences, which can only be explained by considering several other factors in residential living.

In addition to climate factors, another cause of regional differences is the variety of building codes and environmental regulations that exist at the state and local level. California has stronger environmental protection laws than most other states. Its cities also have building codes that require residential construction to meet high standards of seismic resistance and energy efficiency. These factors may contribute to its low per-household energy consumption, which is below that of all other states except Hawaii.

Other factors that influence electrical use per household include:

- Building code standards that require energy-efficient design and construction, thereby reducing demand for energy to heat and/or cool the home.
- Size of individual homes: single-family residences in Houston, Dallas, and Phoenix are, on average, larger than those in New York, San Francisco, or Chicago, where most people live in apartment units or townhouses.
- Building design: multiple-unit buildings are inherently more energy efficient because only the outer walls of these structures are subject to heat transfer (loss or gain). The interior walls between units are heat-neutral barriers, whereas all four sides and the roof of a single-family dwelling are subject to heat transfer.
- The population density of a city: the more residential units are placed in a given area, the less heating and cooling energy per unit is wasted. Economies

of scale apply in several ways: a single furnace or air-conditioning system may serve a hundred residences and do so with half or less of the ductwork needed for an equal number of single-family homes with individual heating and cooling.

Regardless of where your home is located, the amount of electrical energy it uses will probably continue to increase as new electrical and electronic devices are added to the ones currently operating. In 2007, the total U.S. summer peak demand for electricity was about 783 gigawatts (GW) and winter peak demand was about 640 GW. By 2017, the North American Electric Reliability Corporation (NERC) projects summer peak demand will be about 925 GW and winter peak demand about 756 GW.

Cost per kWh for electricity is also expected to rise, so even if your residential use remains unchanged, you will probably pay more for your electric power in the future. The only way to prevent or at least slow the pace of this increasing cost is to make your home more energy efficient—that is, make a smaller amount of electric power do the same amount of work in your home. To do this, you must first learn how power is currently being used in your home and then develop ways to reduce the amount being used while not sacrificing any of the benefits that electric power provides.

INSTALLING AND SETTING UP AN IHD

Once your home has a smart meter installed or you have installed current transformers (CTs) and a measurement transmission unit (MTU) in your electric service box, you are ready to install and set up an in-home display (IHD) to read and analyze the data these devices supply.

The following steps describe the installation and setup of the TED 5000 IHD that is manufactured specifically to receive data from the CTs and MTU, whose installation and setup were described in the previous chapter. Another brand of IHD may be required to work correctly with the smart meter or CTs installed in your home, but the basic installation procedure is similar for most of them. Before beginning installation, be sure you have the correct IHD to receive the transmission of data from your smart meter or MTU.

The CTs described in the previous chapter send data to an MTU that uses powerline technology to transmit its data to a receiving IHD. The CTs are wired directly to the MTU, which in turn is connected by wire to the home's AC electric wiring. Two wires from the MTU connect through a circuit breaker to each of the two 110-volt sides of the 240-volt electric service in the home. A connection to each side of the AC power is necessary in order to be certain that the signal traveling on the AC wiring is available at any outlet in the home.

The MTU receives the raw information from the CTs and converts it into data packets, sets of binary digits that contain the data from the CTs and can be read by an IHD or a computer. These data are sent from the MTU to the home's AC wiring and are distributed throughout the home. An IHD can then be set up anywhere in the home and, when connected to the home's AC wiring, can read the MTU's data.

To install and set up the IHD, follow these steps:

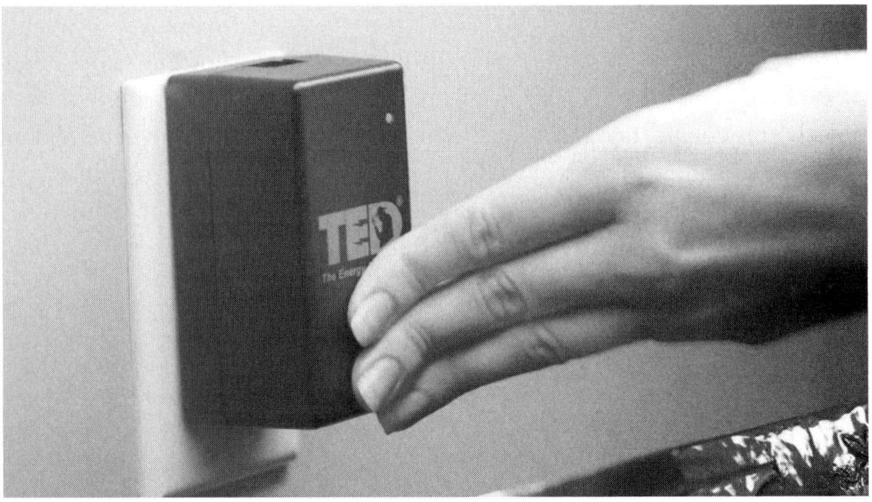

1. Plug the IHD's gateway (the device that reads the data packets transmitted on the AC wiring) into an AC outlet. To avoid interference with its reception, no other device should be plugged into the same outlet as the gateway and it should not be plugged into a power strip.

2. Plug the network cable into the gateway's RJ-45 port and connect the other end of the cable to a port in the home's wireless LAN router.

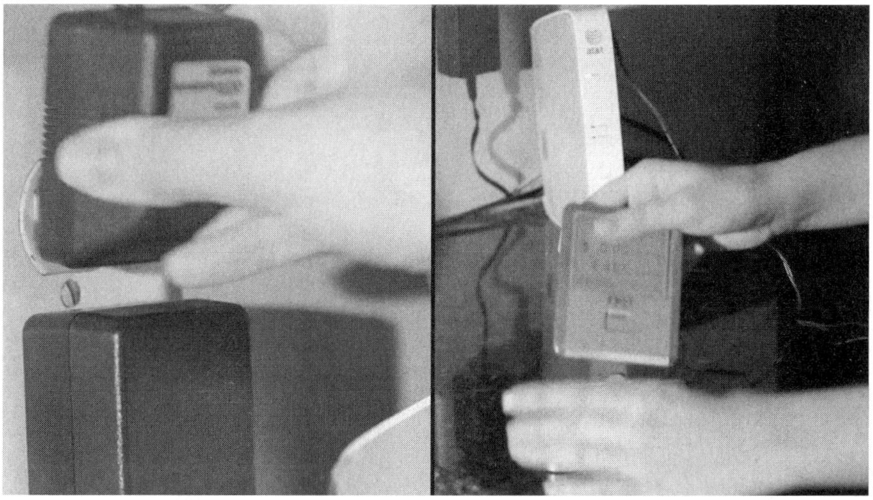

3. Plug the power adaptor of the wireless IHD's base unit into a convenient outlet. Place the IHD into the base unit and it will begin operating.

4. The setup software for the IHD can be found on a disk supplied with the unit or downloaded from The Energy Detective website at http://www.theenergydetective.com. Load the CD on your PC or download the setup program (for Mac or PC) from the website.

© Cengage Learning 2013

5. One you have the setup software installed on your computer, click the TED5000 icon on the desktop to open it. The program will display an opening window and immediately begin searching for the gateway. A few minutes may be needed to complete the search.

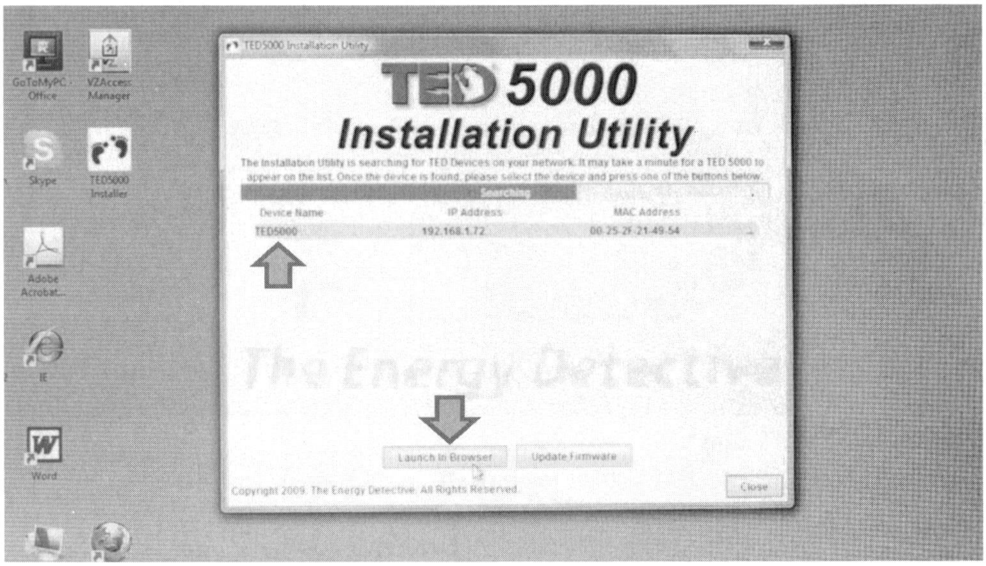

© Cengage Learning 2013

6. Once the program has found the browser, it will display the device on the screen. Click on the device to select it and then click on the "Launch in Browser" button at the bottom of the screen.

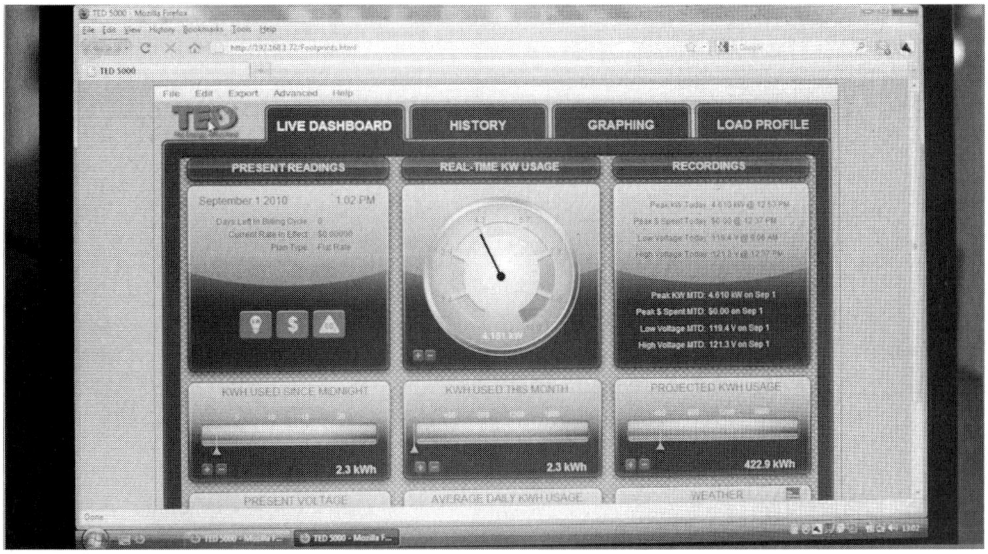

7. When the program is launched, the live dashboard display will appear on screen. Select the System Settings Wizard at the top left of the screen and then select each button at the top of the Wizard screen in order to set up the IHD.

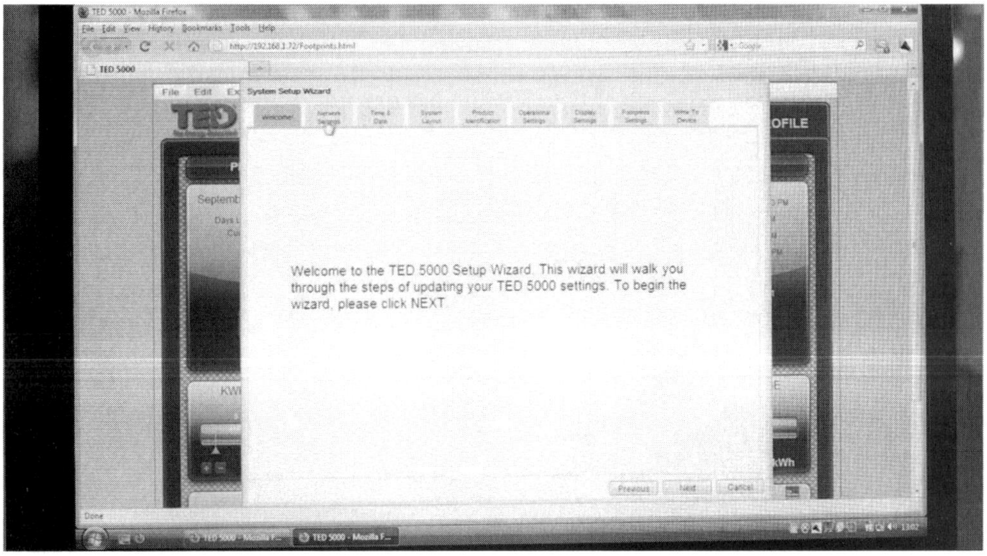

8. As the Wizard guides you through the setup process, you will find some data already entered on the screen. You should check any entered data to be sure they are correct and enter any additional data that are needed for the setup.
9. Among the data you will be asked to enter, or that will be displayed on the screen for you to confirm, are:
   a. The network settings
   b. The correct date and time
   c. The number of MTUs that are supplying data to the IHD
   d. The six-digit product ID numbers for the gateway, the IHD, and the MTU
   e. Display settings

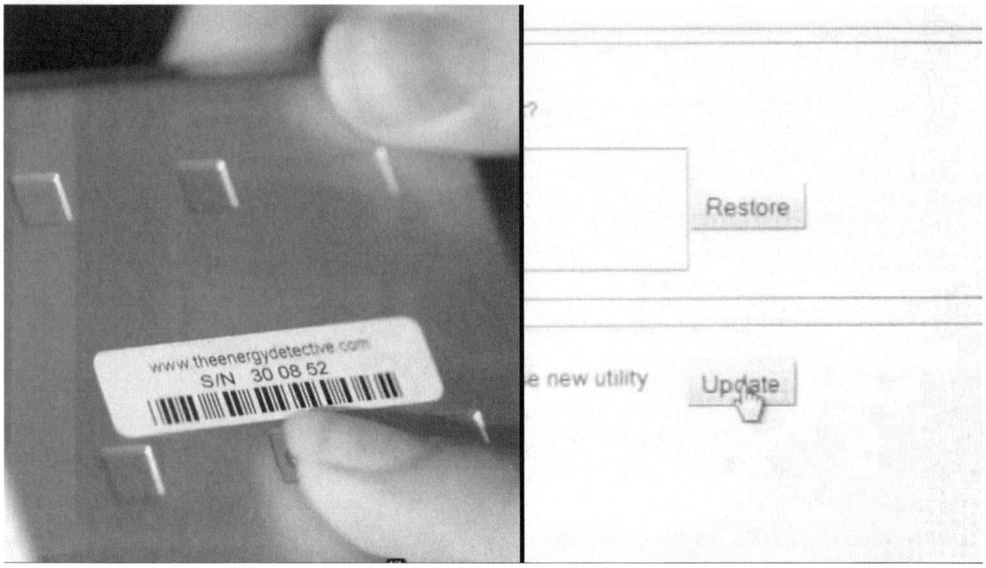

10. If the correct product ID numbers for the gateway, IHD, and MTU are not already entered in the settings, enter them manually from the labels attached to each product.
11. When all setup data have been entered, click the Update button and wait until the program has updated all data before clicking the Finish button to complete the setup.

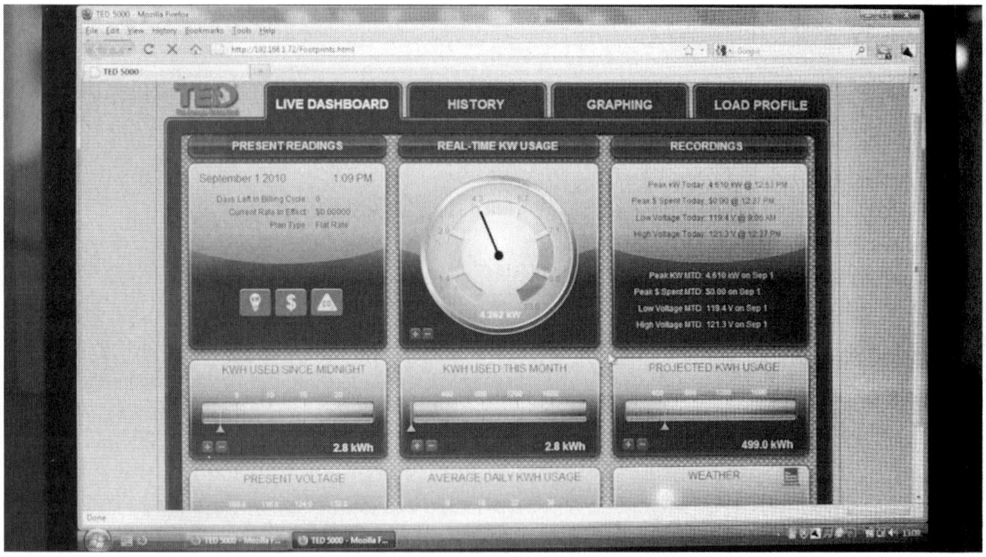

12. When setup is complete, the dashboard screen will be populated with data that updates every few seconds as new information is received from the MTU through the gateway.

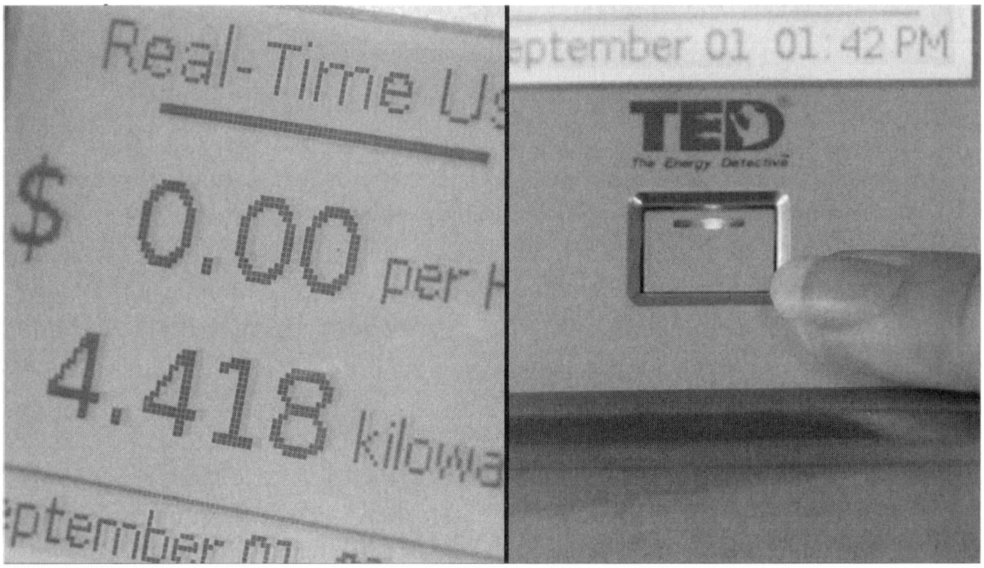

13. Real-time information will be displayed and frequently updated on the IHD. The blinking light at the bottom of the display shows it is receiving data from the gateway.

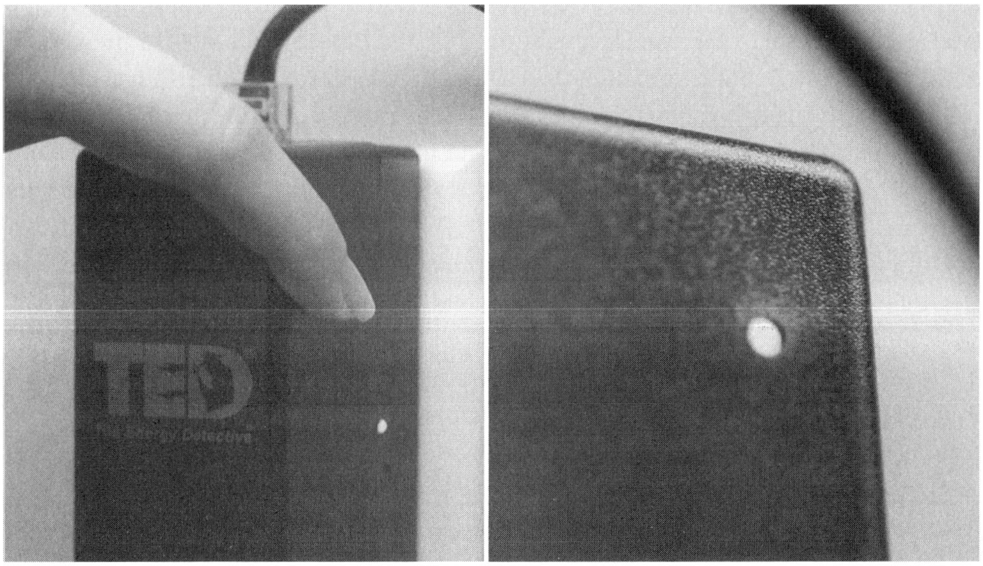

14. The light on the side of the gateway will blink green when it is receiving data from the MTU. The light on the front of the gateway will blink green if it is properly communicating with the wireless display.

READING AND ANALYZING DATA ON AN IHD

The IHD you use to display the data from an electric smart meter or from converter transformers (CTs) installed in the home's service panel will be either a unit made to function with a specific brand of smart meter or CT or a universal display made to function with a variety of smart meter brands and CT installations. Not all IHDs display their data in graph form; some will only display current usage and cost-per-hour figures that the IHD calculates from meter data using its built-in software. To realize the most efficiency from smart meter data, you need to see and analyze electrical use over short and long time periods (days, weeks, and months).

Data from many types of smart meters and CTs, including some that don't display on other IHDs in graph format, can be uploaded on the Internet to the Google PowerMeter website, where users can establish a free account and use the online software to analyze their home energy use. Google PowerMeter is currently the most widely used graphic display for energy use analysis, and we will use data displays from this site to illustrate IHD analysis in this chapter.

Utility Partnerships with Google PowerMeter

As of January 2011, four major utility companies had formed partnerships with Google PowerMeter to display their smart meter data on the PowerMeter site. In the United States, these electric utilities include San Diego Gas & Electric, Blue Ridge Electric, TXU Energy, Wisconsin Public Service, and White River Valley Electric Cooperative, and in Canada, Toronto Hydro-Electric System Limited. In Germany, Yello Strom is on board the PowerMeter site, and in the United Kingdom, first:utility (sic) smart meters can connect to it.

Other utilities are negotiating to make their smart meter output data available on Google PowerMeter. If your utility is not currently listed on the PowerMeter website, you should check periodically to see if it has been added.

To connect any smart meter installed by one of the above utilities to the Google PowerMeter site, the user must proceed by logging on to his or her residential electrical account with the utility and then following the directions on that site to link the individual smart meter's data output to display on Google PowerMeter. The service is free to any customer who has a smart meter installed.

Power Meter Partnerships with Google PowerMeter

Google PowerMeter also displays data from a number of other power-measuring devices. Devices made by the following companies for use in the United States and Canada are currently supported: Blueline, Brultech, Current Cost, Digi, eGauge, PowerHouse Dynamics, The Energy Detective (TED), WattsUp?, and Wattvision.

These companies' products all connect to PowerMeter through a USB connection from an individual power-measuring device to a PC or Mac and from the computer via the Internet to the Google PowerMeter website. Users of power-measuring devices (as opposed to utility-supplied smart meters) must connect to PowerMeter by establishing a free account on the website and then linking their equipment to the account for display on their Google homepage.

Figure 3-3 shows a Google PowerMeter-style graph of residential power use over a 48-hour period. The darker shaded portions at the bottom of the graph represent base-load power use for each of the 2 days. The lighter, varied portions of the graph show the increased amount of electricity used at various times over the 2-day period. These portions of power use are referred to as variable-load power. The highest points in the graph are the peak loads for each of the 2 days. This graph represents only 2 days of energy data and cannot be taken to represent the true base load or the highest peak load for the residence over a longer period of time. For that information, more data and some analysis is needed, as noted in the next section.

The graph in Figure 3-3 shows low electric power usage during the nighttime hours, with use rising in the daylight hours to peaks around 9 p.m. on Friday and 6 p.m. on Saturday. This is a typical daily graph for a home with central air conditioning located in a warm climate during the summer months. In winter, electric power use would be somewhat less because the air conditioning would not be in use and the electric fan motor used to circulate heat in the home during that season would not consume as much power by itself as would the air-conditioning compressor and fan motor combined.

Other seasonal variations in electric power use may show up in graphs for homes located in different climate zones. Many factors may affect both base-load and peak-load readings on the graph of a home's power use over a year's time. To get an accurate

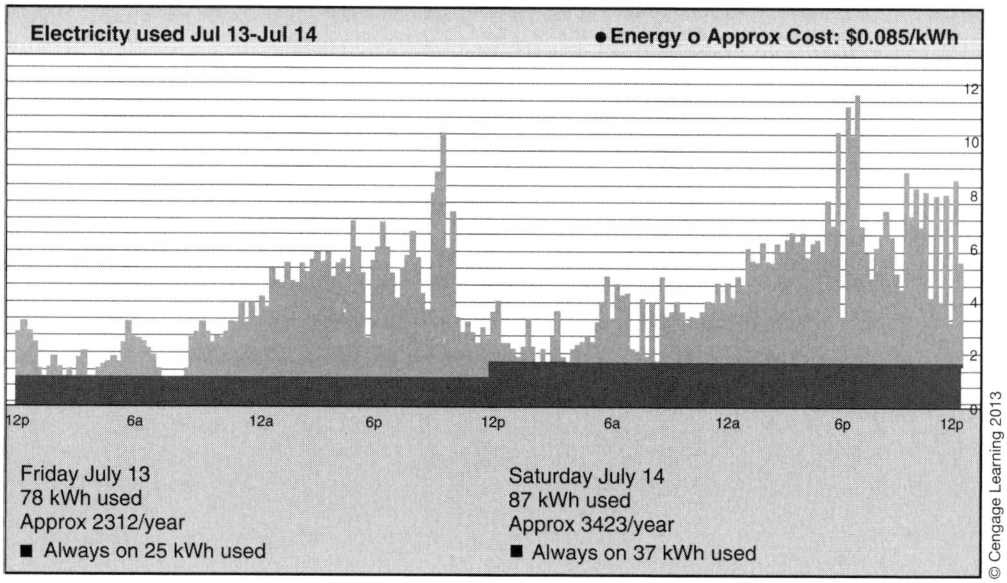

FIGURE 3-3

A GoogleMeter-style chart shows electrical power use over a 2-day period. The darker shading represents base load and the lighter shading variable and peak load.

picture of power use in your home, you should analyze graphs for at least a 30-day period. You should also review use graphs for days during different months throughout the year, especially if your area is subject to wide variations in climate as the seasons change.

Some things to look for in your home electricity use graphs include the following:

- What is the base-load requirement for the home? Does it vary much from day to day? Week to week?
- What times of the day is electric use at or near the base-load level?
- What is happening in the home during base-load times?
- What is the highest peak-load requirement during any day for a month?
- What time of day does peak load occur? Is it the same time each day?
- Is there more than one peak load or near-peak load occurrence each day?
- What is happening in the home during peak-load times?

With at least a month's worth of data about the above questions, you can start to develop ways of improving your home's energy-use efficiency. First, however, you should know a little more about what base load and peak load really mean.

BASE-LOAD POWER

The base load of a home's power requirement is the amount of electricity that is used continuously day and night every day of the year. Power demand almost never drops below the base load unless there is a power interruption, and even then, the base-load demand still exists; it just can't be met while the power is off.

Base-load demand may be only 20% of the peak-load demand in a home, but it can easily account for 50% of the home's total power use. This seeming contradiction is explained by looking at the power graph in Figure 3-4.

In the graph, the base power load is 2 kW and the peak power load goes as high as 12 kW, though only once during the day and for a short period of time. The 2-kW base load means that electric appliances within the home are using 2000 watts (2 kW) of electricity at all times. During the day some appliances may be turned off and others turned on, but the minimum demand for electricity at all times during the day is 2000 watts. This constant base-load power demand of 2000 watts consumes 2 kilowatt hours (kWh) of electricity each hour (2000 watts × 1 hour = 2 kWh). A constant 2 kWh used each hour of the day consumes 48 kWh of power in 24 hours.

Varying levels of additional demand during the day range from a low of about 250 watts (0.25 kW) during the early morning hours to a high of 10,000 watts (10 kW) at the 6 p.m. hour of the evening. The average additional demand over the entire day's time is only about 2.4 kW, and over the 24-hour period an additional 2.4 kilowatt hours (kWh) of power are required each hour to meet this varying demand. Total power consumption during the day to meet varying demand is 2.4 kWh × 24 hours, or 52.5 kWh.

Because the base load is continuous, the amount of power it consumes over time is large even though the amount consumed at any given time may be only a quarter or less of peak demand consumption. For this reason, any reduction in base load in a

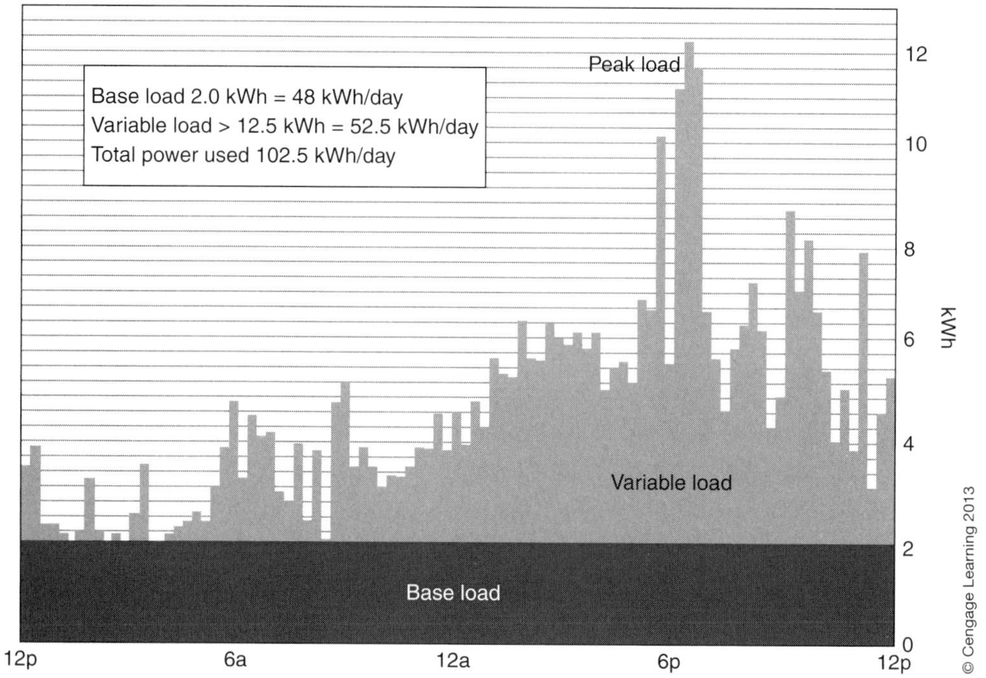

FIGURE 3-4

A day's power use in a typical home shows that almost 50% of the power used was base-load power.

home's power use will have a more significant impact on power costs than a similar-sized reduction in peak demand. This is true even though the charge for peak-demand power may be double or even triple that for base-load power. Unfortunately, because base-load power is used for essential tasks, it is by far the most difficult portion of power demand to reduce.

The base load of a home's power use graph is composed of two main groups of power-consuming devices: electrical appliances that run continuously around the clock every day of the year, and appliances that operate regularly but not continuously.

Continuously Operating Appliances

Not very many electrical appliances in the average home run continuously every day of the year, and most of those that do are not individually large power consumers. As a group, however, they can easily add up to 1000 watts of power being used every hour of the year. Continuously running appliances include:

- Clocks and timing devices that run on AC power
- Transformers that alter voltage and are left plugged in when not in use
- Converters that change AC power to DC for electronics appliances and are left plugged in when not in use
- Communications equipment such as rechargeable phones, routers, and so on
- Internal warmers that keep television sets and other displays ready for instant use

- Sensor circuits such as security systems, garage door openers, and so on
- Electronic equipment with "sleep" mode circuitry that allows the unit to mostly shut down, but still "awaken" almost instantly

To this group must be added some other electric items that we often choose to run continuously for convenience or because starting them each time we want to use them is too time consuming. These include:

- Computers and peripheral equipment such as printers, scanners, and so on
- Entertainment systems such as game players, stereos, and so on
- Decorative lights and other lights left on continuously by choice both inside and outside the home

Each of these devices adds only a trickle to the base load of electric power used in the home, but together they make up a significant stream. Their importance is increased in the energy-efficiency equation because they are almost never reduced in number, but rather increased as the home acquires new technology.

Regularly Operating Appliances

The rest of the home's base load is made up of appliances that run regularly but not continuously. These are usually larger devices in terms of the power they require and they are usually activated by thermostats or other sensors when needed. They include:

- Air-conditioning systems
- Refrigerator and freezer compressors
- Furnace motors and ventilation fans
- Water heaters
- Heat pumps
- Automated lighting
- Sump pumps

These do not operate simultaneously, but on average, at least one or more of them will be running at any given time during a day. The average continuous power required to keep them functioning on their various schedules forms the balance of the residence's base power load.

VARIABLE-LOAD POWER

Simply stated, all electrical demand that isn't base-load demand (that is, power required in addition to base load) is variable-load power. This includes all electrical devices that are used intermittently and controlled by switches. It also includes some of the power used by appliances listed under base-load power, to the extent that these may at times require power above the base-load minimum.

The list of electrical appliances that contribute to variable load in the home is long. We are becoming ever more dependent on the convenience and utility of electricity to make our lives easier and more pleasant. The major electrical devices that demand large amounts of electric power include:

- Electric ranges, cookers, and microwave ovens
- Water heaters
- Space heaters

- Air conditioners
- Electric grills
- Toasters
- Clothes dryers
- Dishwashers
- Refrigerators and freezers

You'll notice that all of these appliances have heating elements or cooling compressors. Heating and cooling functions are the most intense users of electric power in the home. Some of the appliances listed above will draw well over 1000 watts of current when operating, and even the smaller ones frequently require 800 watts of power or more. All of them perform essential services in the home and several of them operating at the same time will usually account for the peak-load demand during any given day.

To perform their individual functions, several major heating and cooling appliances may work in opposition to one another, thus making the home less energy efficient. Take, for example, a refrigerator and a freezer operating in an air-conditioned home during the hot days of summer. Both appliances use refrigeration units driven by electric motors to extract heat from their interiors and keep them at the required low temperatures. But the excess heat removed is then vented directly into the kitchen, where the appliances are located. This heat warms the room and makes the air-conditioning system work harder (consuming more power) to keep the room temperature at a comfortable level. A cooking appliance such as a stove, grill, or toaster also adds heat to the kitchen that the air-conditioning system must then overcome. Similarly, a hair dryer can easily raise the temperature in a bathroom by 15 degrees or more, again making the air conditioning work harder to overcome the excess heat.

Appliances working in opposition to one another is only one of the inefficiencies in home electric power use. In later chapters you will learn specific ways to reduce the amount of electric energy needed for residential power and especially how to bring down the home's peak demand for power. Peak-demand power is the most expensive segment of power for utilities to produce, and they are increasingly charging more for it. Reducing peak demand is therefore not only an energy-saving objective, but a money-saving option as well.

INCREASING HOME ENERGY USE EFFICIENCY—A TWO-PART APPROACH

Once you have an IHD graphic picture of your home's electric power use, you can develop a plan to make your home more energy efficient.

Reduce the Total Amount of Energy Used

Reducing energy use means using less electric power over a given time period than has been used over a similar time period in the past. There are three ways to do this:

- Turn off some electrical appliances that have formerly run continuously for all or part of each day.

- Replace electric appliances with new ones that do the same job but require less electricity to do it.
- Reduce the amount of energy required to heat and cool the home by reducing heat transfer in an out of the home, and by managing excess heat and cold within the home.

Turn Off When Not in Use

Turning off electrical devices that formerly ran continuously means, in many cases, accepting a little less convenience in the speed at which these appliances operate at startup. Switching computers and peripherals completely off when not in use requires connecting the system through a power strip or outlet that has a cutoff switch. The switch cuts off all power to each unit, including the "sleep" circuit of the computer and the warming circuits of printers, scanners, and other peripherals that otherwise keep them ready for instant use. When power is turned back on (and each unit restarted, if necessary) anywhere from 10 seconds to a minute may be required for the system to reboot and be ready for use.

Power savings from cutoff switches on continuously running equipment are not large, but they add up over time. Small devices such as hard drives, routers, modems, and the like usually draw a constant 4 or 5 watts when not in use. Printers, fax machines, copiers, and scanners, which have warming circuits, consume 5 to 15 watts in sleep mode. Flat screens, televisions, and other displays can use more than 50 watts to maintain instant display readiness, depending on their type and size.

An energy audit of all the electrical devices in the home, as described in a later chapter, will tell you exactly how much energy you can save by turning each appliance off, but a typical example illustrates the possible power savings. A computer system consisting of a desktop processor, flat screen, printer, scanner/copier, external hard drive, wireless router, and gateway modem draws about 55 watts of power when in sleep mode. If the system is shut down during the night hours of 11:00 p.m. to 6:00 a.m. and during the weekday hours of 9:00 a.m. to 3:00 p.m. when family members are away from home at school or work, its sleep mode power use will be cut by over 50% (82 hours in sleep mode each week compared to the "always on" mode of 168 hours). Instead of consuming 8.4 kWh of electricity per week, it will only consume 4.1 kWh, a savings at $0.104 per kWh of about $0.43 per week or $1.85 per month.

Similar power-use reductions and cost savings can be realized by switching off other "always on" appliances, but only to the extent that those who use them are willing to wait a few moments at startup time for the equipment to reach full function.

Replace Inefficient Appliances

As the cost of energy has gone up, manufacturers have responded by making their products that use energy more efficient. This trend is taking place in all types of products that use energy, but it is especially evident in cars. A 1955 Chrysler New Yorker sedan weighed more that 5900 pounds and typically traveled about 16 miles on a gallon of fuel in city driving. A 2010 Chrysler Sebring sedan weighs a little over 3500 pounds and travels about 24 miles on a gallon of fuel in city driving. The improved efficiency of the newer car means that it requires 33% less fuel to power the newer car than the older one. Despite the increased price of fuel between 1955 and 2010, fuel costs to drive the

1955 car were about 1.5 times higher than fuel costs to drive the 2010 car ($0.15 per mile versus $0.10 per mile, both figures in 2010 dollars).

People rarely drive 1955 cars today, but there are a large number of refrigerators, freezers, and other electrically powered appliances from that era still in use. And they're relatively just as inefficient compared to today's models as the 1955 car is to its present version.

Refrigeration systems, electronic circuits, drive units, and visual displays have all benefited from new technology developments in recent years, and much of the improved design of these devices has resulted in better performance using less electricity. Any electrical device more than 15 years old is probably less efficient than a current model. Such appliances may require anywhere from 10% to 50% more electricity to operate than a new model.

Replacing major appliances is expensive and will usually need to be done one item at a time as funds are available. During the energy audit described in a later chapter, you will be able to determine which older appliances are the most costly to operate in comparison with new ones. The least-efficient equipment should be replaced first.

Manage the Timing of Energy Use

As previously noted, peak-load power is the most expensive to produce, and consequently costs the most for consumers to use. Lowering a home's peak-load demand is a sure means of reducing overall energy costs.

Peak-load demand can best be lowered by changing the times for activities that use large amounts of power. If these activities can be moved to off-peak hours, a substantial savings can be realized. Activities that require large amounts of electricity that can be moved to off-peak hours include:

- Washing and drying clothes
- Dishwashing
- Cooking
- Refrigerator/freezer operation
- Air conditioning
- Heating

If the timing of any of these activities is moved to off-peak hours, peak demand can be reduced by 1000 watts or more. Most easily moved are the first two activities on the list. Moving these to early morning or late night requires nothing more than a little advance planning. Most washers, dryers, and dishwashers have timers that can be set to start their operational cycles at any time. This allows the user to load the washer or dryer when convenient and then start their operation during off-peak hours. Limiting washing and drying to the hours after 9:00 p.m. and before 10:00 a.m. is a good way to drop peak demand, which usually occurs between 2:00 p.m. and 8:00 p.m.

Changing cooking times to late evening or early morning hours may be more difficult for some meals, but an alternate solution can help. Most cooking requires only an hour or less for a given meal. If it needs to be done during peak demand time, the same reduction in demand can be achieved by turning off the air conditioning while the stove or other cooking appliance is operating. This can be done manually or by means of a controller that automatically blocks the air conditioning or furnace from operating while the stove is on.

Finally, refrigerators and freezers can be timed so that their compressors do not operate during peak demand times. Turning off a freezer for up to 6 hours at a time and a refrigerator for up to 3 hours will not significantly raise the inside temperature in either provided they are not opened frequently during the off time. Again, a little advance planning will enable the user to remove needed items from refrigerator units so as to avoid opening them when they are not operating.

Timing heavy-power-use activities for off-peak hours will definitely save money in areas where utilities use demand pricing to bill for electricity, but it won't reduce the home's overall power use. To do that requires increasing the home's energy efficiency, and the beginning step in that process is an energy audit, as described in the next chapter.

SUMMARY

Analog electric meters measure the flow of current over a given period of time. Analog gas and water meters do the same for those commodities. These measurements are cumulative and utility companies can only use them to charge for the total quantity of energy or water used. Smart meters measure electric gas or water use continuously and send their measuring data in real time to the utilities. This enables them to charge for the amount used and for when it was used, in peak- or low-demand times. An in-home display (IHD) enables a consumer to read and analyze smart meter data to reduce electric, gas, or water consumption and make a home more energy efficient. Some IHDs are made to work with specific smart meters that use proprietary communication systems. Others can be programmed to work with a variety of smart meters. Still others do not require a smart meter at all, but can receive their measurement data from current transformers (CTs) placed inside a home or business electric service box and connected to a measurement transmitting unit (MTU). Installing CTs for an in-home display can be done by a homeowner in areas where local codes permit, but otherwise a licensed electrician should be used. Some MTUs can also connect to Internet sites that will display and analyze their data.

KEY TERMS

Always-on mode The state of an electric appliance in which it always has current running through its circuitry so that it can commence operation instantly without any startup delay.

Base-load power The portion of home or business electric power consumption that is used continuously at all hours of the day and night; base-load demand is the minimum power required at any given time.

Building code The laws or ordinances that govern the construction of residential and commercial buildings so that they meet standards of safety, quality, and energy efficiency.

Continuously operating appliance An electric device that is always working, such as an electric clock or timer; not the same as a device in always-on mode, which may have power connected but not be actually operating.

Cutoff switch A switch used to shut off electricity to continuously running devices or those in always-on mode.

Demand pricing A method of charging for electricity in which power used during high-demand times is billed at a higher rate than that used during low-demand times.

Economies of scale The axiom that larger and more complex buildings and structures are more energy efficient on a per-unit basis than smaller ones. This rule also applies in many other areas.

Energy consumption The total energy in any form (electricity, fuel, gravity, wind, etc.) used in a home or building.

Energy efficiency The ability of a building or device to perform its intended function using the minimum amount of energy; the less energy consumed over time, the more energy efficient the building or device is.

Energy used per household The total amount of energy in all forms used by a household over a given period of time.

Gigawatt A measure of electric power equal to 1000 megawatts (mW). A megawatt is equal to 1000 kilowatts (1000 kW).

Google PowerMeter portal/website A free website where consumers can register to have their electric smart meter data or their monitored electric power use sent for display and analysis using a graphic display on a home computer.

Heat loss The loss of heat energy from a home or other building caused by radiation from the exterior walls and leakage of warm air from the interior.

Heat transfer The passage of heat through a material from its warmer surface to its colder surface or from one material to another; also the passage of heat through the walls, roof, and floors of a home or other building in either direction.

Household energy consumption The total amount of energy in any form used by a home during a given time period.

Kilowatt hour (kWh) A measure of electricity used; equal to 1000 watts of electricity used for 1 hour. It is the usual unit on which electric charges are based (for example, $0.08 per kWh).

Megawatt (mW) A measure of electric power equal to 1000 kilowatts (kW). A kilowatt is equal to 1000 watts (1000 W).

Off-peak hours Those hours during a day when electric power demand is not as high as during the 2-hour period of highest demand for the day.

Peak demand The period during a given day when the largest amount of electric power is required to serve a home or community.

Per capita electrical consumption A measure of the average amount of electricity consumed by each person in a given community; calculated by dividing total energy used by the number of people in the community.

Per-household energy consumption A measure of the average amount of energy consumed by each household in a given community; calculated by dividing total energy used by the number of homes in the community.

Power meter partnerships with Google PowerMeter A means of displaying electric power use using data from a home power meter that is transmitted to the Google PowerMeter site and then displayed on a home computer.

Power strip A multiple-outlet power extension cord that is frequently used to connect components of electronic systems to AC power. Can be used as a cutoff switch for the system, if equipped with a master switch for all outlets.

Regularly operating appliances An electric or electronic device that operates regularly (at least once per day) but not continuously; its operation is usually controlled by a sensor or timer.

Residential electrical account The account a homeowner sets up with an electric utility company to obtain electric power in return for a specified charge per unit used.

Sleep mode A state of an electric device in which its operation is suspended, but its circuits are still powered and ready for instant use on demand.

Timer A device used to turn an electric appliance on or off at set times during the day or over a longer time period.

USB connection Universal Service Bus; a connection port and/or matching plug used to connect electronic devices so that data or low-voltage power can be transferred between them.

Utility partnerships Agreements made between electric, gas, and water utility companies and manufacturers of smart meters and other similar devices; the agreements allow the partners to work together to provide services to the customers of both.

Variable-load power The portion of electric power used by a home or commercial building that exceeds the base power load for that structure.

REVIEW QUESTIONS

1. What is the largest energy consumer in most homes?
2. Why does a home in the Pacific region of the country (California, Oregon, Washington) use 30% less energy for heating and air conditioning than homes in other areas of the country?
3. After the largest energy user, what are three other major energy users in the typical home?
4. What factors besides climate variations affect the annual amount of energy used per capita?
5. The peak demand for electric power across the country is less in winter than in summer. Why is this true?
6. What two major factors make a home more energy efficient?
7. What is Google PowerMeter?
8. Name three manufacturers of power-measuring devices for use in the United States and Canada that are currently supported by Google PowerMeter.
9. True or False: The PowerMeter Web site can be displayed on a PC or a Mac computer.
10. What is base-load power in a residence?
11. Which of the following contribute to base-load power?
    a. Toaster
    b. Electric clock
    c. Space heater
    d. Television set with instant-on feature
    e. Dishwasher
12. What is peak-power load in a home?
13. When does peak demand for power usually occur during the average day?
14. If electricity is billed by demand pricing, how can the homeowner reduce power cost without reducing total power used?
15. True of False: Reducing base-power load in a home will always reduce cost, regardless of the method of charging for electricity.
16. Name three types of electrical devices that usually contribute to base load in a home.
17. What are the major electrical appliances that make up variable-power demand in a home?
18. What are some ways to reduce electric power use?
19. What are some electrical appliances whose use can be partially or completely moved to off-peak hours?
20. True or false: Moving power consumption to off-peak hours will always save the homeowner money on electricity bills.

CHAPTER

4

CONDUCTING A HOME ENERGY AUDIT

OBJECTIVES

After studying this chapter, you should be able to:

- Conduct an electrical energy audit to determine the amount of electrical energy being used by the electrical devices and appliances in the home.
- Determine the amount of gas, oil, and other fuel energy being used in the home.
- Detect air leaks that permit heat transfer into and out of the home.
- Determine the amount of heat transfer into and out of the building through walls, ceilings, and windows.

OUTLINE

I. Introduction
II. Conduct an Energy Use Audit
  A. Measure the Electrical Energy Used in the Home
  B. Determine the Gas Fuel Energy Used in the Home
  C. Determine the Oil, Coal, or Other Fuel Energy Used in the Home
III. Conduct an Energy Loss Audit
  A. Determine Air Leaks in the Home
  B. Determine Heat Transfer into Building from Exterior
  C. Determine Heat Loss through Walls and Ceilings
IV. A Professional Energy Audit Report
V. Summary

INTRODUCTION

In this chapter you will learn how to conduct an energy audit of your home to determine where energy is being used and where it is being lost. You will learn how to measure the actual power use of electrical appliances and devices, both when they are operating and when they are "sleeping." You will also learn how to calculate the amount of gas, heating oil, or other fuel energy the home is using. You will use simple techniques to determine the location of air leaks that permit heat to transfer into or out of the home. You will also learn how to calculate the amount of heat transfer that is occurring through the walls, ceilings, and windows of the home and whether that heat loss or gain can be reduced.

CONDUCT AN ENERGY USE AUDIT

IN THE PREVIOUS chapter you learned how to analyze the total electrical power used in your home. An energy audit is a survey that is designed to tell where your home is using energy and where it is losing energy. The first part will tell you how much electricity is being used by each electrical device in your home and how much energy from other fuels such as gas, oil, and coal is being used. The second part will tell you how much of the energy used to heat or cool the home is being wasted through heat transfer in or out of the building.

Measure the Electrical Energy Used in the Home

Start your energy audit by making two charts listing everything electrical in the home. The first chart will be for electrical appliances and devices and the second will list every light fixture inside and outside the home.

Create an Electrical Appliance Chart

On the electrical appliance chart, show the location of each appliance or device and leave spaces for its rated power and its actual power consumption. Include everything electrical, even if you think a device doesn't use much power. You may be surprised when you measure its actual electricity consumption.

If the device has a label that lists its power rating, put that figure in the rated power space. If it doesn't, leave the space blank; you'll measure the actual power use for each appliance later. Your electrical appliance list will look somewhat like the example in Table 4-1 below, but it will be much longer. You'll probably be surprised at the number of electrical devices you have in your home.

TABLE 4-1 Electrical Appliance Chart

LOCATION	ITEM	RATED POWER	ACTUAL POWER
Kitchen	Range		
Kitchen	Blender	200	
Kitchen	Mixer	100	
Kitchen	Refrigerator	300	
Kitchen	Dishwasher	400	

Don't forget to list appliances that are normally stored out of sight: mixers, hair dryers, electric tools, fans, vacuum cleaners, and so on. In addition to appliances that plug into wall outlets, be sure to also include all the devices that are hard-wired into the home's electrical system. These may include furnaces, space heaters, air conditioners, pumps, vent fans, dishwashers, trash compacters, door openers, security systems, and many others. Remember also that oil and gas furnaces and other appliances also use electric power to run motors and other devices that are part of these systems.

Make a Home Lighting Chart

Make a separate chart of all the lighting fixtures inside and outside the home. Include every fixture, its location, the number of light bulbs or tubes it contains, and the watt rating of each one, which is nearly always displayed on the bulb or tube. Don't take the watt rating of the fixture, which might say something like "60 watts maximum rating," or "maximum bulb size 60 watts," but read the watt rating on the actual bulbs that are in the fixture. It could be more or less than the fixture rating. If it is larger, you should replace the bulb with one that has a rating no higher than the fixture's maximum. But don't make the replacement just yet. In a later chapter we will explain how many light bulbs can be replaced with energy-efficient types that will provide more light while consuming less electricity. Your light chart will look like the example in Table 4-2, but like the appliance chart, it will be much longer. The average home contains 75 to 100 light bulbs of various sizes.

TABLE 4-2 ELECTRIC LIGHTING CHART

LOCATION	FIXTURE	NO. OF BULBS	WATTS PER BULB	TOTAL WATTS
Living room	Overhead—south	5	60 watts	300 watts
Living room	Overhead—north	5	60 watts	300 watts
Living room	Floor lamp	1	150 watts	150 watts
Dining room	Overhead—center	12	40	480 watts
Kitchen	Sink Light	2	40	80 watts
Kitchen	Stove light	2	25	50 watts

Light circuits that have manual or remotely operated dimmers built into them usually use the same amount of electricity as those that do not. Most dimmers function by adding resistance to the circuit to absorb some of the current flow so that the amount of current reaching the light bulb is reduced. This causes the bulb to illuminate more dimly, but the amount of current used by the bulb and the dimmer's variable resistor is about the same as the bulb alone would use if the resistor were not in the circuit.

Measure the Power Consumption of Home Appliances

To find the electrical consumption of each appliance in your home, you'll need a power meter, also known as a watt meter, a power analyzer, or an electricity monitor. A watt meter costs anywhere from $35 to $195 or more, depending on its specifications and quality. A popular one for home use is the WattsUp? Power Meter shown in Figure 4-1,

FIGURE 4-1

The WattsUp? Power Meter displays the power consumption and several other facts about an appliance's power use.

which costs about $149. It can accurately measure power use of any 120-volt appliance that uses up to 1800 watts of power.

In addition to measuring power use in an appliance, the WattsUp? Power Meter can show a number of other facts about power use, including:

- Current watts—the electricity currently being consumed by an appliance plugged into the power meter.
- Minimum watts—the minimum power consumed when an appliance is plugged in but not operating.
- Maximum watts—the power consumed when the load on an appliance is highest.
- Cumulative watt hours—the watt hours of electricity used since an appliance was plugged into the power meter.
- Average monthly kWh—the power that will be consumed in 30 days if the appliance plugged into the power meter is left operating at the present level.
- Cumulative cost—the cost to operate an appliance since it was plugged into the power meter.
- Average monthly cost—the cost to operate an appliance for 30 days as determined by the cost to operate it for a shorter time period (an hour, a day, or a week).

Measure the power consumption of each item on your list of electrical appliances. Start with those items that have an on-off switch and can be plugged into the power meter and then turned on. Plug the power meter into a wall outlet and then plug the appliance to be tested into the power meter.

Turn on the appliance for a few seconds and read the amount of power it is using directly from the meter. If the appliance has more than one operating speed, turn it to the highest speed to measure power use. Appliances with electric motors will consume a surge of power as the motor starts to turn, but this surge will only last a second or two until the motor reaches its regular operating speed. Once the device is at full operating speed, its power consumption will level out to a constant figure. Figure 4-2 shows the power being used by a blender as measured by the power meter.

Appliances should be measured when they are actually working rather than simply running. A mixer is doing more work and consumes more power when mixing a bowl of batter than when turning in air alone. A power drill must work harder when penetrating a board than when simply turning a bit. Hair dryers, electric irons, and other appliances with heating elements in them will use more heat, and therefore more power, when performing their intended tasks than when turned on but standing idle. You should measure each appliance under its normal working load whenever possible in order to get an accurate reading of the amount of power it uses.

If your power meter has the capability to calculate additional power use data, you may want to switch modes while each appliance is operating and see the amount of power it will consume over a period of time and the cost of operating it for a specified length of time. Follow the instructions provided with your power meter to display the additional data it is capable of showing.

You should measure the power use of multiple-unit electrical systems such as computer workstations and entertainment centers as a single unit since their individual components are always used together. To do this, connect each of the components in a system to a 6- or 12-outlet power strip and then plug the power strip into the power meter.

FIGURE 4-2

The power meter displays power consumption by an operating blender; the power being used is higher for blending a full cup of liquid than if the appliance were running empty.

Electronic systems should be measured for power use when their components are turned off or in "sleep" mode and again when the system is fully powered and operating. Nearly all electronic systems consume some power continuously, even when turned off. This continuous power use will be a fraction of the power used when the system is in full operation, but it may still add up to a substantial part of your total electric bill. You can find out how much it costs to run a system continuously by querying the power meter for a monthly cost while measuring the system's power use in "sleep" or "off" mode. Figure 4-3 shows a desktop computer system with some components that operate continuously, some that alternate between "sleep" mode and full operation, and some that are turned on only when needed for a specific task.

To measure power use for a refrigerator, freezer, or other appliance that operates in cycles, you will need to connect the appliance to the power meter for a longer period of time, preferably for a full 24-hour period, in order to get an accurate reading. A freezer may use almost no power when its refrigeration unit is not operating. The unit is controlled by a temperature sensor and cycles on whenever the temperature inside the freezer rises above the set minimum. It continues to run until the inside temperature has been brought down to the set level and then cycles off. Measuring the power use over a full day will give a good average of the unit's power consumption through several cooling cycles and dormant periods.

You will not be able to easily measure the power use of some electric appliances, either because they operate on 240-volt power, which most power meters cannot monitor, or because they are hard-wired into the home's electrical system and cannot be plugged into the power meter. Most of these appliances will carry a label that includes a maximum power rating. If you cannot find the label on the appliance, the power consumption information will also be included in the operating instructions that come

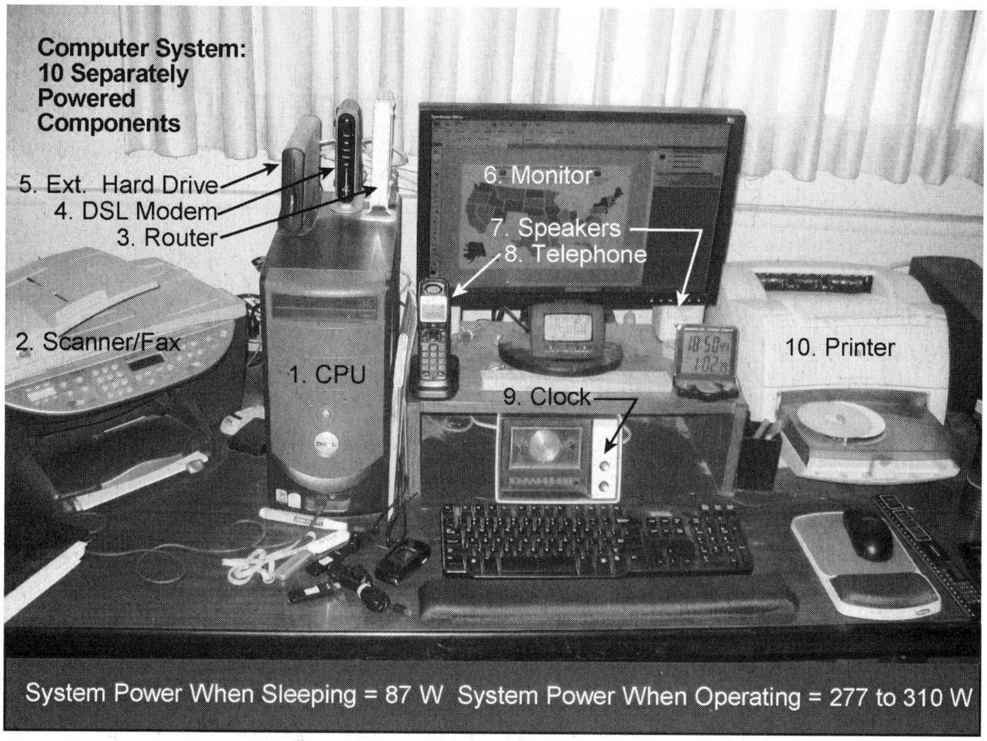

FIGURE 4-3

This composite photo shows the power used by the computer system's 10 independently powered components when in use and when in "sleep" mode.

with it. If these are unavailable, you can find the same information online by searching for the brand and model of the appliance.

Figure 4-4 shows the labels on a washer that draws 10.5 amps of current and an electric dryer that draws 24 to 26 amps of current. These figures can be converted into power consumption using Ohm's law and the Power Law.

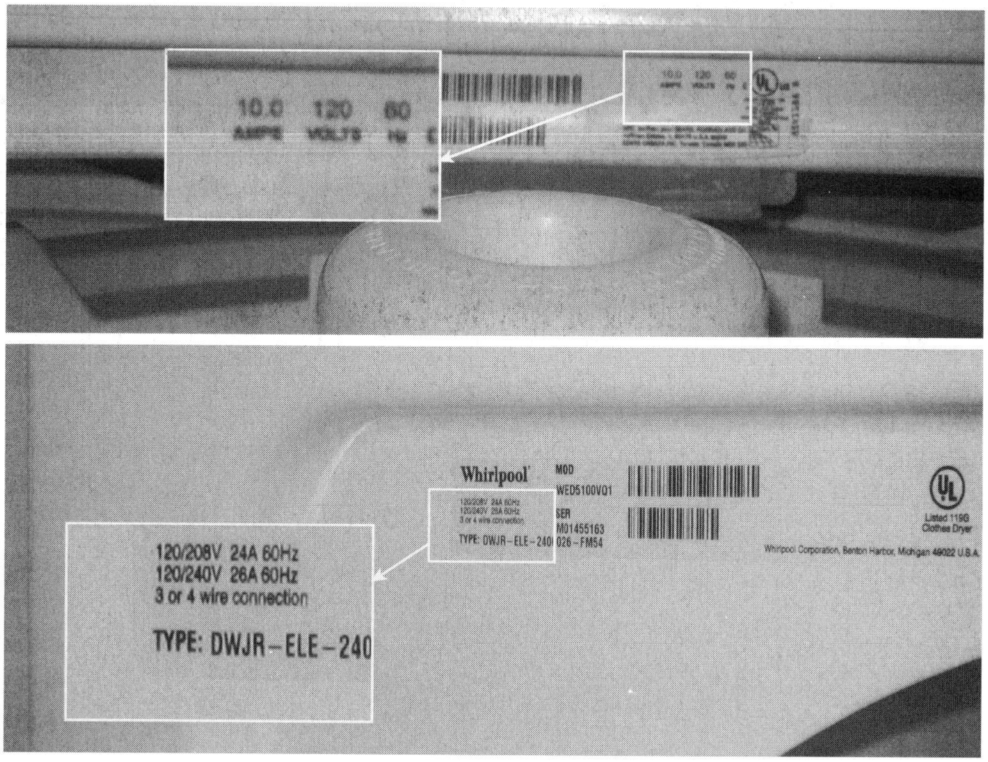

FIGURE 4-4

The labels on a typical washer (top) and dryer (bottom) showing the voltage and amperes each appliance uses. From this information power consumption can be easily calculated.

Ohm's law states that in any circuit, voltage (V) is equal to amperes (I) times the resistance (R) or

$$V = I \times R$$

The Power Law states that in any circuit, power (expressed in watts [W]) is equal to current (expressed in amperes [I]) times the voltage (expressed in volts [V]) or

$$W = I \times V$$

The washer label shows that it operates at 120 volts AC and draws 10.0 amps of current. Using the power formula above, we multiply the voltage times the current to obtain the power in watts that the appliance uses.

$$120 \text{ volts} \times 10.0 \text{ amps} = 1200 \text{ watts power}$$

When operating, this washer uses up to 1200 watts of power. If its operation cycle is 45 minutes in length, it will consume up to 0.90 kWh during each cycle. In actual operation, the washer will not use 1200 watts of power throughout its cycle, but only when its agitator and spin motor is running along with its pump motor. Over a complete cycle, its power consumption will average somewhat over half of the maximum rating if it is washing a full load and less if the load is smaller. Actual power use could vary still more depending on the settings used for individual wash loads.

Even so, a washer is an expensive appliance to run frequently. At an average power draw of 700 watts, it will consume 0.52 kWh of power during each 45-minute wash cycle. If it runs only 3 cycles per week, or 12 cycles per month, it will consume 6.24 kWh of power each month. At a cost of $0.119 per kWh (the U.S. average in 2010), it will cost $0.74 per month and $8.91 per year to operate.

Electric rates vary widely across the United States and are rising as the cost of producing electricity increases. Figure 4-5 shows the average residential rates in the various states as of October 2008. You can use the rate shown for your state to more accurately calculate the cost of using various electric appliances, but be aware that rates in your area may have risen again since these figures were compiled.

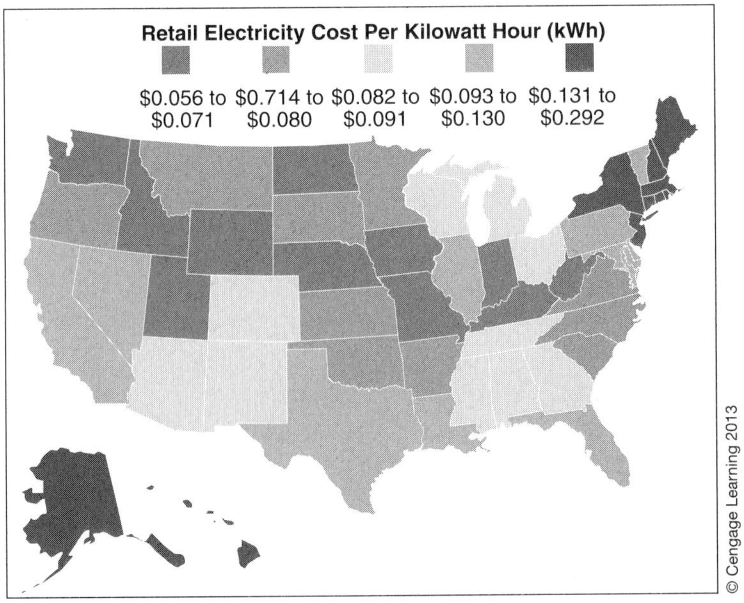

FIGURE 4-5

The various rates for retail electric power across the United States as noted by the U.S. Department of Energy for the year 2008.

The electric clothes dryer consumes much more power than the washer and is consequently more expensive to operate. The label shows that it requires 240 volts and up to 26 amps to operate at maximum capacity. Using the power formula, this calculates to a power draw of up to 6250 watts. Assuming the power draw over a 45-minute drying cycle to be only half of that amount or 3125 watts (the dryer's spin motor will run constantly, but the heating element cycles on and off as needed), the dryer will consume 2.34 kWh per 45-minute cycle. At $0.119 per kWh, that means that each load dried will cost about $0.28; 12 loads per month will cost $3.36, and over a year the cost will be $40.32.

Some typical label ratings for various appliances are listed below. Like the washer and dryer, the actual power consumption of these appliances will be less than the label maximum. Electric appliances that draw large amounts of current and run for extended time periods account for the majority of power used in the home. In the list they are noted in **bold** type, and methods of reducing the total amount of power each uses will be discussed in a later chapter.

Label or nameplate wattages for various household appliances:

- Aquarium = 50–1210 watts
- Clock radio = 10 watts
- Coffee maker = 900–1200 watts
- **Clothes washer = 350–500 watts**
- **Clothes dryer = 1800–5000 watts**
- **Dishwasher = 1200–2400 watts (using the drying feature greatly increases energy consumption)**
- **Dehumidifier = 785 watts**
- Electric blanket, single/double = 60/100 watts
- Fans
  - Ceiling = 65–175 watts
  - Window = 55–250 watts
  - **Furnace = 750 watts**
  - **Whole house = 240–750 watts**
- Hair dryer = 1200–1875 watts
- **Heater (portable) = 750–1500 watts**
- Clothes iron = 1000–1800 watts
- Microwave oven = 750–1100 watts
- Personal computer
  - CPU, awake/asleep = 120/30 watts or less
  - Monitor, awake/asleep = 150/30 watts or less
  - Laptop = 50 watts
- Radio (stereo) = 70–400 watts
- **Refrigerator (frost-free, 16 cubic feet) = 725 watts**
- Televisions
  - 19″ = 65–110 watts
  - 27″ = 113 watts
  - 36″ = 133 watts
  - 53″–61″ projection = 170 watts
  - Flat screen = 120 watts
- Toaster = 800–1400 watts

- Toaster oven = 1225 watts
- VCR/DVD = 17–21/20–25 watts
- Vacuum cleaner = 1000–1440 watts
- **Water heater (40 gallon) = 4500–5500 watts**
- **Water pump (deep well) = 250–1100 watts**
- Water bed (with heater, no cover) = 120–380 watts

When measuring the current drawn by an appliance with a motor, note that the meter will show about three times more current in the first second that the motor starts than when it is running smoothly.

Determine the Gas Fuel Energy Used in the Home

Natural gas is used in many homes to provide fuel for the furnace, water heater, stove, and clothes dryer. Residential gas use is measured in cubic feet by a meter and billed monthly to customers. Billing is usually based on a set cost per 100 cubic feet (CCF) or 1000 cubic feet (MCF) of gas used.

The CCF or MCF used may be converted into decatherms (DTH), a unit of heat measure that is equal to 1 million BTUs (British thermal units). Charges for the gas are then based on the number of DTH of heat value provided. The amount of heat produced by burning a cubic foot of gas varies according to the composition of the gas and any impurities it may contain. For this reason the BTUs produced by gas used in an area are determined periodically by burning tests. Figure 4-6 shows a monthly gas bill based on CCF used during a winter month. The 267 CCF used are converted into DTH using

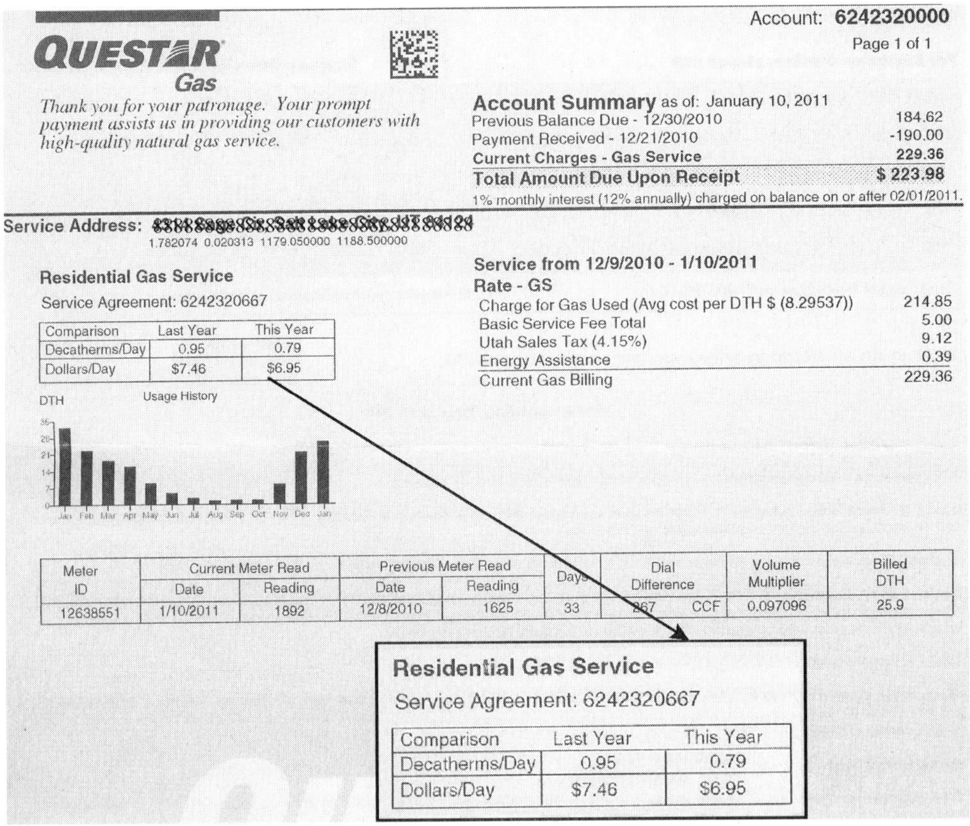

QUESTAR Gas

Account: 6242320000
Page 1 of 1

Thank you for your patronage. Your prompt payment assists us in providing our customers with high-quality natural gas service.

Account Summary as of: January 10, 2011

Previous Balance Due - 12/30/2010	184.62
Payment Received - 12/21/2010	-190.00
Current Charges - Gas Service	229.36
Total Amount Due Upon Receipt	**$ 223.98**

1% monthly interest (12% annually) charged on balance on or after 02/01/2011.

Service Address: [illegible]
1.782074 0.020313 1179.050000 1188.500000

Residential Gas Service
Service Agreement: 6242320667

Comparison	Last Year	This Year
Decatherms/Day	0.95	0.79
Dollars/Day	$7.46	$6.95

Service from 12/9/2010 - 1/10/2011
Rate - GS

Charge for Gas Used (Avg cost per DTH $ (8.29537))	214.85
Basic Service Fee Total	5.00
Utah Sales Tax (4.15%)	9.12
Energy Assistance	0.39
Current Gas Billing	229.36

DTH Usage History
Jan Feb Mar Apr May Jun Jul Aug Sep Oct Nov Dec Jan

Meter ID	Current Meter Read Date	Current Meter Read Reading	Previous Meter Read Date	Previous Meter Read Reading	Days	Dial Difference	Volume Multiplier	Billed DTH
12638551	1/10/2011	1892	12/8/2010	1625	33	267 CCF	0.097096	25.9

Residential Gas Service
Service Agreement: 6242320667

Comparison	Last Year	This Year
Decatherms/Day	0.95	0.79
Dollars/Day	$7.46	$6.95

FIGURE 4-6
A typical gas bill shows how natural gas is charged for based on the amount of heat it provides.

a multiplier that is based on the known BTU heat value of the gas. The bill is based on a charge of $8.29537 per DTH.

This bill shows a decrease in gas used compared to the previous year, even though weather during the current month was colder than the same month during the previous year. The decrease in gas used and cost is attributable to increased insulation installed in the home during the previous summer. The amount of gas used was 16.8% less during the current month, but the cost savings were only 6.8% over the previous year due to the increasing price of natural gas. This is an example of how increasing the energy efficiency of a home can more than offset rising energy costs. It also demonstrates that investment in smart grid technology is a wise practice.

Determine the Oil, Coal, or Other Fuel Energy Used in the Home

About 8.1 million U.S. homes are heated with oil. Over three-quarters of these are in the northeast part of the country (New England and Central Atlantic states). Oil-heated homes have fuel storage tanks and are not connected to a supplier by a pipeline. Homeowners buy heating oil by the gallon as needed, but the price for heating oil can vary widely within a single year depending on supply-and-demand factors.

Some homes are also heated with coal- or wood-burning furnaces, although the number of these is declining. Coal and wood are less expensive than heating oil, but also less convenient and reliable to use as a residential heating fuel. Some rough numbers will enable you to calculate the relative cost of heating with oil or coal as compared to natural gas.

One ton of coal is very roughly equal to 100 gallons of heating oil. A ton of coal usually produces about 15,000,000 BTU or 15 DTH of heat when burned, whereas 100 gallons of oil produces about 12,000,000 BTU or 12 DTH. Comparing an oil furnace with 80% heating efficiency and a coal-burning furnace with 70% heating efficiency, 100 gallons of oil for heating equals 0.91 tons of coal for heating. The cost of coal per ton is far less than that of heating oil, but both are typically more expensive than natural gas.

Homeowners with oil or coal heating systems can determine the annual operating cost of these systems from their fuel bills. Comparing them to natural gas heating costs is only a little more difficult. The gas-heated home just cited used about 26 DTH of heat during a midwinter month; 267 CCF of gas was used to produce that amount of heat. About 271 gallons of oil or 2.47 tons of coal would be required to produce the same amount of heat. At $2.50 per gallon, the heating oil would cost $677.50. At $150.00 per ton, the coal would cost $370.50. Comparing these costs with the cost of natural gas of $230.00 for heating the same home is revealing.

CONDUCT AN ENERGY LOSS AUDIT

The second part of your energy audit is to determine where energy is being wasted through heat transfer in and out of the home. The energy efficiency of a home is determined by how well it is air sealed (defined as the pressure boundary) and how well it is insulated (defined as the thermal boundary). These boundaries are the barriers between inside and outside air. Heat transfer occurs through these boundaries. Transfer via the pressure boundary is by air infiltrating into or out of the home. Transfer via the thermal boundary is by heat being conducted through wall, doors, windows, and other barriers.

During cold weather, heat can be lost from the home through air leaks that allow warm air to escape and cold air to enter the home. It can also be lost by transfer of heat through the solid walls, doors, and windows of the home to the outside. In summer months, cool air chilled by the air-conditioning (HVAC) system can also be lost through leaks. Heat transfer in summer moves heat in the opposite direction through the walls of the home into the cooled interior. This makes the HVAC system work harder and consume more energy to keep the home comfortable.

Determine Air Leaks in the Home

Make a list of obvious air leaks (drafts). The potential energy savings from reducing drafts in a home can range from 5% to 30% per year, and eliminating air leaks will also make the home more comfortable during both warm and cold weather. Look for indoor air leaks, such as gaps along the baseboard or edge of the flooring and at junctures of the walls and ceiling. Check to see if air can flow through or around these places:

- Electrical outlets
- Switch plates
- Window frames
- Baseboards
- Weather stripping around doors
- Fireplace dampers
- Attic hatches
- Wall- or window-mounted air conditioners

Small openings where air can leak out may be difficult to see, but they can be easily detected by means of a simple building pressurization test. To perform this test, you need to slightly increase or decrease the air pressure in your home relative to the outside air. This can be done using a fan that pushes air into the home, thus raising the inside pressure, or one that pulls air out of the home, thus lowering the inside pressure. The slight difference in air pressure in the home (above or below) compared to the outside will cause air to flow through any leaks.

To conduct a pressurization test, you must first close all exterior doors, windows, and fireplace flues. Next, turn off all combustion appliances such as gas-burning furnaces and water heaters, and clothes dryers.

Decreasing air pressure in the home temporarily is the easiest way to do the pressurization test and there are several ways to accomplish this. Try turning on all the exhaust fans that are located in the kitchen and bathrooms. If these fans are not enough by themselves to move enough air for leak detection, you can also turn on the clothes dryer on a setting without heat. If these fans still aren't enough, use a large fan placed in an open window to suck more air out of the house. Be sure to block any open portions of the window that the fan doesn't occupy with cardboard sheets and duct tape so that air pulled out by the fan isn't replaced by outside air coming in through the open window.

The decreased interior air pressure will increase infiltration of outside air through cracks and leaks, making them easier to detect. Use smoke from an incense stick or the flame of a lighted candle to detect points where air is coming into the home. Air flowing through openings will make the smoke or flame waver and move away from the wall as it passes. Wherever you see this movement, air is leaking into the home from outside.

If you have trouble finding the exact places where leaks are present, do the pressurization test in reverse and increase air pressure in the interior. Turn off all fans and the dryer. Turn the window fan around so that it becomes an inflow fan and blows outside air into the home to increase the inside air pressure. This will cause air to pass out of the house through any existing leaks. Use the incense stick again to search for these leaks. The smoke from the stick will be drawn directly into whatever leak is closest to it, as shown in Figure 4-7, thus pinpointing the leak's location.

FIGURE 4-7

Smoke from an incense stick is drawn into an air leak in the frame of a window.

Combustion furnaces, water heaters, and dryers, whether fueled by natural gas, oil or coal, all require a source of air to burn fuel efficiently and produce heat. The air for combustion should not come from inside the home, but from the exterior through a vent or flue that connects directly to the outside. If the vent for combustion air is not large enough to provide adequate fresh air for the appliance, the deficiency may cause incomplete burning of fuel. This is wasteful and can also result in poisonous carbon monoxide gas being produced by the furnace or water heater.

As a general rule, the vent for a combustion appliance should be at least 1 square inch in cross section for every 1000 BTUs that the appliance is rated to produce each hour it operates. An 80,000-BTU furnace, for example, should have an 80-square-inch vent, measuring 10 inches wide by 8 inches high.

Even with a correctly sized combustion air vent, a gas or oil furnace may pull some combustion air from inside the home rather than through the vent. This wastes energy by drawing heated air from the interior out of the home just as if the furnace flue were an outside leak. To prevent this, the furnace should be sealed off from the rest of the home's interior in the same manner as the interior is sealed from the exterior. All leaks in the walls and doorway of the furnace room should be plugged to prevent warm air from leaking out through the furnace and to prevent cold air coming into the furnace through the combustion vent from leaking into the home's interior. Gas-fired water heaters and dryers should also be closed off from the rest of the home so that their combustion air will come only from the outside, not from the home's interior.

When not in use, a fireplace can also remove warm air from the home's interior. If the damper does not close tightly, warm air will flow upward through the chimney in cold weather and wind may drive air downward through it into the home during any season. Even when a fire is burning in a standard fireplace, much of the heat produced by the flames is lost up the chimney and a great deal of warm air is also pulled from the room. The net heating effect of an open fireplace may be as little as 25% of the energy value of the fuel consumed.

A tight-fitting tempered-glass fireplace screen, such as the one shown in Figure 4-8, prevents warm air from passing up the chimney whether the fireplace is in use or idle. The screen still allows about 80% of the heat radiated out from the fire to pass through, thus warming the interior.

FIGURE 4-8

A tempered-glass fireplace screen prevents heat loss from warm air flowing up the chimney when the fireplace is in use or idle. A blower at the right circulates air through the hot tubular grate and back into the room.

Determine Heat Transfer into Building from Exterior

Once you've noted all the possible air leak points inside the home, move to the outside and repeat your inspection there. Air leaks allow heat to move in both directions: into or out of the building. The same leak that causes heat loss from the interior in the winter will allow hot summer air from outside to leak into the air-conditioned interior in the summer. Both directions of heat transfer waste energy by making the home's HVAC system work harder to maintain a comfortable inside temperature.

Look for gaps around pipes and wires that pass through exterior walls, outside electrical outlets, foundation seals, and mail slots (Figure 4-9). Check to see if outside caulking and weather stripping are applied properly, leaving no gaps or cracks, and are still in good condition. Weathering and age will eventually cause nearly every form of caulk to harden and crack (Figure 4-10). When it does, the seal it provides against air leakage is usually lost.

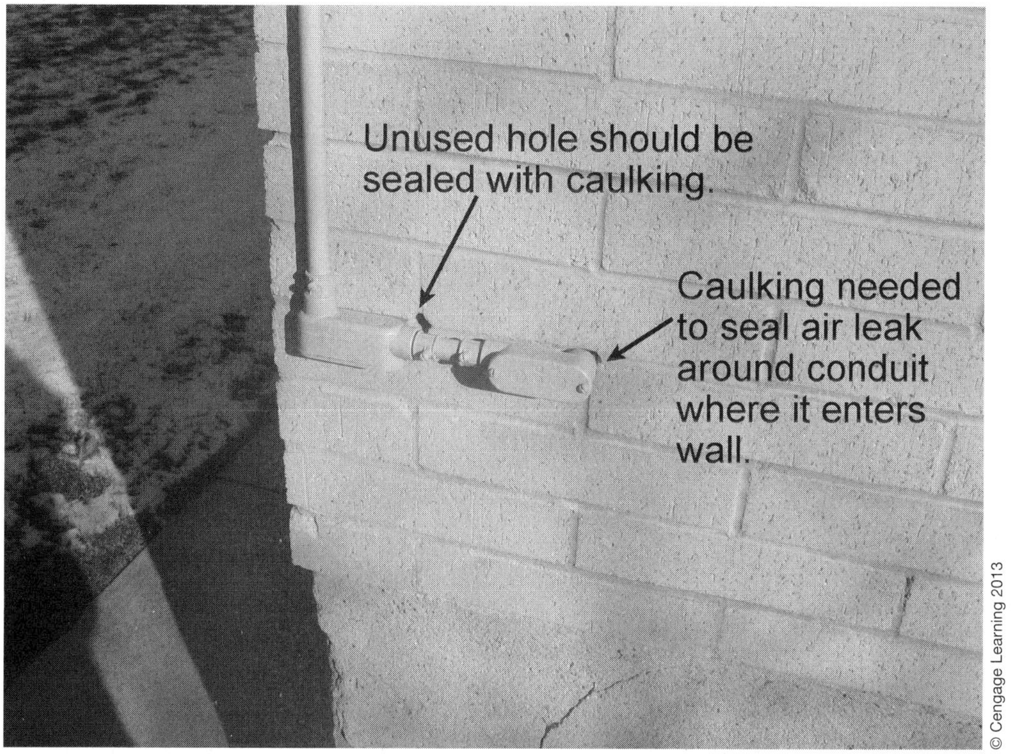

FIGURE 4-9

An electric wire in metal conduit needs caulking where it enters the brick wall to prevent air leakage.

FIGURE 4-10

Old caulking eventually hardens and cracks due to the different expansion coefficients of various materials.

On the outside of your house, inspect all areas where two different building materials meet, including:

- All exterior corners
- Where siding and chimneys meet
- Window casings where they join to brick or siding
- Door jambs where they meet the exterior walls
- Areas where the foundation and the lower edge of exterior brick or siding meet

Inspect the fit of windows and doors for air leaks. If you can see daylight anywhere around a door or window frame in its casing or jamb, then the door or window leaks air. If a closed and latched door fits loosely enough in the jamb that it can be rattled, it will leak air even if light is not visible around the edges. Closed windows that can be rattled in their casings also leak. The older windows frames are, particularly if they're made of wood, the more likely they will shrink or deform enough to leak air around their edges.

Determine Heat Loss through Walls and Ceilings

A large amount of heat can be lost through the ceiling and walls of your home if the insulation levels in these areas are less than the currently recommended minimum. When your house was built, the amount of insulation recommended at that time was probably installed. Some builders may have skimped on the amount of insulation installed, particularly in locations where building codes did not actually require a specific level of insulation, but only recommended it. Rising energy prices have now made increased insulation levels more cost effective and building codes have been updated in most areas to require more insulation in homes.

Depending on a home's age and the quality of its initial construction, the originally installed insulation may be inadequate to reduce heat loss to an acceptable level. If your home is more than 10 years old, you should definitely check the insulation levels in walls and ceilings. Even in a newer home, you may be able to save enough in energy costs by adding additional insulation to more than pay for the cost of installing it.

The most accurate means of measuring heat loss through walls and ceilings is by means of a thermographic inspection. Thermography measures surface temperatures using infrared video and still cameras. These tools record heat as light so that images on the video or film record the temperature variations of a building's skin. Colors range from white for warm regions to black for cooler areas. The resulting images tell an auditor whether additional insulation is needed. They also serve as a quality control tool, to ensure that insulation has been installed correctly. Figure 4-11 shows a thermographic image of the outside of a home. Light shaded areas show the lack of insulation in some parts of the walls, whereas the dark roof is a better heat-loss barrier.

A thermographic inspection can be either an interior or exterior survey. Interior scans are more common and usually more accurate because there is less air movement inside the home than outside and hot or cold spots are more easily located. Heat loss detected in one area of the outside wall might originate at a different location on the inside of the wall. Figure 4-12 shows an interior thermograph that indicates an almost total lack of ceiling insulation in the room.

Although a thermographic inspection is very helpful, a visual inspection may be all that is needed to determine if your home has excessive heat transfer. To check the house's insulation levels, start with the ceilings. Open the attic hatch and inspect the

FIGURE 4-11

An exterior thermograph shows areas of heat loss (lighter shaded areas) caused by poor insulation in the walls of a building. The roof is shaded darker, indicating better insulation there.

FIGURE 4-12

An interior thermograph shows a well-insulated ceiling (light shaded), but wall and corner areas of high heat loss (darker shaded) caused by poor insulation. The scale at right shows temperature differences between dark (cold) and light (warm) areas.

insulation covering the ceilings. Use a flashlight or other illumination to be sure that the entire attic area is insulated. Measure the depth (thickness) of the insulation and observe the type of insulation.

Four basic types of insulation are used in American homes. They are:

- Rolls and batts—flexible pads or material composed of fibers, such as fiberglass
- Loose-fill material—loose fibers or pellets that can be blown into a space
- Rigid foam—sheets of rigid Styrofoam or polyurethane plastic
- Foam-in-place—expanding foam material that can be injected into small areas to control air leaks

In the U.S. building industry, the insulating quality of a material is given in the form of an R-value. The R-value of an insulating material is a measure of its thermal (heat) resistance, that is, its ability to prevent the passage of heat energy through the material from one side to the other. The R-value of any insulation material depends on several factors, including the thickness of the material, the presence or absence of a moisture barrier on the insulation material, and how the insulation is installed (covered by wall board and siding or loose and exposed to moving air currents).

The U.S. Department of Energy has established recommended R-values of insulation in homes located in given areas of the United States based on the general local energy costs for heating and cooling, as well as the climate of an area. Figure 4-13 is a map of the country with areas of lowest insulation requirements (area 1) to highest (area 8) marked in various shades of gray.

Table 4-3 lists the recommended R-value for attics, cathedral ceilings, walls, and floors for each zone of the map in Figure 4-13.

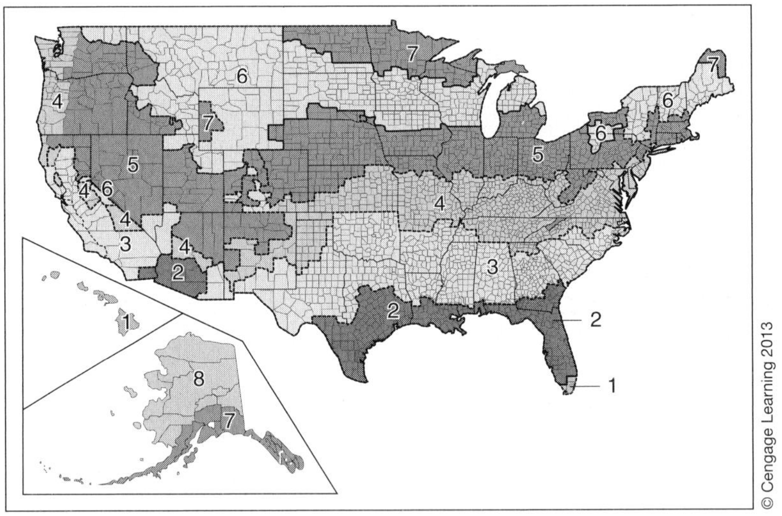

FIGURE 4-13

This map of the United States has eight areas of increasing insulation for homes as recommended by the U.S. Department of Energy. Table 4-3 lists the R-value of insulation recommended for buildings in each area.

To determine the R-value of your home's present attic insulation, measure the thickness of the insulation and note the type of insulation you have. Refer to Table 4-4 to determine the R-value for the type of insulation in your attic and multiply the per-inch R-value by the number of inches of insulation presently installed. Add to the insulation figure the R-value of the sheathing material that encloses the insulation, but only for the thickness installed. Half-inch drywall has an R-value of only 0.45. The 0.9 R-value shown in Table 4-4 is for a full inch of drywall (double thickness of half-inch gypsum board).

If your attic does not have the recommended amount of insulation, you can make your home more energy efficient and save money by adding insulation to bring the R-value up to at least the recommended level for your area.

If the attic hatch is located above a heated or air-conditioned space, check to see if it is at least as well insulated as the attic and closes tightly. It should also be weather-stripped to prevent any air leaks. In the attic, determine whether openings for items such as pipes, ductwork, and chimneys are sealed at the point where they enter the attic space. Note any gaps to be sealed later.

TABLE 4-3 Recommended R-Values

ZONE	FURNACE TYPE	ATTIC	CATHEDRAL CEILING	WALL CAVITIES	INSULATION SHEATHING	FLOOR
1	All Types	R30–R49	R22–R38	R13–R15	None	R13
2	Gas, Heat Pump, Fuel Oil	R30–R60	R22–R38	R13–R15	None	R13
2	Electric	R30–R60	R22–R38	R13–R15	None	R19–R25
3	Gas, Heat Pump, Fuel Oil	R30–R60	R22–R38	R13–R15	None	R25
3	Electric	R30–R60	R22–R38	R13–R15	R2.5–R5	R25
4	Gas, Heat Pump, Fuel Oil	R38–R60	R30–R38	R13–R15	R2.5–R6	R25–R30
4	Electric	R38–R60	R30–R38	R13–R15	R5–R6	R25–R30
5	Gas, Heat Pump, Fuel Oil	R38–R60	R30–R38	R13–R15	R2.5–R6	R25–R30
5	Electric	R38–R60	R30–R60	R13–R21	R5–R6	R25–R30
6	All Types	R49–R60	R30––R60	R13–R21	R5–R6	R25–R30
7	All Types	R49–R60	R30–R60	R13–R21	R5–R6	R25–R30
8	All Types	R49–R60	R30–R60	R13–R21	R5–R6	R25–R30

TABLE 4-4 R-Value of Common Building Materials

INSULATION MATERIALS	R-VALUE PER INCH OF THICKNESS
Fiberglass batt	3.14
Fiberglass blown	2.2
Rock wool batt	3.14
Rock wool blown	3.03
Cellulose blown (attic)	3.13
Vermiculite	2.13
Air-entrained concrete	3.9
Urea polymer foam	4.48
Rigid fiberglass (> 4 lb/ft^3)	4
Expanded polystyrene (beadboard)	4
Extruded polystyrene	5
Polyurethane (foamed-in-place)	6.25
Polyisocyanurate (foil-faced)	7.2

SHEATHING MATERIALS	R-VALUE PER INCH OF THICKNESS
Plywood	1.25
Fiberboard	2.64
Fiberglass	4
Extruded polystyrene	5
Foil-faced polyisocyanurate	7.2
Particle board	1.31
Gypsum drywall	0.9
Hardwood flooring	0.91

When you inspect the attic insulation, check to see if there is a vapor barrier under it. The barrier might be a plastic sheet, tarpaper, or Kraft paper attached to insulation batts. If there is no vapor barrier, you can correct this deficiency by painting the interior ceilings with vapor-barrier paint. This reduces the amount of water vapor that can pass through the ceiling. Large amounts of moisture passing into the attic can reduce the effectiveness of insulation and contribute to structural damage.

Make sure that the attic vents are not blocked by insulation. You also should seal any electrical boxes in the ceiling with flexible caulk (from the living room side or attic side) and cover the entire attic floor with at least the current recommended amount of insulation.

Checking insulation in a wall is more difficult. Select an outlet on an exterior wall and turn off the circuit breaker or remove the fuse for that outlet. Test the outlet by plugging in a functioning lamp or portable radio. When you are sure the outlet is safe, remove the cover plate from it and gently probe into the wall with a long, thin stick or screwdriver. If you encounter some resistance, the wall has at least some insulation. You can also check other exterior walls by making a small hole and inserting a stick to feel for insulation. Choose an unobtrusive spot such as a closet or behind furniture to see what, if anything, the wall cavity is filled with. Probing through a hole cannot tell you if the entire wall cavity is filled with insulation, or if the insulation has settled. Only a thermographic inspection can do this. But a spot check can tell you if the wall is empty and has no insulation, a situation that should be rectified as soon as possible.

If the home has an attached garage, solarium, playroom, or other area that is unheated, walls between the home's interior and the unheated room should be treated as exterior walls and insulated in the same manner. Even though the glass walls and windows of such areas serve to passively heat them, the temperature within them rarely rises to equal the home's interior in winter and will be substantially above it in summer. In either case, significant heat transfer will occur into or out of the home unless the walls between unheated rooms and heated areas are properly insulated.

If the home's basement is unheated or it rests on an above-ground foundation with space beneath the floor, determine whether there is insulation under the living area flooring. Table 4-3 gives the R-value recommended for floors in various zones of the country. If the basement is heated, the foundation walls should be insulated to at least R-19.

A PROFESSIONAL ENERGY AUDIT REPORT

Although all homeowners can perform an adequate energy audit on their own homes, a professional audit, using infrared imaging equipment and pressure testing gear, will provide a more thorough and complete review of the energy efficiency and needed upgrades for the home. Expect to pay from $600 to $1000 dollars for such an audit, but in most cases this cost, as well as the cost of recommended energy-efficiency upgrades, can be recovered over time by reducing the cost of energy used in the home.

A typical energy audit report with a few of the report's accompanying images is reproduced below. This report was made for a home in a Level 5 climate area of the country. This type of energy audit is being performed with increasing frequency, especially on homes that are 10 years old or older. Many new job opportunities are opening for technicians who perform energy audits and install the insulation and other energy upgrades the audits recommend.

Certified Energy Audit Report
March 17, 2009

The quality of your home's heating and cooling performance was determined by how well your home is air sealed, defined as the pressure boundary and how well it is insulated, defined as the thermal boundary.

Energy Use Calculations

The home was measured at 2680 square feet for the first and second floors combined.

Air Leakage

Your home has a medium level of air infiltration/exfiltration. The rate is 2370 cfm (cubic feet per minute) compared with 1588 cfm, which is the recommended maximum rate of air exchange based on your home's measured volume.

Your home is exchanging air to the outside at a rate 1.48 times higher than the recommended amount.

If the cross-section areas of the home's air leaks that exceed the recommended level of air exchange were combined into a single opening, they would be equivalent to an opening 11 inches square.

Scope of Work

The most critical deficiency in the home's performance is the amount and quality of insulation and air sealing in the attic. Heat from the interior rises and passes through the ceiling into the attic. Warm air also leaks into the attic through unsealed openings. This heat loss must be continually replaced by the home's furnace. This is known as the stack effect and emphasizes the need to put a tight lid on the home to reduce it.

The following paragraphs detail our findings and make prioritized recommendations to improve the home's heating and cooling performance.

1. Seal with spray foam along the entire garage ceiling perimeter and seal any evident openings in the ceiling and walls to the basement.
2. Air seal and insulate the two attic doors. These doors must be considered exterior doors and weather-stripped on all four edges as well as insulated by affixing a piece of 1" minimum rigid insulation to the backside.
3. Air seal with spray foam and/or caulk any openings that are visible and/or accessible between the outside and living spaces such as the chimney to sheetrock seams, around all windows and doors, around any outlet boxes in the walls, and around any wire and pipe penetrations that travel through the exterior walls and attic spaces.
4. Replace the 14 existing attic recessed lights with ICAT (insulation contact air sealed) type.
5. Blow in and pack attic ceiling flat with cellulose insulation.
6. Air seal in basement. Spray foam around windows, vertical pipe penetrations.

(Continues)

Certified Energy Audit Report (Continued)

7. Install new weather stripping on the bottoms and sides of the exterior doors, the garage to basement door, and the garage-to-kitchen door.
8. Remove the baseboard trim around the tile floor and caulk/seal where the sheetrock meets the tile flooring.
9. Seal with caulk every crack or gap found between the skylight and drywall and at wall and ceiling intersections.
10. Examine the infrared photos and infrared videos of the home to determine and correct any additional insulation or air leak deficiencies.

Summary

This home is not out of line with energy consumption and leakage rates compared to other homes in this area that are of similar age and square footage. That said, there is plenty of room for improving the home's heating and cooling performance.

We believe that by following the measures recommended above, the home's energy use can be reduced by 20% to 28%.

Selected images of insulation and leakage deficiencies.

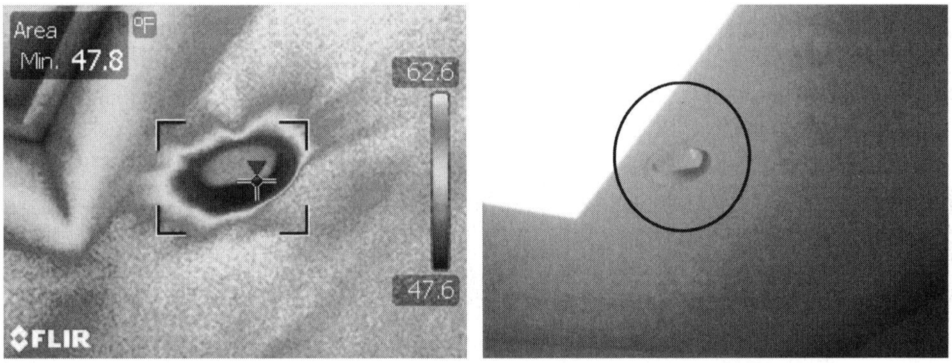

Air leaking in around non-ICAT ceiling lights (dark areas are cold air).

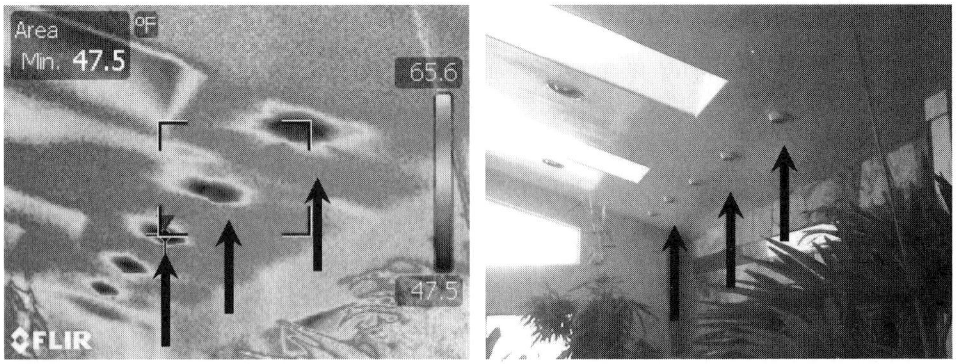

Air leaking in around non-ICAT ceiling lights (dark areas are cold air).

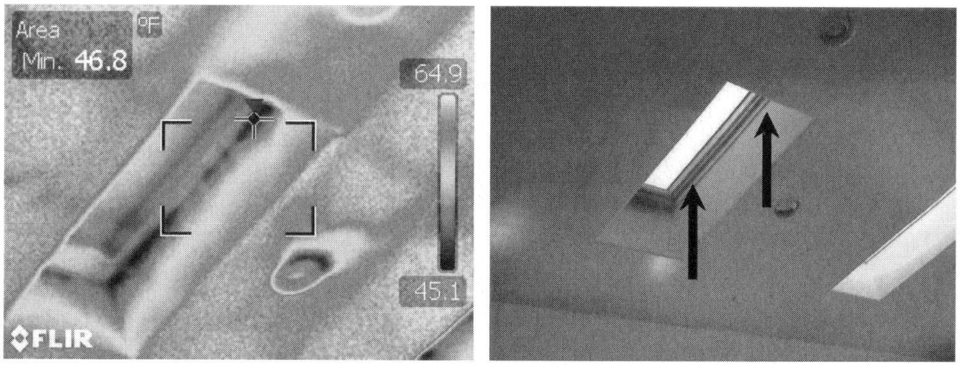

Gaps between skylight frame and sheetrock need to be sealed/caulked (dark areas are outside air leaking inside).

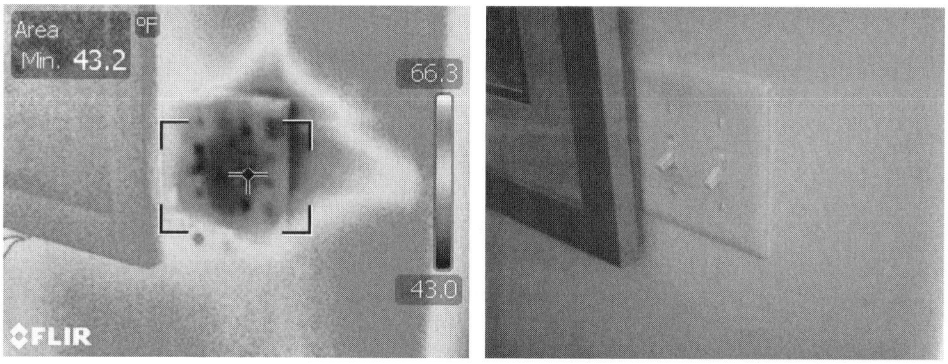

All receptacles and switch boxes mounted on an exterior wall need to be air sealed and a foam gasket installed under the cover plate (dark areas are cold air).

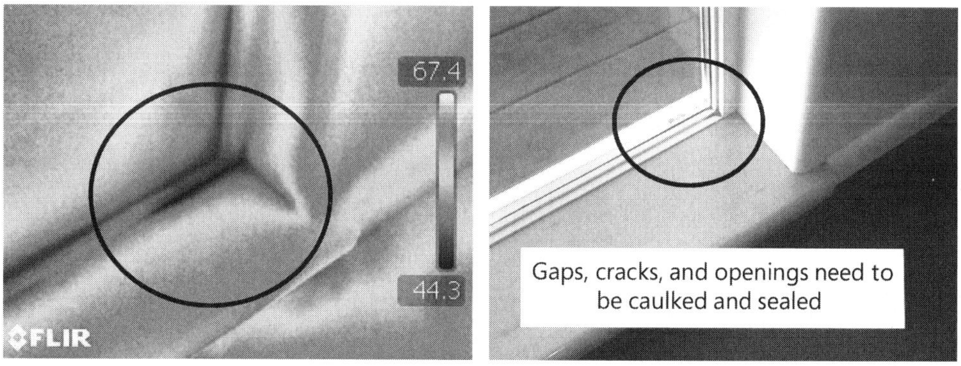

Gaps between window frames and sheetrock needs to be sealed/caulked (dark areas are outside air leaking inside).

(Continues)

Certified Energy Audit Report (Continued)

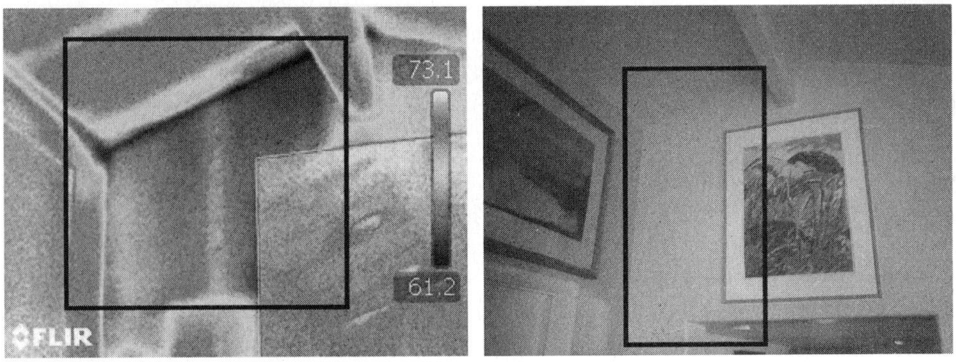

© Cengage Learning 2013

These two stud bays do not have any insulation (cold area is dark). We recommend holes be drilled and insulation blown inside these cavities.

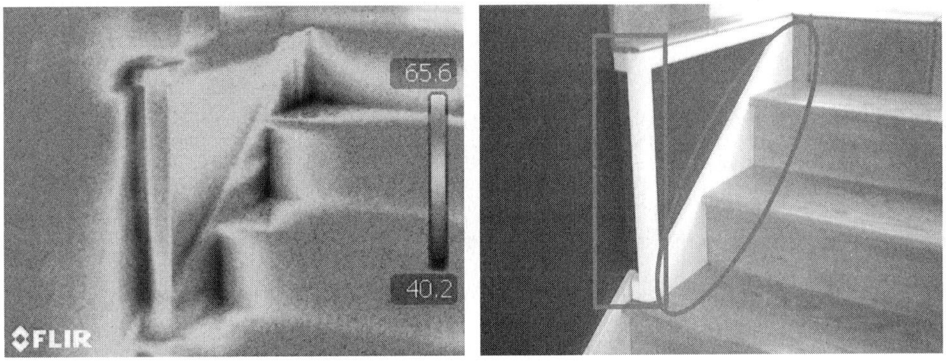

© Cengage Learning 2013

Gaps between steps, wall, and vertical trim (dark areas are outside air leaking inside) need to be caulked and sealed.

SUMMARY

An energy audit is a survey that is designed to tell where a home is using energy and where it is losing energy. An energy audit lists all the electrical appliances and devices in the home, as well as every light fixture inside and outside. To complete the audit, measure the power consumption of each electrical appliance, including those that operate in cycles, those that operate on 240-volt power, and those that are hard-wired into the home's electrical system. For those that cannot be measured, convert their label power ratings into power consumption using Ohm's law and the Power Law. Residential gas use is measured in cubic feet and billed monthly at a set cost per 100 cubic feet (CCF) or 1000 cubic feet (MCF) of gas used. Some homes are also heated with coal- or wood-burning furnaces, which are less expensive than heating oil, but also less convenient and reliable. Reducing air leaks in a home can save from 5% to 30% of energy costs per year. Inside the home, check electrical outlets, switch plates, window frames, baseboards, weather stripping around doors, and fireplace dampers to see if air can flow through these places. Leaks can also be detected by a building pressurization test that slightly increases or decreases the air pressure in the home relative to the outside air. On the outside of the home, inspect exterior corners, where siding and chimneys meet, window casings, door jambs, and areas where the foundation and brick or siding meet. The most accurate means of measuring heat loss through walls and ceilings is by an interior or exterior thermographic inspection using infrared video and still cameras, but a visual inspection may be adequate to determine if a home has excessive heat transfer. There are four basic types of building insulation. They are rolls and batts, loose-fill material, rigid foam, and foam-in-place. Determine the R-value of a home's attic and wall insulation by measuring its thickness and multiplying by the R-value of the kind of insulation used.

KEY TERMS

Average monthly cost The cost to operate an appliance for 30 days.

Average monthly kWh The power consumed in 30 days by an appliance operating at its present level.

BTU (British thermal unit) A measure of heat equal to the amount of energy needed to heat 1 pound (0.454 kg) of water (0.1198 U.S. gallons) from 39° to 40°F (3.8° to 4.4°C).

CCF 100 cubic feet of gas.

Combustion vent A vent to the outside that allows a fuel-burning appliance to draw its combustion air from outside the building where it is located.

Cumulative cost The cost to operate an appliance since it was plugged into a power meter.

Cumulative watt hours The watt hours of electricity used since an appliance was plugged into a power meter.

Current watts The amount of electric current in watts that an electrical device is using.

DTH A unit of heat measure that is equal to one million BTUs.

Electricity monitor Same as a power meter.

Energy audit A survey that is designed to tell where your home is using energy and where it is losing energy.

Foam-in-place insulation Foam material that can be injected into small areas to control air leaks; same as expanding foam insulation.

Hard-wired Permanently connected to a building's electrical system, as opposed to a plug-in appliance.

Heat loss The loss of heat energy from a home or other building caused by radiation from the exterior walls and leakage of warm air from the interior.

Heat transfer The passage of heat through a material from its warmer surface to its colder surface or from one material to another; also the passage of heat through the walls, roof, and floors of a home or other building in either direction.

Insulation Any material intended to impede heat transfer; materials through which heat passes slowly or not at all.

Kilowatt hour (kWh) A measure of electricity used; equal to 1000 watts of electricity used for 1 hour. It is the usual unit on which electric charges are based (for example, $0.08 per kWh).

Loose-fill insulation Insulation made up of small bits of material that can be dumped into place; this type of insulation can also be blown into place.

Maximum watts The power consumed when the load on an appliance is highest.

MCF 1000 cubic feet of gas.

Minimum watts The minimum power consumed when an appliance is plugged in but not operating.

Ohm's law $V = I \times R$ or voltage (V) is equal to amperes (I) times the resistance (R).

Operating cycle The complete programmed set of functions performed by an electrical device in a timed sequence or in response to preset conditions.

Operating mode The full-functioning operating status of an electrical system or device.

Power analyzer Same as a power meter.

Power consumption The amount of electricity used by a device when operating normally.

Power Law $W = I \times V$; power (expressed in watts [W]) is equal to current (expressed in amperes [I]) times the voltage (expressed in volts [V]).

Power meter A device for measuring the amount of electricity used by an appliance and calculating other information from that measurement.

Power rating The maximum amount of power an electrical appliance is designed to use safely.

Power surge A sudden, momentary rise in electric current flow or voltage; electric motors create a current surge when starting.

Pressure boundary The physical barrier of walls, windows, doors, and so forth that prevents air from passing into or out of a building.

Pressurization test A method of testing a building for air leaks by slightly raising or lowering the inside pressure to cause air flow through the leaks.

Resistance The opposition of any material to the passage of electricity through it; measured in ohms.

Rigid foam insulation Sheets of rigid Styrofoam or polyurethane plastic used to prevent heat transfer.

Rolls and batts Insulation made in flexible pads of material composed of fibers, such as fiberglass.

R-value A measure of an insulation material's thermal (heat) resistance, that is, its ability to prevent the passage of heat energy through the material from one side to the other.

Sleep mode A state of an electric device in which its operation is suspended, but its circuits are still powered and ready for instant use on demand.

Thermal boundary The insulation barrier that reduces or prevents heat transfer by convection through the walls floors and ceilings of a building.

Thermographic inspection A means of measuring heat loss through walls and ceilings by measuring surface temperatures using infrared video and still cameras.

Vapor barrier A waterproof sheet of material placed under attic or wall insulation to impede the passage of moisture through the insulation.

Watt meter Same as a power meter.

Watt rating The amount of current that will flow through a light bulb or other electric device when operating.

REVIEW QUESTIONS

1. What is an energy audit?

2. True or False: A light fixture connected to a dimmer switch uses less electricity than one connected to an ordinary on-off switch.

3. What do a watt meter, a power analyzer, and an electricity monitor have in common?

4. In addition to measuring current watts—the power being used by an appliance—what other data can a power meter measure or calculate?

5. Does a washing machine use more power, less power, or the same amount of power when run through an operating cycle empty than it does when run through a cycle with a full load of clothes?

6. What is "sleep" mode for an electric appliance or system?

7. If the label of a hard-wired appliance says that its power rating is 15 amps, will it always draw that much power when in operation?

8. What is Ohm's law?

9. If an entertainment system draws 287 watts of power when operating, how many kilowatt hours of electricity will it consume in 3 hours of operation?

10. What do the terms *CCF* and *MCF* mean in relation to natural gas?

11. What do the terms *heat transfer* and *heat loss* mean?

12. What are some places to check for air leaks in a home's interior?

13. What is a pressurization test?

14. How does a thermographic inspection identify heat loss?

15. What are the four types of insulation typically used in homes?

16. What is the meaning of R-value in relation to insulation?

17. What is the Power Law?

18. True or False: Additional insulation added to an attic area would usually be the foam-in-place type.

19. True or False: When insulating a home's walls, the wall between an unheated attached garage and the home's interior should be treated as an exterior wall.

20. What is a vapor barrier in relation to insulation?

CHAPTER

5 UPGRADE HOME ENERGY EFFICIENCY BY REDUCING HEAT TRANSFER

OBJECTIVES

After studying this chapter, you should be able to:

- Detect and seal air leaks inside and outside a home to reduce heat loss from the interior.
- Determine if the existing insulation in a home is adequate to meet current energy-efficiency standards.
- Install additional insulation to bring a home up to minimum current R-level standards.
- Rate the energy efficiency of a home's windows, doors, and skylights and upgrade or replace them as necessary to improve their performance.

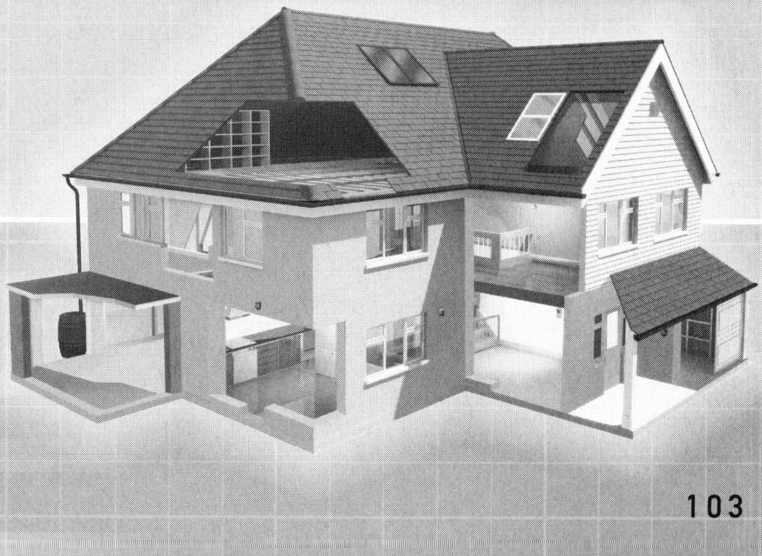

OUTLINE

I. Introduction
II. Reduce Heat Transfer in and out of the Home
  A. Seal Leaks
  B. Insulate
  C. Energy-Efficient Windows, Doors, and Skylights
III. Energy Efficiency Outside the Home
  A. Awnings and Screens
  B. Landscaping for Your Location
IV. Summary

INTRODUCTION

In this chapter you will learn how heat transfer strongly affects the energy efficiency of a home. You will learn methods of eliminating sources of heat loss from the home, such as interior and exterior air leaks, inadequate insulation, and inefficient windows and doors. You will also learn how to reduce heat transfer into the home that occurs through walls, windows, doors, and skylights when the outside temperature is substantially above the inside room temperature. Finally, you will become familiar with techniques that can be implemented outside the home to further improve its energy efficiency and reduce its power and fuel consumption.

REDUCE HEAT TRANSFER IN AND OUT OF THE HOME

ONCE YOU'VE IDENTIFIED the causes of heat loss in your home, you can greatly reduce the home's energy consumption and save the cost of wasted energy you are now paying for by eliminating every cause of heat loss to the maximum extent possible. Start with repairing the air leaks. They should be addressed first, so they won't be covered by any new insulation or other construction changes that are made later.

Seal Leaks

Seal the outside leaks you have detected first. Any cold air that gets into the wall cavities of a building will find its way into the interior or it will absorb heat from the structure. Either way it causes heat loss on the inside of the building. After every outside leak is sealed, then proceed to the inside and seal every point where air is still entering the interior rooms.

Seal Outside Leaks

Any unsealed opening through which pipes, conduit, wiring, or cables pass from the outside through an exterior wall should be sealed airtight with caulking or similar material that provides as much insulating R-value as possible. Large leaks should be stuffed with rock wool or a similar material and then caulked over the insulation material to create an airtight seal. Caulking alone is not sufficient to reduce heat loss from this type of leak, especially when a metal pipe or conduit, both excellent heat conductors, passes through the opening. If accessible, these leaks should be sealed on both sides of the exterior wall to avoid the kind of damaging heat transfer illustrated in Figure 5-1.

FIGURE 5-1

Old caulking on this air-conditioning chiller line has hardened and shrunk, leaving cracks and openings that allow air to pass through in both directions.

Other outside air leaks require varying types of seals to block air and prevent heat loss. Among the most common problem areas and possible solutions for them are the following:

- Exterior brick corners—Outside brick corners rarely crack and can usually be caulked if they do. Inside corners crack more often and sometimes the bricks of the two walls separate as much as half an inch. Caulking alone will not do for an opening this wide. Foam-in-place insulation should be injected into the crack to fill it as full as possible. Once the foam has set, caulk a smooth layer over it that seals to both edges of the crack. If you can't find caulk in a matching color to the brick, paint over the seal.
- Exterior siding corners—Check for lapped siding, whether of wood, aluminum, vinyl, or other material, that does not meet in a tight seal at outside or inside corners. If outside corners are capped, caulk any edges of the caps that appear unsealed. If edge boards cover the corner, inject foam-in-place insulation wherever the siding does not meet flush with the edge board. The same procedure should be used for inside corners that are capped or finished with edge boards. If neither is present, try to caulk a seal along the entire corner where the siding boards meet.

FIGURE 5-2

This window casing needs new caulking all the way around; the top opening is so wide that a filler must be inserted before caulking can be used to finish the repair.

- Joints where siding and chimneys meet—If the crack between siding and chimney is wide, fill it as full as possible with foam-in-place insulation and seal the set insulation with an outer layer of caulking. If the crack is narrow, caulking alone may be sufficient, but it will be made easier by stuffing the crack first with fiber insulation. Use a wide putty knife to force the insulation in as far as it will go, then cover the opening with caulk sealed to both the siding and the chimney.
- Window casings where they join to brick or siding—Window casings should be carefully caulked all the way around (Figure 5-2). Stuff any cracks that are wide enough to admit a putty knife with insulation before covering with a caulk seal. If the cracks around a window casing are wide, replacing the window is a better option than filling the cracks with foam-in-place insulation. Insulation can be used, however, if budget constraints rule out new windows.
- Door casings where they meet the exterior walls—First, be sure the door frame is not loose in its opening. If the frame is not rigidly set in place, adjust its position to true vertical and add fasteners as needed to keep it in place. Caulk around the entire frame, inside and outside, using stuffed insulation if necessary. Make sure the door fits correctly in the jamb and install good-quality weather stripping around all sides. Be sure the bottom of the door is also weather-stripped so it mates tightly to the threshold when the door is closed.
- Areas where the foundation and the lower edge of exterior brick or siding meet—Cracks where brick and concrete meet, as shown in Figure 5-3, can be caulked or filled with fine sand mortar if they are wide. In either case, fill the crack as fully as possible with the sealing material. Siding may overlap the foundation, but not be sealed to it at all. If an opening exists, it must be filled with a sealing material all around the house. Foam-in-place insulation can

FIGURE 5-3

Large temperature variations have caused this caulking to crack and pull away from one side of a brick/concrete joint. New caulking should be of a type that remains permanently flexible.

be used for narrow openings up to three-quarters of an inch. Wider cracks should be closed with wood stringers wide enough to fill the opening. A bead of caulk laid on both sides of the stringer before it is set in place will seal the crack effectively without the need for additional caulking later. Figure 5-4 shows a cross section of siding filled with caulked stringers.

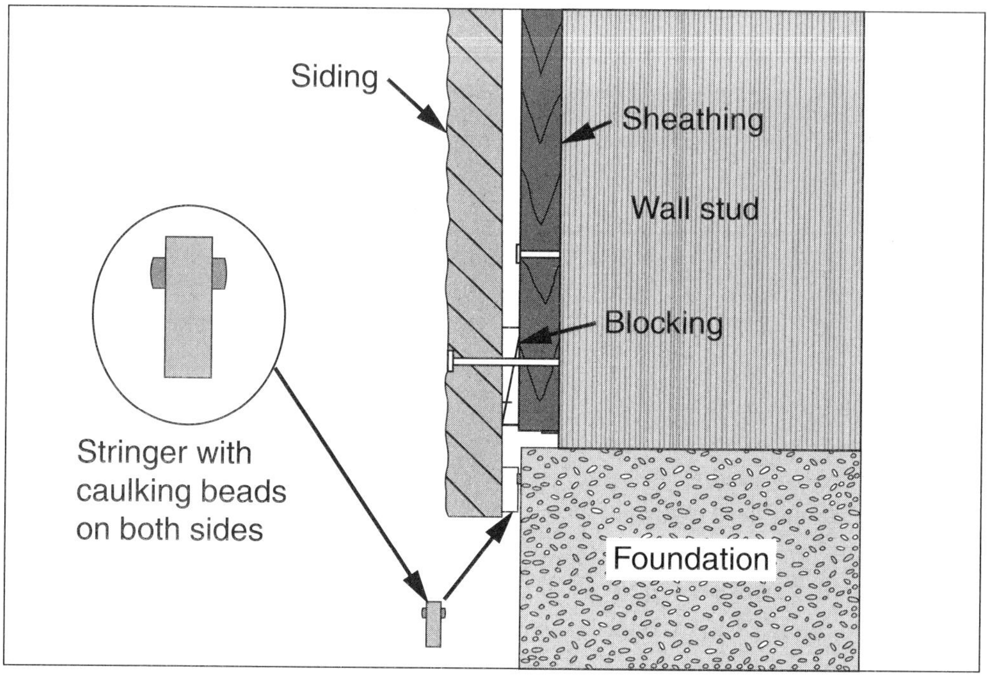

FIGURE 5-4

A correctly sized stringer with caulking on both sides can be used to seal the opening between siding and the home's foundation. All sides of the home must be sealed to prevent cold air from leaking into wall cavities.

Seal Inside Leaks

Air leaks around pipes and wiring, and holes in other unfinished areas that are accessible from inside the house, should be sealed airtight with caulking or similar material that provides as much insulating R-value as possible. For some leaks, especially large ones, this may mean first filling the opening with rock wool or a similar material and then spreading an airtight seal of caulking over the insulation material. Insulation material alone is not enough to eliminate this type of leak. The material must be covered and sealed to prevent any air from passing through it in either direction. Figure 5-5 shows an air leak in a brick wall that should be filled with insulation and then caulked for a complete seal.

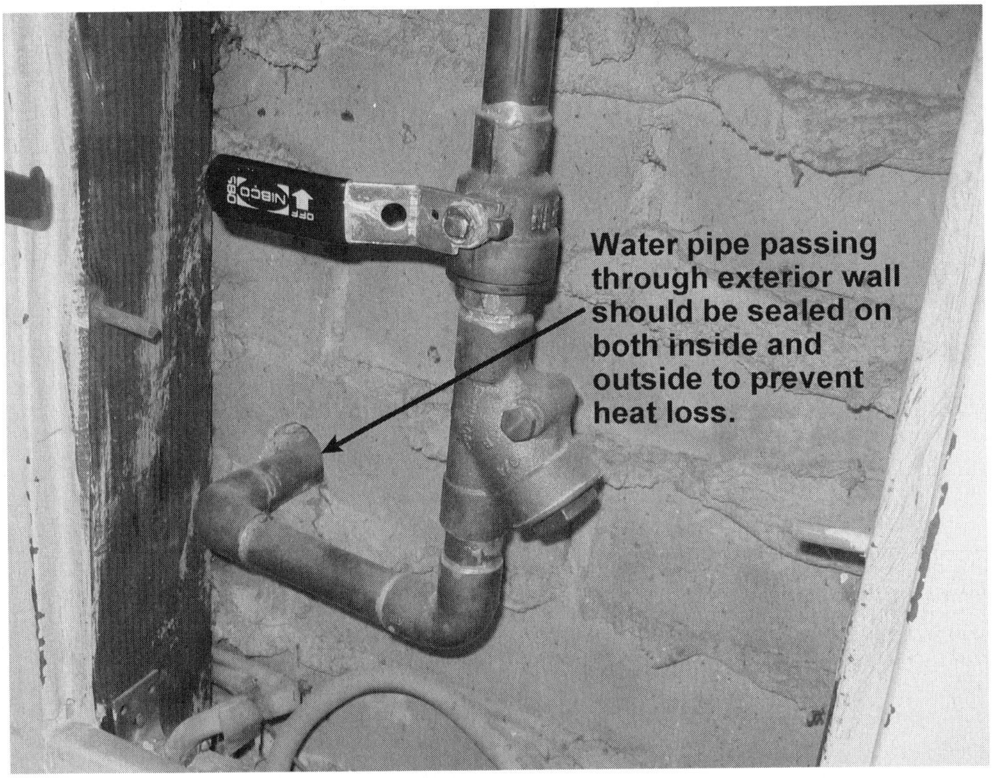

FIGURE 5-5

A water pipe passing through an external wall should be packed with insulation and sealed with caulking on the inside and outside of the wall.

Other inside leaks can be more difficult to correct. The points where most inside air leaks are detected, such as wall outlets or baseboards, do not connect directly to the outside of the house. Air leaking from these points indicates that air is flowing into and out of the wall cavities through leaks on the outside of the home. Stopping the outside leaks to prevent cold or hot air from getting into the walls will probably do more to stop the inside leaks than anything you can do at the point of entry into the interior.

When you've done all you can to stop outside leaks, some inside leaks may still show air flow. Following are some suggestions for dealing with specific leak points:

- Electrical outlets on exterior walls—Remove the faceplate and put a thin layer of foam insulation around the outlet box. Use only insulation approved for placement near electrical wiring. You may also be able to use foam to seal the inside of the box at the points where the wires enter. Replace the faceplate over the foam insulation
- Switch plates on exterior walls—Use the same procedure as for outlets. Do not get foam insulation on the switch mechanism.

- Window frames—If replacement is not an option, weather-strip the frame on the inside around the edges that interface with the casing. For added protection, you can also weather-strip the outside. If you don't need to ever open the window, you can caulk it on the outside.
- Exterior wall baseboards—Remove the baseboard and refit it more tightly to the wall and floor. Before replacing the baseboard, run a bead of caulking along the back about one-half inch from the top of the board. Run another bead along the bottom of the board. Replace the board, making sure that the caulking seals it against the floor and the wall.
- Weather stripping around doors—Replace the weather stripping and fit the new material carefully so that it seals the closed door effectively. Use high-quality weather stripping. Its cost will be repaid many times over in energy savings.
- Fireplace dampers—If the damper is badly heat warped, it should be replaced. If this is not an option, little else can be done about the damper if it is to remain useable with the fireplace. If the fireplace is not used, an airtight decorative faceplate can be fitted over it.
- Attic hatches—Fit the hatch as tightly as possible within its frame. Weather-strip all sides where they meet the frame. Insulate the top of the hatch with as much insulating R-value material as the rest of the attic. Use rigid Styrofoam or similar material.
- Wall- or window-mounted air conditioners—Seal all four sides of the unit in the window. Use fiberboard or plywood to fill large openings. Caulk small openings so an airtight seal is created all the way around the unit and to the window frame.

Before attempting to seal leaks around windows, decide whether they should be replaced with more energy-efficient designs. Correctly installed replacement windows will be airtight both between the frame and casing and between the casing and wall material.

If the home has an unheated basement or crawl space, these should be treated as outside exposures and the floors above them sealed to prevent air leakage. Most openings can be more easily sealed from below than from inside the house, but every opening should be made as airtight as possible. In areas with severe climate, the floors should also be insulated, as noted in the following section.

Insulation added to the floors of a home, especially in below-grade basements or crawl spaces, should not be sprayed foam, as this type will trap moisture against the floor joists and flooring, causing rot to develop in both. The best way to block heat transfer through floors above basements or crawl spaces is to seal the lower level airtight. If additional insulation is needed, it should be Styrofoam panels or other water-resistant material.

Insulate

If you've determined that your home needs additional insulation, the next step is to find out if any government or utility company grant programs are available in your area to help you pay for installing it.

The federal government has a Weatherization Assistance Program (WAP) that can help pay for insulation upgrades as well as other efficiency upgrades. Not all homeowners qualify to receive this assistance but most do. To find out if you qualify for a WAP grant, visit the U.S. Department of Energy WAP website at http://www1.eere.energy.gov/wip/wap_apply.html.

You should also check whether your state government or even your local county or city government has any weatherization or energy-efficiency program that will provide grant funds, rebate all or part of the costs of weatherization upgrades, or provide a tax credit to the homeowner for such upgrades. If you find such a program, be sure to read the requirements and qualification limits carefully to be sure you do all that is required to obtain the assistance. Some programs may require that work be done only by professional contractors. Others may permit do-it-yourself installation of upgrades, but impose conditions on its quality or documentation to receive the benefit.

In addition to government programs, many utility companies also offer grant or rebate programs to help homeowners upgrade their home energy efficiency. To find out about programs in your area, contact your local electric and/or gas utility. As with government programs, be sure you know and follow the requirements for participating in a utility-sponsored program before you start your upgrade.

Government and utility programs may also apply to other types of energy-efficient upgrades to your home, including such projects as window and door replacement, upgrading of furnaces and air-conditioning systems, and replacement of water heaters with more efficient models. Be sure you explore all the possibilities for obtaining assistance for increasing your home's energy efficiency.

Increasing the R-value of insulation in your attic can be done with either batting or blown-in insulation, and several materials are available for each method. Professional equipment is required to blow insulation into place. The homeowner can install batts or hire a commercial contractor for the work to be done. Figure 5-6 shows blown fiberglass insulation being blown into place over a layer of originally installed loose insulation.

Either batts or blown insulation will deliver excellent results, but both require careful installation to perform well. Both types depend on their ability to trap and hold small pockets of air so that warm or cold air cannot pass through the insulation. Blown-in insulation creates a seamless blanket in an attic or within wall cavities. Careful installation is required to achieve this same result with batts.

Before deciding which insulation method to use, compare the cost. The cost of fiberglass insulation blown in by professionals was about $0.40 per square foot in 2011. Fiberglass batts cost about $0.45 per square foot for the material alone. Installation costs will be in addition to that. You will only save money by using batts and doing the work yourself if you can find fiberglass batts deeply discounted.

When working with fiberglass or other batt insulation, wear a respirator to minimize throat irritation from particles and to prevent insulation particles from being inhaled into your lungs. Gloves, a long-sleeved shirt, and goggles are also recommended.

If the existing insulation is batts placed between the ceiling joists, install the new batts across the joists at a right angle to the existing material. This will help cover any gaps in the original installation. Don't worry if the original insulation has settled and doesn't come up to the tops of the joists. Lay the new insulation across the joists and leave an air space between the old and new insulation. It won't affect the R-value of either one.

FIGURE 5-6

Blown insulation consisting of small pieces can easily fill all the irregular cavities of an attic space.

Check the existing insulation for dark spots that may indicate air leaks from the interior living space. The insulation traps dirt from air as it passes up into the attic. Locate and plug any leaks before installing the new insulation. Also, be sure to use unfaced batt insulation when covering existing insulation. Any facing on the new insulation will act as a moisture trap and possibly cause stains on the ceiling below (Figure 5-7).

When installing batts around roof framing members, cut the batts so they fit tightly around each piece of wood. Work from a two-and-a-half-foot square plywood piece set across the joists so you don't press down the existing insulation or fall through the ceiling. Do not block the air vents between the exterior soffits (eaves) that permit air flow into the attic with insulation. Attic ventilation from the eaves to the roof vents must be maintained. Purchase foam or cardboard baffles to place between the joists as they pass over exterior walls. These baffles will keep a pathway open for air to enter the attic. Figure 5-8 shows batt insulation being installed across attic joist that are already filled with insulation.

Do not press down batts to fit under or around a pipe or electrical cable. The compressed batt loses its R-value because its dead-air space is reduced. Split the batt to half its thickness, slide half behind the pipe or cable, and lay the remaining piece on top. Insulate around an electrical box by cutting the insulation to go around the box and loosely tucking as much as you can behind the box.

The split-and-cut technique is also used wherever insulation must go around an obstacle or the available depth changes, such as where a ceiling joist joins a rafter. No gaps should be left at the top or bottom of bays or where batts butt together. Do not

FIGURE 5-7

This batt insulation was so badly deteriorated by moisture condensed from leaking air that it needed to be replaced.

FIGURE 5-8

New insulation being installed across the insulation-filled joists of an attic. Note the installer's plywood working platform to avoid damaging the existing insulation.

insulate above ceiling-mounted light fixtures unless they are designed for insulation contact (rated IC). Doing so can cause a buildup of trapped heat that could start a fire. For best insulation efficiency, ceiling-mounted light fixtures should be insulation contact air sealed rated (ICAT) so that air from the home's interior does not pass through them into the attic.

Loose-fill or sprayed foam insulation is the best choice for adding insulation to existing finished walls. These two types of insulation can be added without much disturbance to finished areas of your home. Professional equipment is required for foam installation, and many homeowners will prefer to have blown insulation installed by a contractor. But you can do the job yourself using cellulose blown-in insulation. This material is composed of small particles that fill in the nooks, crannies, and irregular areas of wall space quite well. It is mainly made up of recycled newsprint paper and does not contain any resins made with formaldehyde, which can release harmful vapors into the air.

Blower equipment for installing cellulose can be rented at many home improvement centers. When working with cellulose or any blown-in material, always wear safety goggles and a mask or respirator. Installing the insulation in walls is a five-step process that requires two people.

1. Locate the studs in the wall using a stud finder. Cut a 3-inch hole between two studs about 6 inches below the ceiling. The hole should be cut with angled sides, as shown in Figure 5-9, so that the cutout piece can be easily replaced in the hole later. Repeat this step between each pair of studs along the length of the wall. Number the holes and the matching cutout pieces and set them aside.
2. Insert the blower hose as far as possible into one of the holes with the nozzle pointed down into the wall cavity. Seal the hole around the hose with a rag to prevent air and insulation from blowing back into the room.

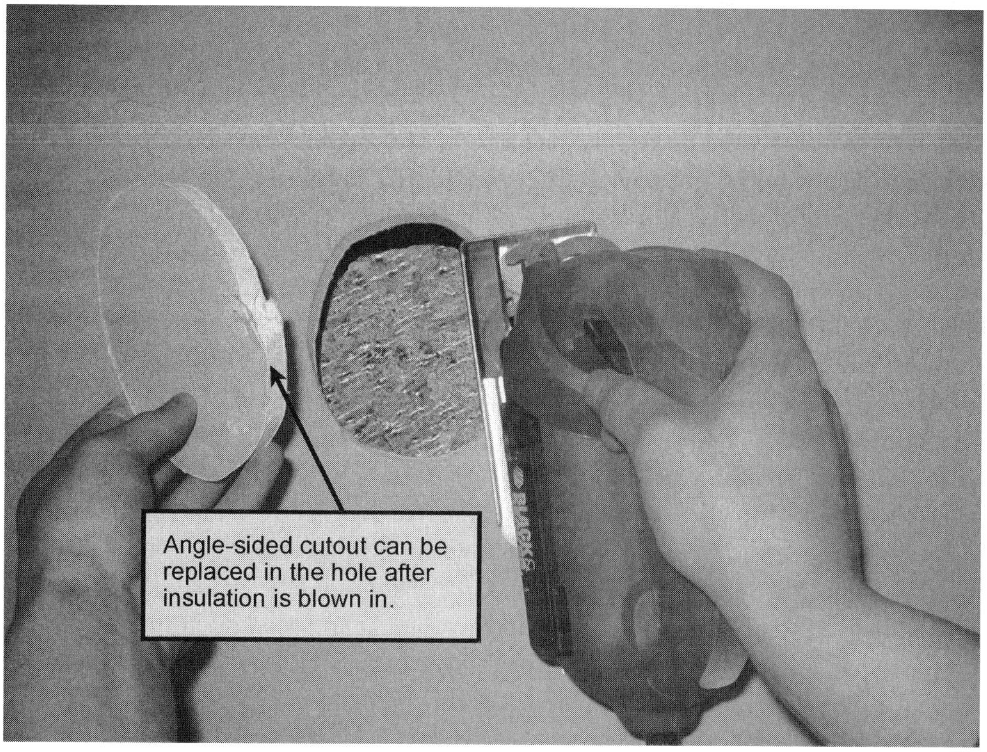

FIGURE 5-9

A small angle-sided hole cut between studs allows insulation to be blown into wall cavities. The cutout piece can then be refitted in the hole and faced smooth with spackling paste.

3. Hold on to the hose while a helper turns on the blower and feeds insulation into it. Pull the hose back as the cavity fills. Stop the machine when you note some resistance and insulation no longer blows into the cavity.
4. Repeat steps 2 and 3 for each wall cavity where you have made a hole. Fill the entire wall with as much insulation as possible.
5. When all insulation is in place, insert the numbered drywall cutouts back into their matching holes using spackling compound to seat them firmly in place. Spackle the hole area to a smooth surface and paint to match the wall surface.

Energy-Efficient Windows, Doors, and Skylights

Energy performance ratings for windows, doors, and skylights are different than those for insulation. This is because glass, the main component of windows and skylights, and a significant component of doors, transfers heat in two ways: by direct conduction of heat through the glass from its warmer side to its cooler side, and by radiation of heat through the glass into a house from the sun or out of a house from the air and other objects in the home's heated rooms.

Energy-Efficient Windows

The rate at which a window, door, or skylight conducts heat flow is called its U-factor. A U-factor may refer to only the glass in a window, skylight, or door or it may represent the entire unit's performance, including frame and spacer material. In either case, the more energy efficient a window, door, or skylight is, the lower its U-factor rating will be.

The amount of solar radiation admitted through a window, door, or skylight, and subsequently released as heat inside a home, is expressed as its solar heat gain coefficient (SHGC). The lower the SHGC, the less solar heat a glass panel transmits. A window with a low SHGC rating will block more heat from the sun from entering the home and reduce the load on the air-conditioning system in summer. Conversely, a window with a high SHGC rating will permit more solar heat gain during the winter, thus reducing the load on the home's heating system during that season. The SHGC rating you need for a window, door, or skylight should be determined by whether you want your windows to allow less heat transfer out of the home year-round to reduce the load on the air-conditioning system in summer, or more heat transfer into the home year-round to reduce the load on the heating system in winter. With uncoated glass, you can have one or the other, but not both.

Another method of describing the heat radiation of glass and other building materials is by their emissivity or E-value. Emissivity is a measure of how much radiation (of all wavelengths) a material absorbs and then emits again as radiant energy. Emissivity values can range between 0 (for a material that doesn't absorb any radiation but reflects 100% of the energy that strikes it) and 1 (for a perfectly black material that absorbs 100% of the energy that strikes it and then emits it again as heat). The lower a material's E-value, the better it is as a building material because it reflects radiation away from a home or other building rather than absorbing it and then re-emitting it into the interior. Table 5-1 shows the emissivity or E-value of common building materials.

Flat window glass has a high E-value (.92 to .95), which means that it absorbs and re-emits nearly all of the radiant energy striking it. This makes it a poor building material

TABLE 5-1 Common E-Values

EMISSIVITY VALUES FOR COMMON MATERIALS			
MATERIAL	E-VALUE	MATERIAL	E-VALUE
Asbestos: board	.96	Granite	.96
Brick: common	.81–.86	Gypsum	.85
Brick: common, red	.93	Paint: aluminum	.45
Brick: facing, red	.92	Paint: light blue	.95
Brick: facing, yellow	.72	Paint: flat white	.99
Carbon: candle soot	.95	Paint: ultra flat black	.97
Cement	.54	Plasterboard	.90
Charcoal: powder	.96	Plastic: acrylic, clear	.94
Concrete: dry	.95	Plywood: untreated	.83
Glass	.92	Skin, human	.98
Glass: polished plate	.95	Snow	.80

for energy efficiency. Silver, on the other hand, has an E-value of .02 and aluminum .03 to .05, making them excellent reflectors of radiant energy. Adding very thin coatings of silver, aluminum, tin, or other materials with low E-values to the exterior side of window glass can raise the E-value of the coated glass dramatically (as high as .85).

A voluntary program of the National Fenestration Rating Council (NFRC) tests, certifies, and labels windows, doors, and skylights based on their energy performance ratings. The NFRC label provides a reliable way to determine a window's energy properties and to compare products. This NFRC label can be found on all ENERGY STAR® qualified window, door, and skylight products, but ENERGY STAR bases its qualification only on U-factor and SHGC ratings.

If your home's windows are single-pane glass, double-pane replacements will improve their insulation R-value, although not as much as professional window contractors might suggest. Single-pane glass windows have an R-value of just .91. Double-pane windows with a quarter-inch air space between the panes have an R-value of 1.69, and triple-pane windows with one-quarter-inch air space between each pane have an R-value of 2.56. Widening the air space between panes can further increase the R-value to 3.23 (for triple-pane windows), but a lot of heat will still be lost through the glass.

Glass coated with a metal oxide has a higher R-value than ordinary glass. This type of glass has a semiconducting coating of tin oxide doped with fluorine atoms 0.3-0.4 microns thick. The coating has the unusual property of transmitting almost all visible light, while reflecting heat waves (infrared) at room temperature. It is deposited in a few seconds by a chemical vapor deposition (CVD) process applied to the hot glass while it moves through the manufacturing process. Metal-oxide-coated window glass can have an R rating of up to 5, making it more than twice as efficient as ordinary glass for insulating against heat transfer into or out of the home.

An increased R-value for glass windows can also be achieved by adding a film overlay to the inside surface of the glass. Window film interrupts the transfer of radiant heat energy by reflecting some of the energy back instead of transmitting it or absorbing it. Transmission is the percentage of solar energy (all wavelengths) that passes through window glass. Absorption is the percentage of solar energy absorbed to that of the total solar energy. Absorption and transmission together represent the e-value of the glass. Reflection is the percentage of solar energy reflected back from the surface of the glass.

With solar window film installed on the inside of a window, the percentages of energy reflected, transmitted, and absorbed are changed dramatically. Reflection of energy and absorption of energy increase and transmission decreases. This results in a reduction of solar heat transferred inside the home. Window film increases the R-value of glass to about the same level as does a metal oxide coating. Figure 5-10 shows how heat transfer is reduced by the addition of film.

FIGURE 5-10

Adding solar film to window glass increases its reflectivity from 5% to 50%; absorption is also increased but transmission is reduced from 87% to 12%, effectively doubling the R-value of the glass.

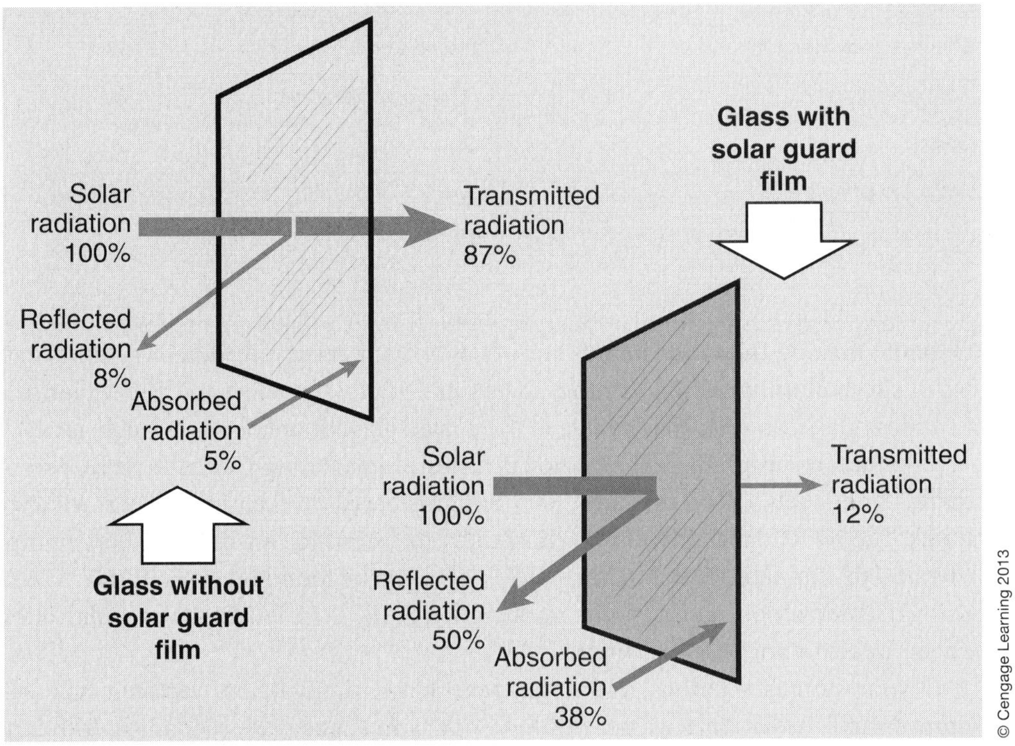

Replacement windows are expensive, especially if they must be custom made to fit odd-sized openings in older homes. If your budget won't allow for new windows, you can still improve the R-value of the ones you have.

Adding storm windows on the outside of the existing window glass is the best way to upgrade older windows. Storm windows raise the R-value of single-pane windows to about 2.00. Storm windows added on the inside of the existing glass pane work as well and are less costly than external windows, but are not as durable. Storm window types range from plastic sheets or films designed to last only for one heating season, to glass multiple-pane units that will last for many years. Storm windows made with glass or plastic panels can be further enhanced by the addition of film layers or special coatings, but these features can easily bring the cost of storm windows up to or above the cost of multiple-pane replacement windows.

For the do-it-yourself homeowner, interior storm windows are easier to install and to remove than the exterior type. They won't last as long, but they usually require less maintenance and are easier to seal tightly to the primary window. Interior storm windows also are often the best choice for apartments and houses with more than one floor. If you can afford exterior storm windows, you can probably afford some newer, more energy-efficient windows, which will be a better investment.

Glass-pane storm window types offer better visibility and longer life than plastic-pane types, but glass is heavy and fragile. Glass windows are more difficult for the homeowner to install and remove. Plastic-pane windows are inexpensive and relatively easy to install, but also easy to scratch or otherwise damage. Some plastics may turn yellow or cloudy over time due to long sun exposure.

Storm window frames can be made of wood, aluminum, or vinyl, each of which has advantages and disadvantages. Aluminum is very strong, light, and almost maintenance free, but it conducts heat rapidly, making it a poor choice for window frames. Even if the home's original windows have aluminum frames, vinyl storm window frames will be better than matching the metal ones.

Wood storm window frames have the same qualities as original wood window frames. They insulate well, but weather with age. They also expand and contract with varying weather conditions. This can make them hard to remove or close in summer and cause air leaks around them in winter. Wood frames are thicker than aluminum or plastic ones and so don't look attractive if used with either of the other types of original windows. Also, like original wood frames, they require more maintenance than metal or plastic ones.

Vinyl frames are usually made of polyvinyl chloride (PVC) with ultraviolet (UV) light stabilizers to reduce the damaging effects of sunlight on the material. These help, but long exposure to intense sunlight will still degrade vinyl substantially. It will also expand and warp at high temperatures, and crack in extremely low temperatures.

Whether hung inside or outside the original windows, storm windows must be hung parallel to the primary window and sealed within the window opening on all sides. The original window should also be carefully sealed and weather-stripped before the storm windows are hung. Weather stripping rather than caulking should be used to seal storm windows so they can be easily removed.

Exterior-mounted storm windows must have "weep holes" at the bottom of the frame to drain moisture that collects between the original window and the storm window. These drainage holes reduce the insulation value of the windows, but they are essential in damp climate areas. Without them water will condense and freeze between the windows in winter and cause rot to develop in the original window frame in summer.

Energy-Efficient Doors

Exterior wood doors permit heat loss about equally with windows of the same size. As with windows, replacing doors can increase the R-value by double or more. Wood exterior doors have an R-value of about R-2 if they are hollow core and R-3 if they are solid core. If a door contains a window, its R-value will be less. The R-values of most steel and fiberglass-clad entry doors range from R-5 to R-6 (not including the effects of a window.) A 1½-inch thick steel or glass door without a window provides double the insulating value of a solid wood door of the same size.

Sliding glass doors and glass paneled doors lose heat even faster than other types of doors because glass is a poor insulator. Modern glass doors with metal frames have a thermal break, which is a plastic insulator between the inner and outer parts of the frame. Doors with double- or triple-pane glass and low-conductivity gases between the glass panes can have an R-value up to R-4, almost four times the value of a single-pane door.

When replacing patio doors, keep in mind that hinged doors can be sealed more tightly than sliding types. Even with good weather stripping, it is impossible to stop all the air leakage through the upper and lower tracks of a sliding door and still be able to open and close the door. With use, the weather stripping wears down and air leakage increases as the door ages. Worn weather stripping can be replaced on some sliding doors, but not all.

Replacement doors are usually pre-hung in a steel or wood frame. The door will be properly weather-proofed in its frame, but must be properly installed to preserve its insulation R-value. The existing door frame and door must be removed from the rough opening and the new door must be placed as square as possible in the opening so that the door seals tightly to the jamb and swings properly.

Before replacing the interior trim, apply an expanding foam insulation to seal the new door frame to the rough opening and threshold. On the outside, caulk the new frame to seal it to the exterior siding or brick. The foam and caulking prevent air from leaking around the frame.

If replacement doors are too expensive or the existing doors are still in good condition, you can still increase their R-value by adding storm doors outside the existing door. Storm door frames can be made of aluminum, steel, fiberglass, or wood (painted or not). Metal-framed storm doors may contain foam insulation within their frames. Wooden frames require more maintenance than the other types and are becoming less popular.

High-quality storm doors include low-emissivity and sometimes double-paned glass. Most storm doors have a removable glass panel that can be replaced with a screen to provide ventilation. The screen provides no insulation value and should only be used during times when the HVAC system is not operating. In hot climates, a glass storm door placed over an older door may trap too much heat against the entry door and possibly damage it. In this case an opaque metal or plastic panel should be installed in place of the glass.

Energy-Efficient Skylights

The physical size of a skylight will strongly affect the temperature and the light level of the room below it. A skylight should not be larger in area than 5% of the floor area in rooms with wall windows and no more than 15% of the total floor area of rooms with no other windows.

The skylight's position is also important in determining its light and heat performance. In the northern hemisphere, north-facing skylights provide constant illumination and relatively low heat transfer. South-facing skylights are the opposite, providing maximum solar heat gain in all seasons and strong illumination. Those on east- or west-facing roofs provide a half day of strong heat gain and illumination (morning for east facing, afternoon for west facing) and a half day of medium light and moderate heat gain. As with storm doors and windows, the heat transfer characteristics you choose are active year-round, although some mitigation for skylights is possible from outside, as noted below.

ENERGY EFFICIENCY OUTSIDE THE HOME

Energy efficiency outside a home consists mainly in devising methods to block sunlight from creating heat transfer into the interior during hot seasons, and removing the blocking mechanisms to allow sunlight to assist in heating the interior during cold seasons. Sunlight is a powerful heat and light source. In summer, every square yard of a building's surface area is being bombarded with the equivalent of about 1500 watts of heat and light energy. In winter, the figure is somewhat lower because the sun is lower in the sky and its energy strikes at a more oblique angle.

Awnings and Screens

Awnings and opaque shades are the best way to prevent direct sunlight from shining through windows. Outside shades are far better for controlling sunlight than inside shades, which can block light very well but have little effect on heat transfer.

In all areas of the country where summer heat makes air conditioning a desirable feature in a home, awnings or outside shades on all the home's windows and glass doors will reduce the amount of cooling required by preventing heat from transferring through windows to the interior. Retractable awnings and roll-up shades have increased utility because they can be put out of the way during cold days when transfer of the sun's heat is a welcome supplement to the home heating system (Figure 5-11).

Shades are less expensive than awnings, especially retractable ones, and neither will last for more than a few years in direct sunlight. Ultraviolet rays in sunlight eventually destroy or deteriorate almost any material, but generally cotton, wood slats, or

FIGURE 5-11

A large retractable awning shields double doors with multiple glass panes from overhead sunlight and from solar glare reflected off the surface of the swimming pool.

bamboo fare better than synthetic fabrics or plastic. When buying shades or awnings, note what the retraction gear and cords are made of as well as the awning or shade itself. If these components fail first, the utility of the whole product is lost or greatly reduced.

Remote-controlled, electrically powered shades and retractable awnings are more expensive than those operated by hand, but they may be a good investment. If you need the ability to deploy or retract your home's window coverings often to take advantage of changing sunlight and weather conditions, motorized controls that can be set to a schedule or activated manually with the touch of a button are a real convenience, not to mention the most energy-efficient means of using the shades and awnings.

Landscaping for Your Location

Direct sunlight pours a large amount of heat energy through the roof, walls, and windows of a home in summer that the HVAC system must cool down in order to maintain a comfortable temperature inside the building. Similarly, wind—moving air—that is either much warmer or much colder than the preferred inside air temperature increases the work load of the HVAC system by forcing hot or cold air against the sides of the building and through any openings to the interior. Even if the hot or cold air can't penetrate into the home, it will increase the amount of heat transfer through the walls and roof in direct proportion to its temperature difference and its velocity.

Anything that can be done around a home that blocks sunlight in summer or that deflects wind in any season increases the home's energy efficiency by reducing heat loss or heat gain. Landscaping and outside structures can both contribute to this facet of energy efficiency.

Shade

Deciduous trees are natural sun blockers. Placed on the south side of a home, trees can keep direct sunlight off the south wall, the one most exposed to solar heating, for almost the entire day during the late spring, summer, and fall. Mature trees, with a height of 20 feet or more, can also shade much of the roof. In winter, when sunlight contributes useful heat to the home's interior, the trees shed their foliage and allow it through.

Trees placed on the west and east sides of a home won't do as much to block the summer sun as those on the south, but can still contribute a lot of useful shade. West-side plants are especially helpful in taking the edge off late afternoon sunlight coming through windows and beating against walls. North-side trees will contribute little to energy efficiency in the home because their shadows can only shade the home during the early mornings and late evenings of summer. Figure 5-12 shows a home with trees placed strategically on the south side for maximum shade.

Bushes can also block sunlight from a home, if placed near walls on the south, west, and east sides. Even ivy and other climbing plants that attach themselves to walls can absorb sunlight before it strikes the walls of a home and causes heat transfer to the interior. Like shade trees, both bushes and climbing plants should be deciduous so that their foliage won't block winter sunlight that is useful in the home.

FIGURE 5-12

A house filled with windows is made much more energy efficient by shading overhangs and south-side trees.

Windbreaks

Nothing the homeowner can install will stop wind, but there are ways to deflect it away from the home and to reduce its velocity (Figure 5-13). Before installing any windbreak you should first decide whether wind is a serious problem in your area. Homes in nearly all areas of the United States, except perhaps Alaska, will benefit to

FIGURE 5-13

Fence panels and landscaping are combined in this yard to deflect and reduce the velocity of wind; the more densely windbreak elements are concentrated, the more effective they are.

some extent from shade in summer, but only a fraction of the country is subject to persistent strong winds. You are probably well aware of the fact if you live in a windy area, but it is a good idea to check with your local weather bureau to find the number of days that your location experiences per year with wind above 10 miles per hour. If it is less than 100, you might want to skip any windbreaks in your yard. If it is more than 200, windbreaks can be a good investment. If it is in between, you will have to judge whether windbreaks make sense for your budget.

Windbreaks can be created in several ways. Among the most common are:

- Earth berms built up on the side of a home that faces the prevailing wind. A berm produces a lift in the wind that creates slower-moving, swirling air below, thus reducing its effect on the home.
- Walls and fences, solid or slotted, to allow passage of air through them, break up wind, and slow its velocity in the area immediately downwind, again benefiting the home beyond them (Figure 5-14).
- Dense foliage bushes or trees planted in a row also slow the wind and deflect some of its force. They can be planted on top of a berm to increase their effect or along a wall or fence.
- Outbuildings such as workshops, detached garages, storage sheds, and so on, if they can be placed between the home and the prevailing wind direction, are effective in splitting a wind stream to go around or over the home.

If you are planning on installing any of these features in your yard, only a small additional effort may be needed to place them where needed for a windbreak.

FIGURE 5-14

A solid 6-foot wood panel fence provides maximum protection from wind; note the heavy steel posts set in concrete that are required to support the wood panels against heavy wind pressure.

SUMMARY

To reduce heat transfer in and out of the home, all outside and inside leaks should be sealed airtight. Outside leaks should be addressed first because sealing them will prevent cold exterior air from penetrating into wall cavities and from there into the interior through inside leaks. Areas to be particularly checked include corners of brick or siding, joints where siding or roof and chimney meet, window casings and door frames, and the areas where brick or siding meets the foundation. After all outside leaks are corrected, any remaining air flow through interior leaks should also be blocked using expanding foam and caulking. Inside locations that may require attention include wall outlets and switches, window frames, baseboards, fireplace dampers, attic hatches, wall- or window-mounted air conditioners, and weather stripping around doors. If the home needs additional insulation, you should determine if any government or utility company grant programs are available to help pay for installing it. Federal, state, and local assistance as well as utility company incentive programs are available in many areas for energy-assistance upgrades. Insulation in a home's attic can be increased in R-value using either unfaced batts or blown-in fiberglass, rock wool, or cellulose. Wall insulation must be blown in through small holes cut in the top of each wall cavity. Home energy efficiency can be further upgraded by replacing windows, doors, and skylights with higher-R-value models. If replacement is too expensive, storm doors and windows as well as improved weather stripping and sealing can be almost as effective. On the outside of a home, awnings and shades can also help reduce heat transfer into the interior. Further reduction can result from strategically placed shade trees and bushes. Windbreaks such as berms, fences, walls, tree lines, and outbuildings can slow or deflect strong winds and also contribute to reduced heat transfer.

KEY TERMS

Baffle A cardboard or other rigid block across the space between joists or studs used to prevent insulation from obstructing vents on the outside of the baffle.

Batt insulation Fibrous insulation made of fiberglass or similar material formed into thick sheets (batts), sometimes with a facing adhered to one side.

Berm An artificial hill or rise in ground level intended to conceal a view or to deflect wind.

Blown-in insulation Any insulation composed of small bits of material that are blown into place through a hose using an air blower.

Caulking A viscous (thick) semi-liquid material that can be squeezed into place to plug leaks or seal cracks. Good-quality caulking forms a thin skin after placement, but does not harden for many years.

Cellulose insulation Insulation made of cellulose fibers, primarily composed of recycled newsprint and other paper.

Deciduous trees and bushes Plants that shed their leaves in fall and grow new ones in spring; used to provide shade in summer while allowing sunlight to pass in winter.

Door jamb The part of a door casing against which the door fits and seals to prevent air passage.

Double-pane glass Window glass consisting of two panes parallel to one another with an air- or gas-filled space between them.

Emissivity value A measure of how much radiation (of all wavelengths) a material absorbs and then emits again as radiant energy; emissivity values range between 0 (no absorption, 100% reflectivity) and 1 (100% absorption, no reflectivity)

Expanding foam insulation Semi-liquid insulation material that when squeezed out of a container expands to fill a space and seal against air passage.

Foam-in-place insulation Foam material that can be injected into small areas to control air leaks; same as expanding foam insulation.

Inside leak An opening from the interior of a building to the cavities in the walls that permits air to flow in either direction.

Insulation Any material intended to impede heat transfer; materials through which heat passes slowly or not at all.

Joist The cross members that support the floor of a building.

Loose-fill insulation Insulation made up of small bits of material that can be dumped into place; this type of insulation can also be blown into place.

Low-emissivity (Low-E) glass Glass that has been treated to reduce heat transfer; the treatment may be in the composition of the glass or in a film or coating applied to one or both surfaces.

National Fenestration Rating Council (NFRC) An organization that tests, certifies, and labels windows, doors, and skylights based on their energy performance ratings.

Outside leak An opening in the exterior surface of a building that allows air to pass from the outside into the wall cavities.

Rafters The cross members that support the ceiling of a building and form the angled framing of the roof above.

R-value A measure of insulation material's thermal (heat) resistance, that is, its ability to prevent the passage of heat energy through the material from one side to the other.

Single-pane glass Window glass composed of only one pane; usually found in older homes and buildings; not as energy efficient as double- or triple-pane glass.

Solar heat gain coefficient (SHGC) The amount of solar radiation admitted through a window, door, or skylight, and subsequently released as heat inside a home. The lower the SHGC, the less solar heat a glass panel transmits.

Storm door A glass-filled door placed on the outside of an exterior door to improve its resistance to heat transfer.

Storm window A second pane of glass or plastic placed inside or outside of an existing window to increase its resistance to heat transfer; the storm window makes the single-pane window a double-pane window.

Stud The upright frame members in the walls of a building.

Triple-pane glass Window glass consisting of three panes parallel to one another with an air- or gas-filled space between each pair of panes.

U-factor The rate at which a window, door, or skylight conducts heat flow.

Wall cavity The space in a wall between two studs that can be filled with insulation to improve energy efficiency.

Weather stripping Any sealing strip of material placed around the edge of a door or window to prevent air flow; can be made of rubber, fabric, metal, plastic, or any material that will improve the seal between the moving window frame or door and its casing.

Windbreak Any object that blocks, slows, or deflects wind.

Window casing The wood or metal structure that fits in a window opening and contains the movable window frame.

Window frame The wood or metal square or grid that contains the glass of a window and can usually be opened and closed.

REVIEW QUESTIONS

1. Why should outside air leaks be sealed before inside leaks?
2. True or False: Insulation material can be pressed down to a smaller thickness without affecting its R-value.
3. What are some areas to check when searching for outside air leaks?
4. How can a wall outlet or light switch be sealed against air flow?
5. If a fireplace damper leaks and can't be replaced, what alternative can be used to stop the leak?
6. Which is likely to be less expensive to have installed professionally in an attic, batt insulation or blown-in insulation?
7. What safety equipment is required when installing blown-in fiberglass insulation?
8. What do dark spots on existing attic insulation indicate?
9. True or False: Batt insulation added in an attic should be laid crossways to the joists in order to minimize the effect of any missed or settled areas in the existing insulation.
10. When cutting access holes for installing blown-in wall insulation, why should the holes be cut with angled sides?
11. What are the two ways by which glass transfers heat?
12. What is the U-factor?
13. True or False: Changing a window from single-pane to double-pane glass just about doubles its R-value, and changing to triple-pane glass about triples the R-value.
14. Are added storm windows as effective as changing windows from single- to double-pane glass?
15. What are "weep holes," and why are they necessary in outside storm windows?
16. True or False: A fiberglass- or steel-clad insulated door has double the R-value of a solid wood door.
17. Why do sliding doors have a lower insulation value than doors that open on hinges?
18. If a room has no other windows, what is the maximum area that a skylight should have in order not to overheat or overcool the room?
19. Should awnings or shades be placed on south-facing windows or north-facing windows? Why?
20. Will the trees placed on the west side of a home be more effective for shading the building than those placed on the east side?

CHAPTER

6

TAMING THE ENERGY HOGS IN YOUR HOME

OBJECTIVES

After studying this chapter, you should be able to:

- Describe how heating and air-conditioning (HVAC) systems work.
- Increase the energy efficiency of the HVAC system, space heaters, and room air conditioners.
- Define the characteristics of tank-type and tankless water heaters and describe how to obtain maximum energy efficiency from each.
- Describe how clothes washers, dryers, dishwashers, stoves, and pumps use energy to perform home work tasks and how each can be made to operate more efficiently.

OUTLINE

I. Introduction
II. Design and Operation of HVAC Systems
 A. Central Heating Systems
 B. Central Air Conditioning (A/C)
 C. Single- and Multiple-Unit HVAC Systems
 D. Zoned and Non-zoned HVAC Systems
III. Improving the Energy Efficiency of a Home HVAC System
 A. Seal and Insulate Ducts
 B. Maintain the HVAC System
 C. Automate the HVAC System
IV. Space Heaters
V. Room Air Conditioners
VI. Water Heaters
 A. Tempering Tanks
 B. Lowering Water Temperature
 C. Tankless Water Heaters
VII. Clothes Washers, Dryers, Dishwashers, Stoves, and Pumps
 A. Clothes Washers
 B. Clothes Dryers
 C. Dishwashers
 D. Stoves
 E. Pumps
VIII. Summary

INTRODUCTION

In this chapter you will learn that most of the energy consumed in a home is used by the heating, ventilation, and air-conditioning (HVAC) system and a few other high-current electrical devices. You will learn how to reduce the energy consumption and the cost of the HVAC system in two ways: by making the system operate more efficiently and by reducing the amount of time it must operate to perform its job. You will also learn how to get maximum efficiency from the other major electrical appliances in the home and how to determine which of them to use for various functions. Finally, you will learn how to reduce the cost of electric power in areas with demand billing by timing the use of major appliances as efficiently as possible.

DESIGN AND OPERATION OF HVAC SYSTEMS

HEATING, VENTILATION, AND air-conditioning (HVAC) systems in American homes vary as widely as the locations and climates in which they are located. Heating systems for low-altitude southern-latitude homes may be little more than a few space heaters placed in strategic locations, whereas whole-house air conditioning (A/C) is a must. At the other extreme, a home in the mountains of Montana or almost anywhere in the northern tier of states may need a central heating system for 9 months of the year and no A/C at all. And then there are many areas, such as Chicago, Omaha, and Denver (to name just three of hundreds), where central A/C and central heating are both necessities for year-round home comfort.

Maximizing the energy efficiency of an HVAC system means controlling two factors: the area (volume of air) in the home that the system must heat and cool, and the temperature level to which the system must heat or cool that area. Most automated HVAC control systems are designed to work with central heating and A/C rather than with room-size units. We'll discuss these whole-house systems in the next section and deal with the smaller units later.

Central Heating Systems

Central heating systems are divided into two types, forced air and radiant. Forced-air systems all function by drawing air from the home's interior into a furnace, where it is heated and then blown out again into the home through heating registers connected to the furnace by metal ducts. All forced-air heating systems rely on the continual transfer of air from the home's interior to the furnace (where it is heated) and back again. They keep the air temperature in the room comfortable by heating the air itself.

A typical forced-air heating system is shown in Figure 6-1. The system is shown combined with a central A/C system, which uses the same ducts and blower as does the furnace. When heating is required, the furnace provides it, and the A/C unit is idle. When cooling is necessary, the furnace is idle, and the air conditioner takes over.

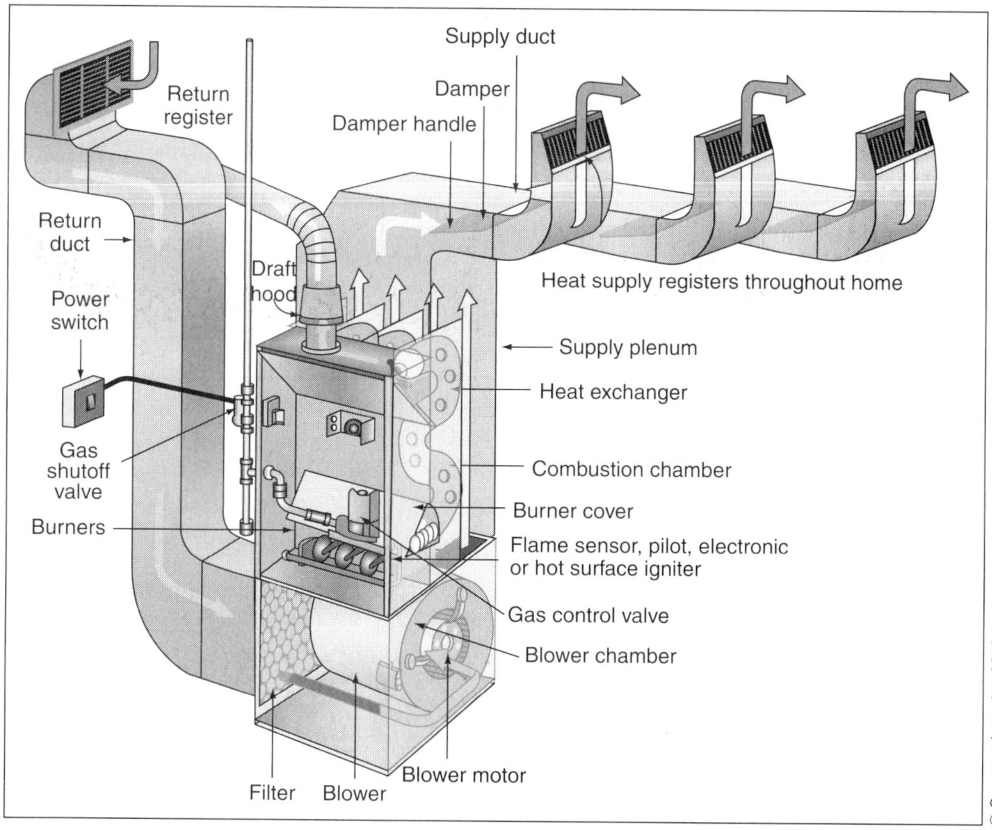

FIGURE 6-1

A forced-air heating system can be fueled by either natural gas or oil and is located in the basement or on the main floor of a home.

Radiant heating systems don't have any air movement system. They work by heating the air in the house through convection or by heating the house structure itself rather than the air inside it. Older radiant systems use circulating hot water or steam from a central boiler to heat standing radiators or baseboard radiators in each room of the house. These units warm the air around them, which then expands and becomes lighter. The heated lighter air rises away from the radiator and is replaced by cooler air, which, in turn, is heated and rises away from the radiator. Radiator heating systems do not heat evenly and are not as efficient as other radiant or forced-air systems. Few, if any, of these systems are being installed in new homes, although millions still function in older homes.

Radiant heating systems now being used in newer homes consist of a large water heater connected to pipes that run at spaced intervals under the floors of the home. The heating system operates by heating water in the central boiler and then circulating the water through all the pipes beneath all the floors of the house. Heat from the pipes is absorbed into the structure of the house and radiates from there to the home's interior air, warming it to a comfortable temperature. A section of a radiant heating system is shown in Figure 6-2.

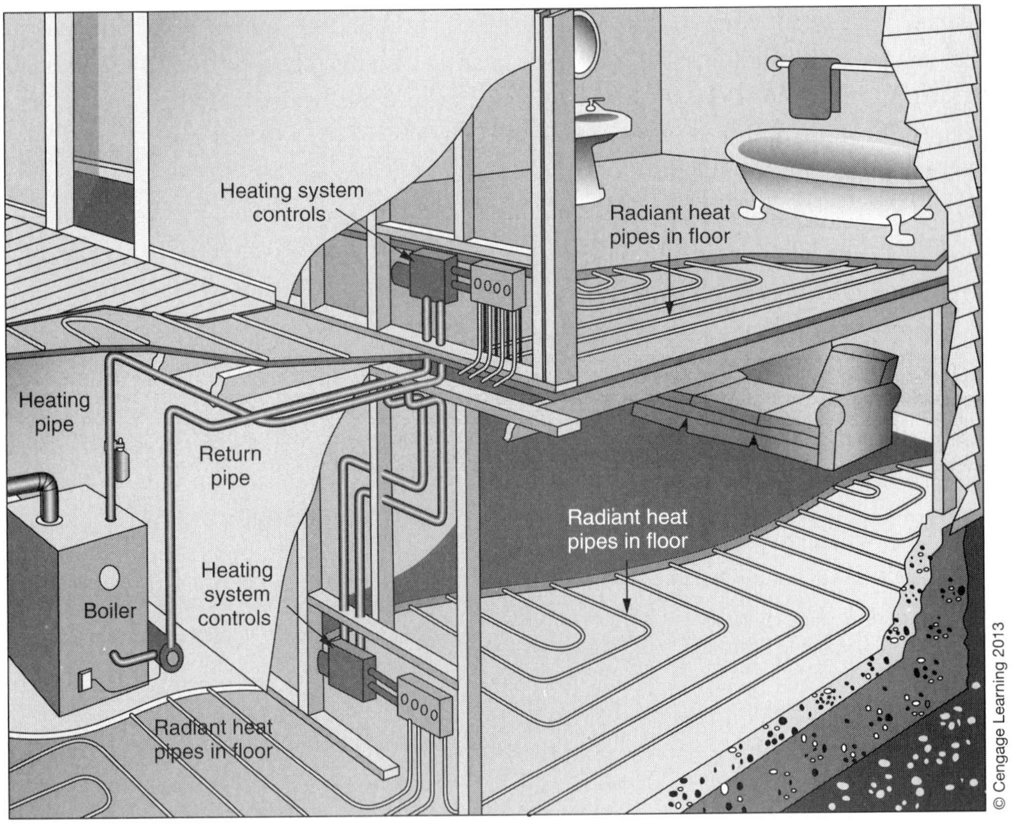

FIGURE 6-2

This radiant heating system uses pipes in the home's floor to heat the structure of the home and radiate warmth into the interior.

Although this heating method sounds less efficient than forced-air heating, it is actually more energy efficient and provides more even heat distribution throughout the home than the other type. For these reasons, radiant heating systems, though more expensive to install initially, are growing in popularity.

Radiant systems do not, however, work well as air-conditioning systems because cold cannot be radiated into the air as heat can. A cooled structure absorbs heat from the surrounding air, but not rapidly enough to meet the requirements of air conditioning in the home. Consequently, radiant heat systems are mostly installed in homes located where air conditioning is not needed. A separate forced-air cooling system can be installed with radiant heating, but this considerably increases the expense of the home's total HVAC system.

Furnace Fuels

Fuel for the combustion chambers of most modern furnaces is either heating oil or natural gas. Many coal-burning furnaces remain in use, but they are a declining minority. Some that were once coal-burning have been converted to burn oil or gas, and others are gradually being replaced with more energy-efficient oil or gas designs. All new residential furnace systems use oil or gas.

As noted in Chapter 4, natural gas is the least expensive fuel for furnace heating, although its cost is rising. Coal is less expensive than oil, but is still declining in use because oil is cleaner and easier to use. Coal-fired furnaces are inherently less efficient than those that use oil or gas.

Both forced-air furnaces and radiant systems can be fueled by either oil or gas. The choice of which to use is usually a matter of which is available. Gas tends to be preferred wherever it is available, both because it is cheaper and because it burns cleaner (the main combustion products are carbon dioxide and water). Gas also leaves no ash or residue in the furnace, which must be removed periodically.

The amount of either oil or gas fuel that flows to a furnace is not controlled as a means of regulating the unit's heat output. Fuel settings are constant in both oil and gas furnaces, as is the amount of heat produced by the units when running. The total amount of heat produced by the system to heat the home is controlled simply by turning the furnace on or off. When enough heat has been added to bring the interior temperature up to a preset level, the furnace turns off. No more heat is added and the temperature eventually falls below the set level, at which point the furnace activates and adds more heat to bring it up again.

Some newer furnaces for use in very cold climates have two operating levels; the first level is used for most heating and operates the furnace at up to 75% of its rated capacity. The second level, which operates the furnace at maximum capacity, is used only if the first level is unable to maintain a comfortable temperature in the home because of extreme cold outside.

Because the fuel consumption and hence the energy cost of operating a furnace is a constant whenever the unit is running, the only way to increase its energy efficiency is to reduce the amount of time that it operates. Before discussing ways to do this, let's look at air-conditioning systems to determine if they are also constant costs when operating.

Central Air Conditioning (A/C)

Central air-conditioning (A/C) systems are all forced-air designs. They move air in much the same way as a furnace heating system. In fact, most A/C systems use the same ducts and vents as the heating system. Instead of heating the air that is drawn into the system from the home's interior, the air conditioner cools it by refrigeration and then blows it back into the home through the duct system.

Refrigeration cooling is provided by the A/C condenser and compressor unit, which is located outside the home. Chilled refrigerant from the condenser unit flows to the A/C unit and cools the air passing through it. Excess humidity condenses out of the cooled air and is carried away by a drain pipe.

A/C units are mainly electric powered and use no other fuel. The compressor in the condenser unit requires a fairly large motor to operate it. The blower unit that moves air over the cooling coil and circulates it through the home is also powered by an electric motor. These motors make air conditioners heavy consumers of electric power. In most homes the A/C system is the largest user of electric power because it requires substantial power and because it operates for long periods of time.

Air conditioners work like a refrigerator to cool air and circulate it. Air conditioners consist of a condensing/compression unit and a cooling coil through which air is blown by a fan or blower. The condensing/compression unit and cooling coil are connected by pipes or tubing so that a refrigerant fluid can circulate between them. The condenser unit is located outside the home so that the heat it extracts from the cooled air can be exhausted to the outside air. The cooling unit is located inside in the main air-flow duct of the HVAC system where it can cool the air passing through it.

The refrigerant, which is a gas when it is at room temperature and pressure, is held inside the sealed circulation system in the A/C unit. As the refrigerant passes through the condenser/compressor unit, it is compressed by a compressor into a liquid under high pressure. Compressing the gas makes the liquid it forms hot and it is cooled by the fan of the condenser/compressor unit as it passes through the condenser coil. From the coil, the cooled liquid, still under high pressure, flows to the cooling coil in the air-conditioner unit. There it is permitted to expand back into a gas at lower pressure and, as it does so, it absorbs a large amount of heat from the cooling coil and the surrounding air. This cools the air, which then is forced into the home by a blower. The expanded gas, which has absorbed heat from the air, flows back to the condenser/compressor unit, where the processes of condensation and compression turn it back into liquid form and its heat is extracted in the condenser coil by the cooling fan of the condenser. Figure 6-3 shows how this cooling cycle in an air conditioner works.

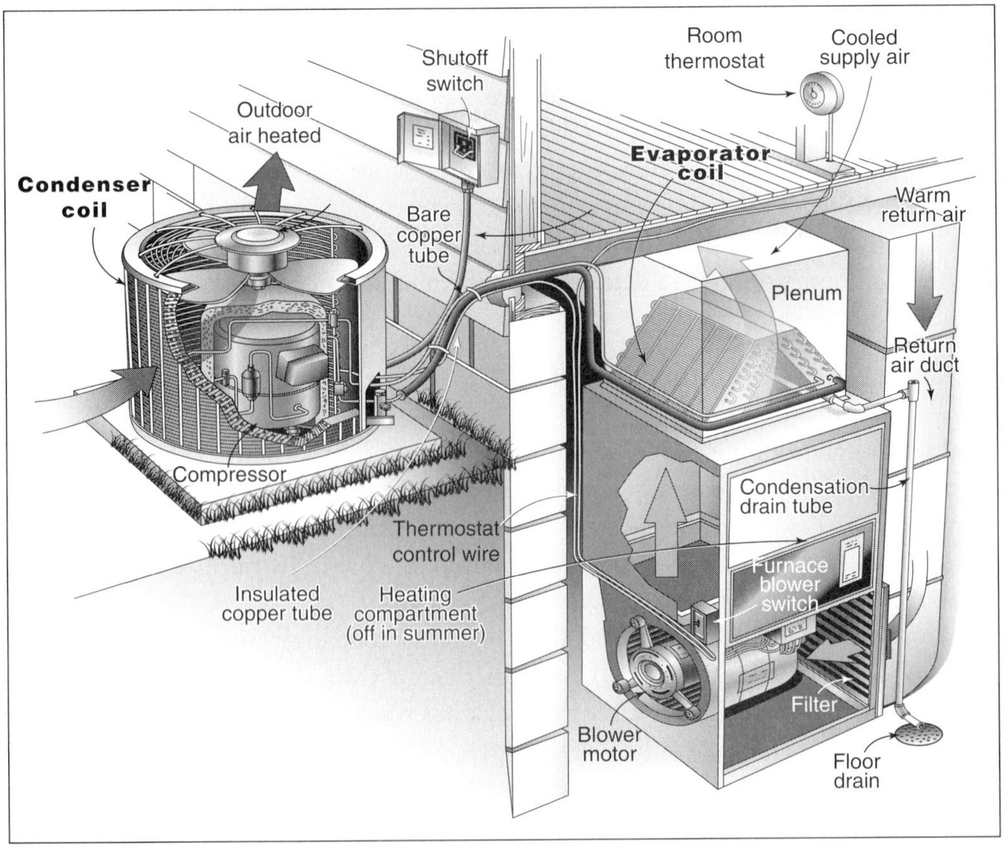

FIGURE 6-3

A combination heating and air-conditioning system has its blower and heating and cooling units inside the home, whereas the condenser, compressor, and heat extraction components are outside for better heat diffusion to the atmosphere.

Single- and Multiple-Unit HVAC Systems

Most homes have only a single forced-air furnace, which serves to heat the entire house. The larger the home is, however, the more difficult it is to heat adequately and evenly with a single furnace. Long duct runs from the centrally located furnace to the furthest rooms in the home dissipate some of the heat in the air traveling through them so that by the time it reaches its destination, its temperature has dropped too low to raise the room temperature to an acceptable level. Similarly, the sheer size of many homes may exceed the heating capacity of a single furnace.

The efficiency of A/C systems is also limited by the distance their cooled air must travel and the size of the blower required to move it. Though blowers run more quietly than fans, they still make some noise and the noise level increases with the size of the blower. As with a single-unit furnace, a point is reached where the size of the home exceeds the ability of one A/C unit to effectively cool it. A second unit is the logical answer.

The Air Conditioning Contractors of America (ACCA) publishes Manual J, which contains the industry standard for calculating heating and air-conditioning loads in residential buildings. To meet or exceed its recommended furnace capacity, many homes have two or more furnaces and A/C systems located in different parts of the structure. Whenever there are two or more furnaces and A/C systems in a home, the HVAC system is by definition divided into zones, each of which is heated by a different furnace and cooled by a separate A/C system. One zone, with one HVAC system handling its heating and cooling, might be, for example, the main floor of the home, while another HVAC system handles the lower floor or the second floor. The controls of each system must be programmed to work together or the efficiency of both is degraded and the home is poorly heated and cooled.

Radiant heating systems are less likely to require two or more separate heating units than forced-air systems. This is because radiant home heating systems can be larger in overall capacity than forced-air units and because radiant systems are more efficient in carrying heat over long distances without significant loss. Finally, radiant heating systems can be more easily divided into zones and their heat distribution adjusted for distance than forced air systems. Few homes with a floor area of less than 5000 square feet have more than one radiant heating unit, and many of these systems can go to even higher capacity.

Homes with radiant heating systems may have a forced-air A/C system for cooling. The heating and A/C systems are completely separate, but they can be controlled by one set of thermostats and a single central controller.

Zoned and Non-zoned HVAC Systems

Heating and cooling zones are not just the result of multiple furnaces or air conditioners. They can be set up in homes with only a single furnace and A/C unit as well. The purpose of heating zones is to more evenly and efficiently distribute heat throughout the home. The two factors that make zones necessary in many home HVAC systems are distances from the central furnace to some parts of the home and the effects of outside climate.

In an HVAC system, the further heated or cooled air travels from its source (the HVAC system), the more heat loss or gain occurs in transit. Warmed air loses heat to the duct system and cooled air absorbs heat from the ducts as it moves. In a home with only one zone and relatively equal heated air distribution to all rooms, this results in the rooms closest to the HVAC system being the warmest in winter weather and the coolest in summer. Those furthest away are the reverse: warmer in summer and colder in winter.

Radiant heating systems with only one zone also have this problem. The heating water cools more by the time it reaches the most distant distribution pipes than it does when traveling to the nearby ones. The result again is uneven heating.

The A/C system also must deal with unequal heat gain happening in different parts of the home. Summer sunlight heats the more exposed southern side of the building more than the northern side. In this situation, equally distributed cool air results in an unequally cooled home, with south-side rooms remaining too warm because they do not receive enough cooled air.

Even with good insulation in the walls, attic, and floors, some heat transfer occurs from the warmer side of these surfaces to the colder one in both summer and winter weather. Figure 6-4 shows the approximate percentage of heat loss or gain through different parts of a home's structure and how these variations can result in unequal heating and cooling if the home has only a single zone.

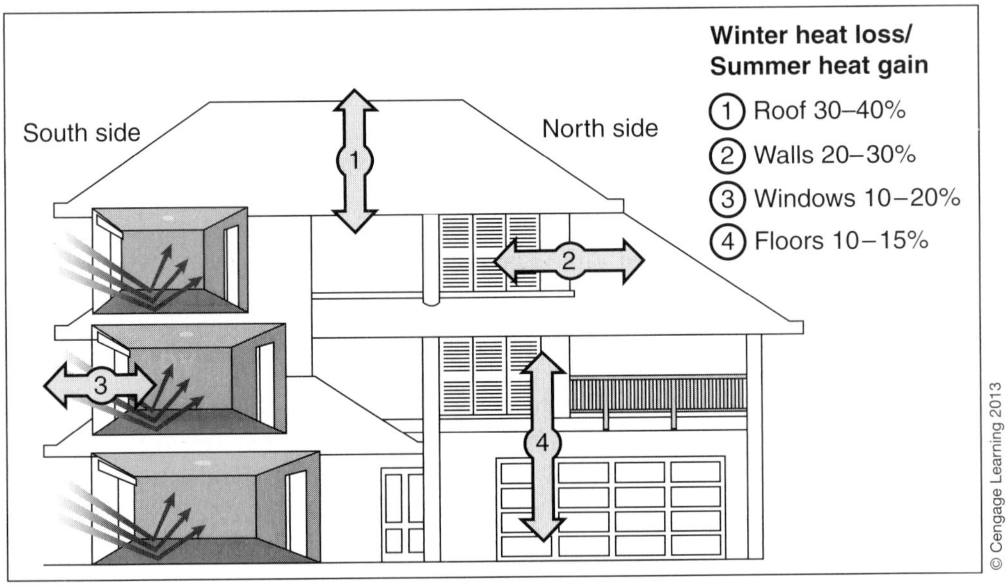

FIGURE 6-4

Heat transfer in and out of a home creates uneven temperatures in different rooms. Greater transfer through the roof makes the upper floor colder in winter and hotter in summer.

Dividing the home's HVAC system into several zones allows the heating and cooling level of each zone to be adjusted independently. Each zone can be set for a comfortable temperature level and the HVAC system can provide varying levels of heating or cooling to the zones to maintain that temperature. In addition to maintaining a more even temperature throughout the home, a multiple-zone HVAC system may use less energy than a single-zone system because it does not need to provide the maximum amount of heating or cooling to every part of the home, but only to the zone with the greatest heat loss or gain.

Figure 6-5 shows a house divided into zones so that each can be controlled independently. In this example, Zones 4, 5, and 6 will require more heating because they

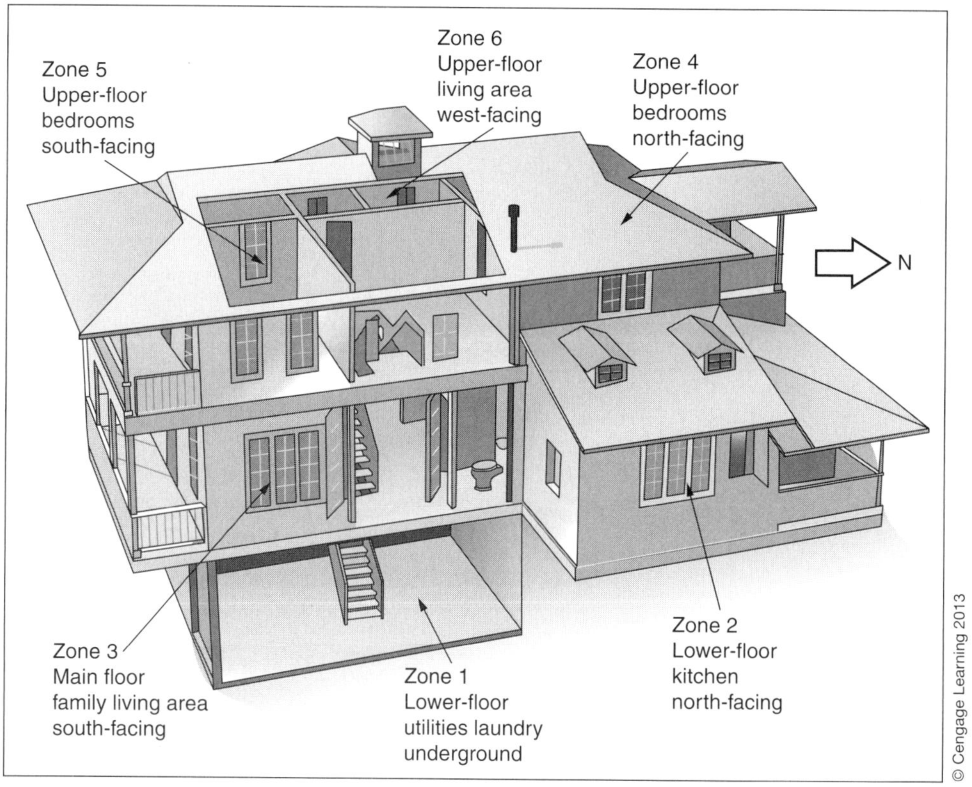

FIGURE 6-5

Dividing a home into zones allows the HVAC system to heat and cool the home more evenly by adjusting the level of each zone independently.

are upper-floor zones and will lose more heat through the roof. Zones 3 and 5 will require more cooling because they are south-facing and have large windows. Zone 5 will require the most heating because it is also upper floor. Zone 2 may require less heating because it contains many heat-producing appliances. For the same reason it may require more cooling. Zone 2 may require little heating or cooling because of its insulated location and because it is infrequently occupied.

Separate temperature controls in each zone allow the system to compensate for inequalities in heating and cooling in the different zones of the home by sending more or less heated or cooled air to needed areas, independent of what is happening in other zones. Each zone is treated like a separate apartment with its own HVAC system and controls. If the controls of each zone had to be continually adjusted for changing conditions by hand, the system would be too cumbersome to use effectively. Automating the HVAC system allows it to be controlled by a preset program based on input from temperature sensors (thermostats) located in each zone.

Figure 6-6 shows a ceiling-installed air-conditioning system such as might be used in a home with radiant heating. Each room in this home is a separate zone with its own thermostat in the room and a remotely operated damper placed in the air duct leading to the ceiling air vent. Cooled air flow to each room is regulated by the damper, which receives its signals from a central controller. The controller receives temperature data from each room thermostat and adjusts all the duct dampers to maintain the programmed temperature levels. Once programmed and put in operation, the system requires little additional input or maintenance.

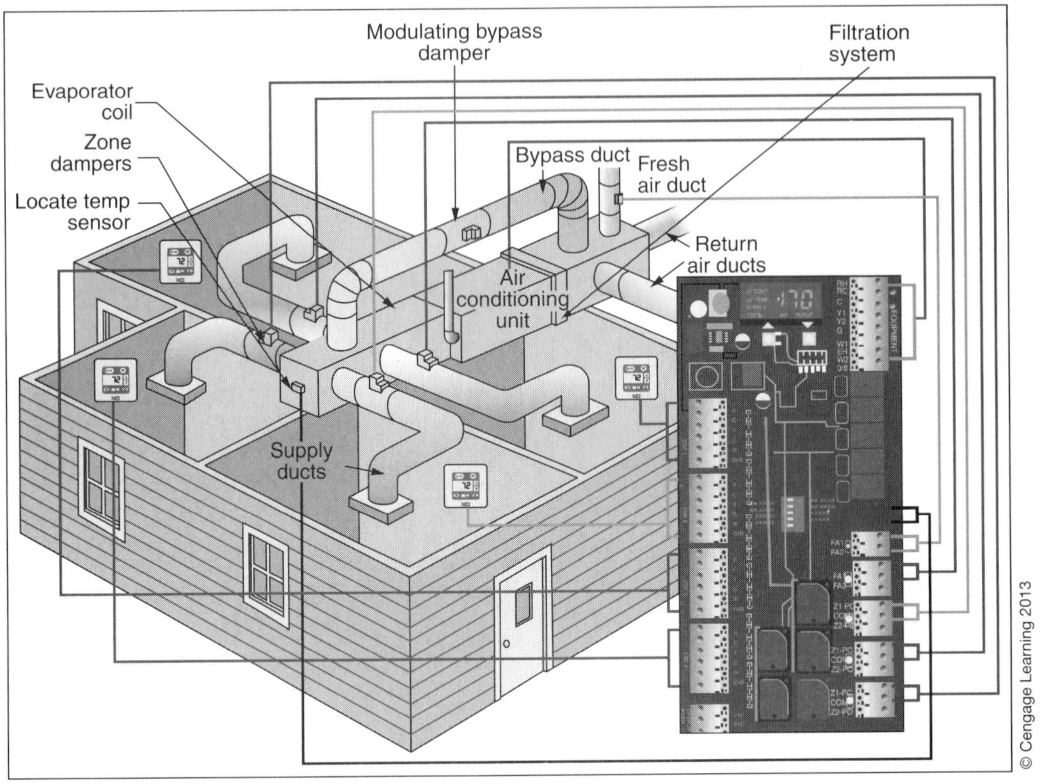

FIGURE 6-6

A programmed controller adjusts dampers to maintain the set temperature in each of four zones. Each room's thermostat can be set to the same or a different temperature.

IMPROVING THE ENERGY EFFICIENCY OF A HOME HVAC SYSTEM

After reducing heat transfer in and out of the home by first sealing leaks and then increasing insulation to the maximum practical R-levels, there are several additional ways to improve the energy efficiency of the home's HVAC system.

Seal and Insulate Ducts

Think of the ductwork in your home as large hoses that deliver warm or cool air instead of water to the interior. Because they're mostly out of sight, ducts go unnoticed and can leak for years without any obvious sign of the defect. All of the ducts that distribute heated and cooled air should be sealed airtight, but frequently they are not. Leaks can be the result of poor-quality original construction or the effects of aging on sealing tapes and other components. Whatever the cause, any conditioned air leaking from ducts before reaching its destination is a direct loss of efficiency for the system and an increased cost to the homeowner.

Field research by the California Energy Commission indicates the average home can reduce heating and cooling costs by 10% by eliminating duct leaks. The California standards that were implemented in 2005 still allow up to 15% air leakage from ducts before repairs are required. That is a high tolerance level, but the fact is that leakage can be reduced close to zero by careful inspection and sealing.

Check for leaks wherever the ducts are exposed and seal them with good-quality duct tape or mastic. If duct sections are not firmly attached in place or fit loosely to one another, they should be fastened together with metal screws and fixed in place with metal straps or hangers to joists or wall studs before any attempt is made to seal leaks. If any leaks are openings large enough for mice or insects to enter, the inside of the ducts should be checked for unwanted residents before they are sealed again.

Duct tape is the easiest material to use in sealing duct leaks, but it has serious disadvantages. Nearly all duct tape adhesive will dry out and harden over time. Hardened tape easily detaches from the duct and allows the leak to resume. If you use duct tape, use only the highest quality with adhesive guaranteed to last a specific length of time under use conditions (alternate heating and cooling). If you can't find long-lasting tape, use mastic that can be applied with a putty knife or squeezed from a tube and then smoothed into place. Mastic forms a skin on its outer surface where it is exposed to air. The skin prevents further contact with air and the interior mastic remains pliable for many years. Duct leaks sealed with mastic should remain tight for the life of the home.

Ducts concealed within finished walls can be checked indirectly for leaks. After outside and inside air leaks in the home's structure have been sealed, turn on the furnace or air conditioner and check the interior locations you previously checked for air leaks again. If you feel warm or cool air entering a room through wall outlets, around baseboards, or through fixtures, a duct leak may be the cause.

Repairing concealed duct leaks is difficult and usually expensive. It can be done by removing the wallboard to expose the leaking duct, or by injecting expanding foam sealant into the wall near the location of the leak in order to fill the opening and block air flow. Both of these procedures involve some damage to the walls with the accompanying risk of failure to find the leak in the predicted area. A compromise solution is to seal the places where leaking air is entering the room tightly enough that the conditioned air will find an easier path flowing through the duct to its destination instead of out through the leak.

Heat loss also occurs by conduction from ducts just as it does from glass, wallboard, or siding. Where the ducts run through interior areas where the surrounding air in conditioned, the loss is reduced, but where they run through unheated basements or other open unconditioned spaces, heat loss can be significant.

Ducts in an unheated basement, crawl space, garage, or attic should always be insulated, but in most older homes they are not. Fortunately, this problem can be corrected using batt insulation that the homeowner can install. Before you start, be sure the ducts are completely sealed airtight. Any leaks that remain will permit air to push under the insulation and possibly break the seal that makes it effective. Also be sure that you wear gloves, a respirator, and long-sleeved clothes when working with any insulation.

Insulation is much easier to handle and attach in place if four hands are available for the job. Find a helper and you'll more than cut the time for the job in half.

Use batts that are faced on one side with aluminum foil or Mylar. Measure the circumference of the rectangular or round duct and cut the batt to that length if the duct is rectangular, an inch or so longer if it is round. Wrap the cut piece around the duct, butt the ends together, and secure it with duct tape. Use insulation in widths of 3 or 4 feet to minimize measuring and cutting. Continue this process the entire length of the duct. As you encounter hangers or other obstacles, cut the batt to go around them, but don't press down its thickness if you can avoid it. Doing so reduces the R-value of the insulation. Once done, go back and tape completely along each seam where insulation batts join with one another.

If you find that some areas are too tight to fit insulation completely around the duct, wrap it as much as you can and tape the batt in place to seal it to the duct along the edge of the area where insulation won't fit. Again, don't press down the insulation. Tape its full thickness to the duct and cover the exposed end with more tape. Figure 6-7 shows ducts in an older home wrapped with foil-faced insulation.

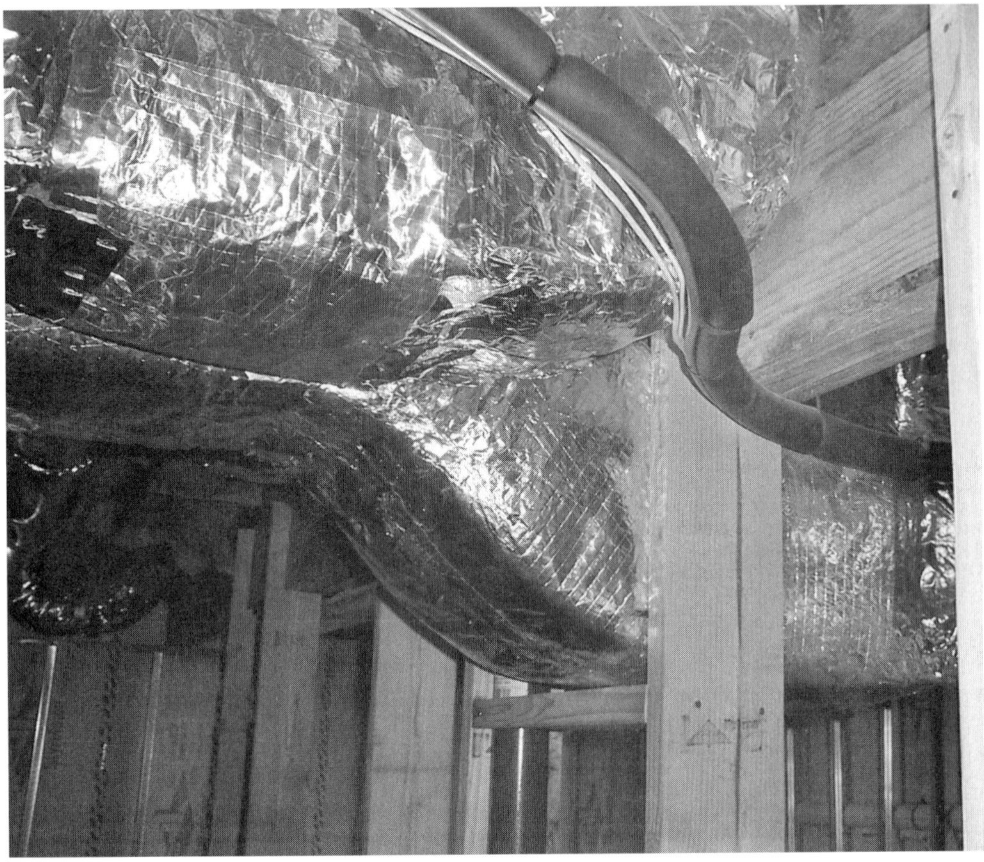

© Cengage Learning 2013

FIGURE 6-7

Insulating ducts, especially in areas where they pass through unheated spaces, will increase HVAC efficiency significantly.

Maintain the HVAC System

HVAC systems are generally very reliable, but they should be checked at least annually to maintain good performance. Among other items, air-conditioner refrigerant level and compressor operation should be checked. In furnaces, combustion chambers should be checked for cracks or corrosion, blower motor belts should be checked for wear, and bearings should be lubricated if necessary.

In all HVAC units the air filters should be checked regularly and replaced as frequently as necessary. A dirty air filter impedes the flow of air through the HVAC system and can impair efficiency more than any other defect. There is no set interval for replacing filters. The schedule depends on how continuously the HVAC system operates and on the level of pollution and dust in the ambient air. A replacement filter is cheaper than the energy the HVAC system will waste trying to drive air through a plugged filter. For this reason, it is better to err in replacing the filter too often than to wait and find that your system has been operating for a month with a filter through which no light, let alone air, can pass.

Automate the HVAC System

After you've done all you can to reduce heat transfer in and out of your home and in and out of the ductwork of your HVAC system, still more efficiency can be achieved by reducing the amount of time the system operates so that it does not condition air in areas where it is not needed or at times when it is not needed.

Adding Zones

If your home currently has a whole-house HVAC system for heating, cooling, or both, you may be able to save money by dividing the system into two or more zones. To determine if a zoned system would save energy, answer the following questions:

- Are large parts of your home (second floor, bedrooms) unoccupied for 2 hours or more during a 24-hour period?
- Are some areas of the home warm or cold while the rest of the house is at a comfortable temperature?

If the answer to either question is yes, adding zones to the HVAC system may be an advantage, but only if it can be done cost effectively. To determine this, first determine which areas in the home should be zoned separately. Decide on only as many zones as are needed to equalize heating and cooling in the home or to permit temperatures in some areas to be raised or lowered when unoccupied. Zoning isn't needed if you just want to raise or lower the temperature in the whole house at night or while the family is away. That can be accomplished with a programmable thermostat, as we will discuss later. You only need zones if different areas of the home need different heating and cooling levels at the same time.

When you have decided whether zones are needed and their location, check the ductwork of the HVAC system to determine if access is available to the ducts leading to the areas you want to zone. If it isn't, installing zone hardware will likely be an expensive project that won't be cost effective. If the ducts are accessible, you can hire a heating contractor to install wireless remote-controlled dampers in the ducts. Local building codes may permit you to do the job yourself, but even with easy access, it is not a simple installation and in most cases is better left to professionals.

Install Programmable Thermostats

Whether or not your home has a zoned HVAC system, a programmable thermostat is a must for the efficiency of any system. A simple programmable thermostat, allows you to adjust the temperature in the whole home or in one zone four times during a 24-hour period. A typical schedule might look like that in Table 6-1.

TABLE 6-1 A Typical Four-Part Schedule for a Programmable Thermostat

TIME OF DAY	TEMPERATURE SETTING	ACTIVITY IN THE HOME
6:30 a.m.	72 degrees	Family awakens, prepares for work and/or school
8:00 a.m.	62 degrees	Family leaves, home unoccupied
4:00 p.m.	72 degrees	Family returns, active in home
11:00 p.m.	60 degrees	Family asleep

In the example in Table 6-1, the HVAC system operates at a reduced level for 15½ hours in every 24-hour period. This will save a minimum of 10%, and perhaps much more, depending on weather conditions, of the energy it normally consumes with a corresponding reduction in operating cost. If the thermostat controls the entire HVAC system, the savings will apply throughout the home. If it controls only one zone, the energy savings may be less, but still substantial.

Installing programmable thermostats in each zone may result in still larger energy savings because some areas of the home, such as bedrooms, may be kept at reduced temperature levels for 22 hours of each 24-hour period. Each zone can be programmed independently and the HVAC system can maintain the set temperatures in each zone without regard to the others in the home. This allows the system to heat and cool each area of the home as its use and its individual energy requirements dictate. The energy and cost savings from operating the HVAC system only when and where it is actually needed can run as high as 25%.

SPACE HEATERS

Residential space heaters are appliances designed to heat a single room or an even smaller area within a room rather than an entire house. Space heaters can be permanently installed in a wall or ceiling, or they can be free-standing portable units that connect to a wall outlet and can be moved from room to room as the need requires. The two main types of heaters are defined by their method of transferring heat.

Convective heaters transfer heat by convection. The heating elements, which can be electric or fuel-burning, warm the air surrounding them or heat an intermediate filler such as oil or stones, which in turn transfers heat to the air. The warm air then warms objects and people in the room or space. Convective heaters provide constant, diffuse heat, but need an airtight and insulated room in which to work efficiently. Air leaks and heat that transfer through poorly insulated walls require them to work harder to maintain a set temperature, just as in a central HVAC system.

Convective space heaters are often used for supplemental heating in bathrooms, entry ways, and other areas in addition to the main heating system. They work well in these applications but should always be thermostat controlled so as not to waste energy

FIGURE 6-8

A 1500-watt European-style electric wall heater can warm the confined space of a bathroom quickly and costs about $0.20 per hour to operate.

by overheating an area. A controlled space heater can serve the same function as zone control in a room that occasionally requires additional heat to maintain a comfortable temperature. The heater can be as energy efficient as the zone and may be easier to install and control.

Radiant heaters, such as those shown in Figure 6-8, usually have tungsten filaments protected by heat-resistant quartz envelopes. These are mounted in front of metal reflectors and operate much like incandescent light bulbs. Instead of visible light, however, they radiate their energy in the infrared heat spectrum, converting up to 85% of their input power into radiant energy and transforming the remainder to conductive and convective heat. Radiant heat passes through air without warming it much and is absorbed directly by solid objects, including clothing and skin. For this reason, radiant heaters can keep people warm even in poorly insulated areas or outside a building. They require more energy to do so, however, because heat transfer from the warmed objects and people to the surrounding cooler air continually lowers their ambient temperature. Like convective heaters, radiant heaters work well to provide supplemental heat in enclosed areas. They are most efficient in applications where heat loss is limited and the heated area is sealed.

All radiant heaters and many convective heaters use electricity as their power source, but convective heaters can also be fueled by natural gas, propane, heating oil, or kerosene. These fuels all consume oxygen from the room in which the heater is located. They also produce as combustion products both carbon dioxide and water vapor. Combustion-type heaters should only be used in ventilated areas lest oxygen levels be depleted or carbon dioxide build up to toxic level while they are operating. A low oxygen level in a room may also cause a fueled space heater to begin producing carbon monoxide, which is even more toxic. Ventilation from the outside reduces the efficiency of the heater by the influx of cold outside air, but it is still a necessary safety precaution for any heater that operates with an open flame.

The best efficiency for all types of heaters is to operate them only when needed to maintain a comfortable temperature in an occupied space. This can be done by the diligent use of manual controls, but will probably be more consistent when control is given to an automated thermostat and a motion sensor to determine if the room is in use.

ROOM AIR CONDITIONERS

Window or wall-mounted air conditioners are the cooling equivalent of space heaters. They work well in a confined, well-insulated space, but multiple units are not effective as a substitute for whole-house air conditioning. Room air conditioners must be mounted so that the heat extracted from the room air can be expelled to the outside and not re-circulated back into the home. Any room air conditioner that does not vent heat to the outside will not cool a room. A stream of cold air may blow from its front, but an equal or greater amount of heat is being discharged from the unit's heat exchanger back into the room. In applications where outside venting is not possible, an ordinary fan or an evaporative cooler (in dry climates) is a better cooling option than an air conditioner.

The best efficiency from a room air conditioner is obtained by sealing it carefully within its window or wall mounting. Doing so keeps the conditioned air inside the room and the extracted heat outside (Figure 6-9). If a room air conditioner is used as a supplement to the home's central air conditioning in an area where additional cooling is needed, it should be controlled with a thermostat to avoid overuse.

FIGURE 6-9

This correctly installed window air conditioner is sealed on all sides so that all the heat it extracts is vented to the outside while the cool air it produces is directed inward.

WATER HEATERS

Most residential water heaters in the United States are the tank type that consist of a cylindrical upright container in which water is kept continuously hot and ready for use. Tanks range in size from 20 to 100 gallons. Natural gas is the most popular fuel for water heaters, but many are also use electricity, propane, or heating oil as energy sources. Gas is the least expensive energy source and, because heating water requires large amounts of energy, the cost differential between gas and other energy sources for water heaters is large.

Tank-type water heaters that store heated water use energy at a slower rate than the other major type of water heater, known as a tankless or on-demand heater. After being heated, water is stored in the tank for later use, but over time it will cool and require reheating. When the heated water in the tank is used, additional incoming water that replaces it must be heated. This continual heating and reheating consumes large amounts of energy and makes water heaters one of the largest energy users in the home. Increasing water heater efficiency can result in considerable energy cost savings, but attaining better efficiency can be difficult.

As noted in a previous chapter, natural gas use is measured in CCF (100 cubic feet), which is converted to a heat content unit called the therm, equal to 100,000 British thermal units (BTUs). A BTU is the heat energy required to increase the temperature of 1 pound of water by 1 degree Fahrenheit. A gallon of water weighs about 8.3 pounds (3.8 kg), and so 8.3 BTUs are required to raise its temperature by 1 degree. To raise a 50-gallon tank of 55°F (13°C) water up to 120°F (41°C) requires 26,975 BTUs of heat ($50 \times 8.3 \times [120 - 55]$), or approximately 0.26975 CCF, used at 100% heating efficiency. A water heater that produces 40,000 BTUs of heat per hour will take 40 minutes to heat the water, again assuming 100% efficiency. At a cost of $1.00 per therm, the gas to fuel the heater for heating the 50 gallons will be about $0.29. If the incoming water is colder, say 38 degrees, as it is in winter in many community water systems, heating 50 gallons to 120 degrees will require 34,030 BTUs (0.34 therm) and will cost $0.34.

In comparison, an electric water heater with a 4500-watt heating element, if 100% efficient, will require about 7960 watts of power and 1.75 hours of time to heat 50 gallons of 55-degree water to 120 degrees. At a cost of $0.119 per kWh (the U.S. average in 2010), the 50 gallons of hot water will cost will cost $0.95, almost three times the amount for a gas-fueled water heater. Cost for an oil or propane heater will be higher than the gas figure, but still considerably lower than for the electric heater.

All of the above figures assume 100% efficiency for the various types of water heaters, but in fact no water heater is that good. Energy efficiencies of water heaters vary greatly. Electric heaters are most efficient, with a transfer rate of heat energy to the water of up to 98%. Gas-fired heaters have maximum transfer rates of only about 85% (the remaining heat goes up the exhaust flue and is lost). The low end of efficiency for water heaters can be 80% for electric and 50% for gas- or oil-fired units. Even at these lower figures, gas water heating remains the least expensive and electric the most costly. Raising the efficiency of either system can save significant amounts of energy and money for the homeowner.

The amount of energy required by tank-type water heaters for heating water, and the resulting cost, can be reduced in several ways:

- Raise the temperature of the incoming cold water so that less energy is required to heat it.
- Lower the required hot water temperature.
- Increase the insulation around the water heater to reduce heat transfer so that less re-heating of water is required.
- Increase the heat transfer efficiency of the heating unit so that it uses energy more effectively.

The last two of these methods can only be accomplished by replacing an inefficient and poorly insulated water heater with a new energy-efficient model. Because water heating is energy intensive, even a modest improvement in a unit's insulation or heat transfer efficiency can save significant amounts of energy. Any water heater more than 10 years old should be considered for replacement, even if it is still functioning normally. New models will almost always be more efficient in both insulation and heat transfer. When looking for a replacement water heater, the most important consideration is its energy-efficiency rating.

The more insulation a water heater has, the better, as it reduces heat loss. New water heaters have insulation ratings ranging from R-6 to R-24. Current-model water heaters have polyurethane foam (PUF) insulation in place of the fiberglass or other batt insulation used in older units. A low-R-rated unit can be somewhat improved by adding an outside layer of insulation around it in the form of a fiberglass "blanket" or jacket. The insulation should have a foil or vinyl film on the outside. The insulation batt is wrapped around the tank and the ends butted together and taped. The insulation blanket should also be taped over the top of the tank so that it cannot slide down and block air flow to the heating unit or cover safety and drainage valves and controls.

Other improvements to look for in a replacement water heater include:

- Check valve devices at the inlet and outlet to control pressure
- Heating cycle timers, especially on electric units, to time heating for off-peak hours
- Electronic ignition in place of pilot light for fuel-fired models
- Sealed combustion air-intake systems in fuel-fired models
- Pipe insulation
- A drain to handle water condensed out of the combustion process
- A flame arrestor screen to resist ignition of flammable vapors near the water heater
- A thermal cutoff switch now required as a safety measure for all heaters

Size should also be a factor. The smaller a water heater's storage capacity is, the less energy is required to keep water in the tank heated. Our culture has conditioned us to the idea that bigger is better, but water heaters can be an exception. Most homes don't need more than a 50-gallon water heater and many smaller ones can get along well with a 40- or even 30-gallon model. This is especially true if the home is occupied by only one or two persons. Seventy-gallon and larger heaters assure that the home always has a ready supply of hot water, but to do so they consume a large amount of energy keeping water hot that is mostly not used.

If a smaller water heater is adequate for your home, be sure to get a fast recovery unit that will replace its hot water quickly. Fast recovery units have larger heating elements that enable them to heat a given volume of water more quickly, but they do not consume more total energy over time than slower recovery units. Fast recovery helps minimize the occasions when a small-capacity water heater will be unable to meet demand for hot water while still reducing the energy consumption for keeping water heated and ready for use.

Tempering Tanks

Raising the temperature of the water coming into a water heater requires an additional holding tank, called a tempering tank, and is only cost effective if the incoming water is well below the average 55- to 60-degree temperature common in the United States. In cold climates the winter temperature of incoming water is typically 38 degrees or lower. A tempering tank can raise that temperature as high as 60 degrees before the water enters the water heater. The tempering tank is a passive heating device and so requires no energy input to raise the water temperature by up to 22 degrees. This can result in energy savings of 30% for water heating.

A tempering tank is simply a holding tank of about half the size of the home water heater. It is connected in series with the water heater so that cold water first enters the tempering tank. There its temperature gradually warms by heat transfer to near that of the surrounding air. When hot water exits from the water heater, the warmed water in the tempering tank flows in to replace it. The warmed water requires considerably less energy to heat than if it were brought directly into the heater at 38 degrees. A tempering tank and water heater system is shown in Figure 6-10.

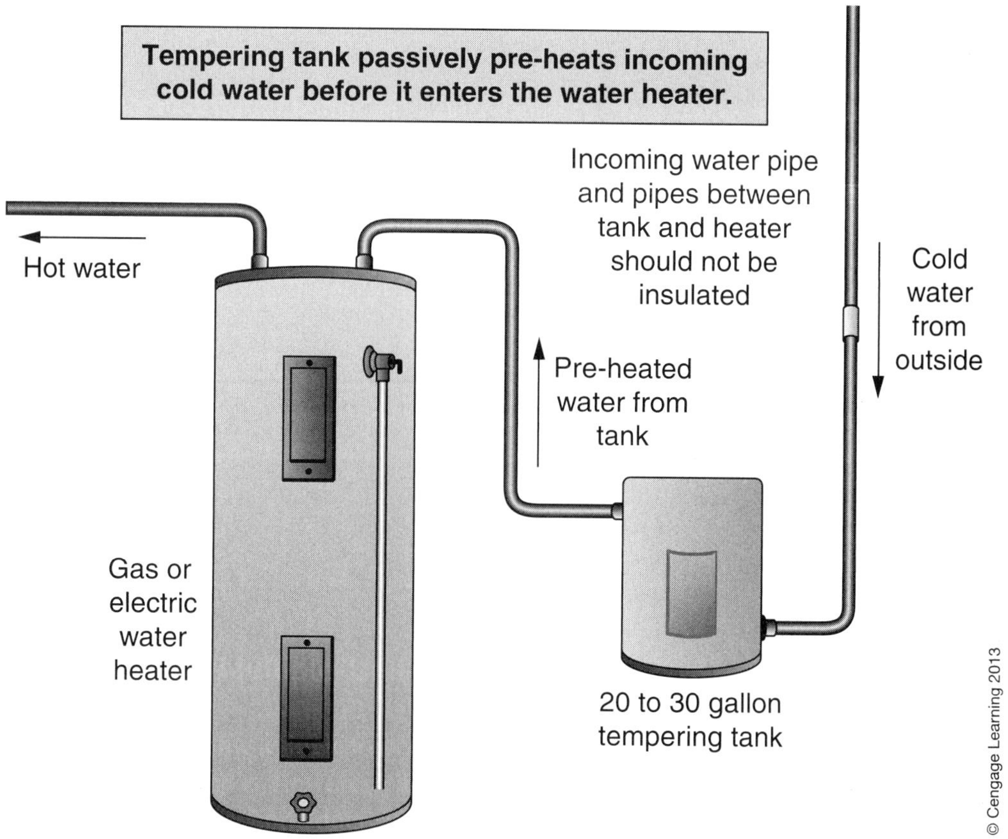

FIGURE 6-10

A tempering tank can reduce energy costs by passively pre-heating water before it goes into the water heater.

The tempering tank should be placed in the warmest area of the home, where heat transfer can warm its contents the most. This area may be in the basement near the furnace (temperature 55–60 degrees) or in a heated area of the home (temperature 70–75 degrees). It should generally not be in the attic. If the temperature there is above 55 degrees, a better energy-saving solution is to increase the insulation to reduce heat loss rather than trying to recapture a fraction of the lost heat with a tempering tank.

The performance of a tempering tank and an associated water heater can be further improved by connecting them to a solar powered warming panel. We'll discuss this application in more detail in Chapter 9.

Lowering Water Temperature

Hot water temperatures of 105–120 degrees Fahrenheit are preferred for dishwashing, laundry, and showering. This requires the water temperature to be raised from 65 to 75 degrees or more. Many older water heaters were set at 140 degrees or even higher, but this is not necessary and wastes energy by heating the water so hot that it must then be mixed with cold water to bring its temperature down to a comfortable level.

Water temperature of 105 degrees is adequate for laundry and for dishwashing, especially if the dishwasher has its own heating element to further raise the water temperature during the wash cycle. Water temperature for showering and bathing is somewhat a matter of personal preference—105-degree water is fine for cleansing, but may be too low for personal comfort for some people. Water heater temperature should be set to the minimum temperature that the home's residents find acceptable so that hot water for showering can be used from the water heater directly and does not have to be mixed with any cold water to get a comfortable temperature.

For every degree the water heater's temperature setting can be lowered, 8.3 BTUs of heat energy will be saved on each gallon of hot water used. The Uniform Plumbing Code reference shower flow rate is 2.5 U.S. gallons per minute. At this flow rate, a 6-minute shower uses 15 gallons of hot water. If the temperature of the water is reduced from 120 to 110 degrees, heat energy consumption will be reduced by 1245 BTUs for every shower taken. For two persons taking daily showers, this will amount to about 75,000 BTUs saved per month. Savings on dishwasher and laundry hot water usage will be similar.

Tankless Water Heaters

Tankless water heaters, also called continuous flow, or on-demand, water heaters, are gaining in popularity. Rather than heating and storing water for later use, this type of water heater heats water almost instantly as it flows through the device. Tankless heaters do not store any water internally except for what is in the heat exchanger coil.

Because they have a limited delivery capacity, more than one tankless heater is usually required in a home. A separate unit may be necessary at each point where hot water is dispensed. In a bathroom this may include separate heaters for the sink and the shower or tub.

The main advantages of a tankless water heater is its ability to provide a continuous flow of hot water and save energy by not having to maintain a supply of stored hot water as conventional tank water heaters do. The main disadvantages of these devices are the limited flow rate of hot water they can supply and water temperature fluctuations due to the varying temperature of the incoming water.

When a hot water tap connected to a tankless water heater is opened, the flow turbine in the heater senses the flow and activates the heating process. The unit's control board calculates the difference between incoming water temperature and the desired water temperature as set on the temperature controller and adjusts the flow of gas to the burner assembly and electronically ignites the heater. If the heater is electric, current flow is adjusted in the heating coil and heating begins. A continuous flow of water is heated to the desired temperature as it circulates through the copper heat exchanger and is delivered through the tap. The water heater shuts down as soon as flow is turned off.

Tankless water heaters located where the hot water is dispensed provide almost instant hot water and also save more energy than larger units that are installed to serve a full bathroom or even a whole home. The point-of-use tankless water heaters don't have to fill a length of pipe with hot water before reaching a tap and leave no hot water in the pipes after the water is shut off. This type of tankless water heater is usually limited to a flow rate of about 1.5 gallons per minute and, for this reason, is usually used in combination with a central water heater. This system provides the benefit of instant hot water along with a higher continuous flow rate when needed. Until recently, point-of-use water heaters were almost always electric, and therefore more expensive to operate per gallon of hot water delivered than central water heaters fueled by natural gas or other fuels.

In ideal circumstances, point-of-use tankless heaters can be more efficient than storage water heaters. The absence of a storage tank saves energy compared to conventional water heaters that have to reheat tank water periodically. Central water heaters of any type also waste energy and water getting hot water through the pipe from the water heater to the faucet and more energy when hot water is left to cool in the pipe when the faucet is turned off. Tankless heaters are more efficient because they heat water only for immediate use and only as much as is currently needed.

Full-on/full-off tankless water heaters do not have a variable heat output level; the unit is either fully on or completely off. Modulated tankless water heaters use a flow sensor and inlet water temperature sensor in combination with an outlet water temperature sensor and choke valve to vary their heat production according to the flow of water passing through the unit. The result is a hot water output stream at a constant output temperature within a range of ±4°F, but a fluctuating flow rate.

At present, the most cost-effective configuration for tankless water heaters in a home is to install a central tankless water heater to supply most of the house, and install one or more point-of-use tankless heaters at any distant faucets or bathrooms outlets. Only electric point-of-use tankless water heaters were available until recently, and they are still by far the most common type. Natural gas and propane central tankless heaters are now readily available. These are less expensive to operate than electric models and should be used wherever the required fuel is available (Figure 6-11).

When buying a whole-house gas-fired tankless water heater, you should purchase a unit that can function continuously when raising the water temperature by 75 to77°F. In a winter climate area, this means finding a unit with the heating capacity to produce 4 gallons of hot water per minute starting with 38–50°F inlet water temperatures. This size unit is referred to as a two-appliance heater, meaning that it can supply two typical outlets

FIGURE 6-11

An electric tankless water heater (left) is about the same size and has about the same heating capacity as a gas-fueled model (right); the gas model is currently more expensive and also uses some electricity to power its combustion fan.

with continuous hot water simultaneously, even in winter. This same unit in summer may produce 6 to 7 gallons per minute when starting with higher inlet water temperature, but the winter capacity should be available to assure adequate hot water year-round.

Tankless water heaters are only slightly more energy efficient than the best-quality tank water heaters, and they are usually more expensive to purchase and install. Other advantages and disadvantages, however, may make one type of water heater or the other preferred by individual homeowners. Among the advantages of tankless water heaters are the following:

- Unlimited hot water is available with a tankless water heater. This can also be a disadvantage, however, as running out of hot water self-limits use, whereas a tankless heater has no limit.
- Less physical space is required for most tankless water heaters.
- There is little risk of water damage from a tankless heater because there is no stored water to leak from a tank failure or rupture.
- Constant water temperature is maintained by tankless heaters, although flow rate will vary; tank heaters maintain constant flow rate but the temperature decreases as the tank is emptied of hot water.
- Tankless water heaters precisely control water temperature, which means no extreme temperature levels or spikes.

Tankless heaters also have some disadvantages, including:

- A startup delay in hot water delivery for central tankless heaters that is about equal to that for tank heaters; point-of-use systems are faster, but still not instantaneous in their hot water delivery.
- Intermittent-use hot/cold delivery caused by the short delay (1–3 seconds) between the time when the water begins flowing and when the tankless heater's flow detector activates the heating elements or gas burner; this can result in a flow of hot water, followed by a small amount of cold water, followed by hot water again.

- Installation cost for a tankless system is higher, especially if it is replacing a tank-type heater, due to the need for higher-capacity power lines or gas pipes and exhaust vents.
- Heat source flexibility is not always as great for tankless heaters, which are usually limited to electric power or natural gas fuel.
- Tankless water heaters have a limited ability to achieve cooler water temperature because they often have minimum flow requirements before the heater is activated that may make precise mixing of hot and cold water difficult.
- Time-of-use metering and peak electrical load charges for power may be greater for tankless electric heaters, which are high-current devices whose hot water use tends to peak during high-demand times of the day.

CLOTHES WASHERS, DRYERS, DISHWASHERS, STOVES, AND PUMPS

Clothes dryers, dishwashers, ranges, ovens, and pumps are all heavy users of electric power, either because they contain electric heating elements or are driven by powerful electric motors, or both. Increasing the energy efficiency of each of these appliances is a matter of timing when and how often they are used and using them in a manner that consumes only the necessary amount of energy to perform their work tasks.

Clothes Washers

Few strategies are available for saving energy when using a washing machine. Because these machines run in programmed cycles, the only adjustment to their operation is usually the length of the wash portion of the cycle. This should be kept to the minimum needed to adequately clean the load.

If the washer has adjustable settings for load size, these should be used as accurately as possible and the washer should not be overloaded for the setting chosen. Small-load settings usually do not alter the length of the wash cycle, but they do reduce the amount of hot water used, and this is an energy-saving effect. Heat energy is also saved by operating the washer on a warm water setting for washing and cold water for rinsing. Hot water is rarely needed for washing or rinsing clothes.

Clothes Dryers

Gas-heated or electric clothes dryers work by blowing heated air through wet clothes in a revolving drum. The warm moving air absorbs moisture from the clothes and is exhausted to the outside through a vent. In all but the most humid climate conditions, using unheated air in the dryer will also dry the clothes, but not as rapidly. Heated air can hold more moisture and dries the clothes more quickly than air at the ambient room temperature.

The operating efficiency of a dryer is reduced by any obstruction blocking the air flow through it. Lint filters should be kept clean and exhaust hoses and vents should be kept clear of any obstructions so that maximum air flow is maintained. Because a dryer is usually connected to its outside vent by a flexible (and collapsible) hose, the most common restriction of air flow is a kink in the hose or a narrowing of a portion of its length by being squeezed between the dryer and the adjacent wall. Both of these conditions constrict air flow and waste energy, but are easily corrected by replacing the hose and adjusting its position to avoid repeating the problem.

Energy consumption by a dryer can be reduced by operating it on a medium- or low-heat setting. The tradeoff is that each load will take longer to dry using the lower heat settings, but in most cases this is not a significant issue. Many people like the dryer cycle to keep pace with the washer's cycle that typically runs between 30 and 45 minutes, but unless such speed is really necessary, a lot of energy can be saved by allowing more time per load and operating the dryer on a low-heat or no-heat setting.

Drying times at these settings will vary depending on humidity levels and climate in your area. Experiment a bit to find the optimum settings to use for drying clothes in the maximum time (rather than the minimum) you can allow. You may find that towels and heavy items require a medium or even high setting, especially in humid weather. Synthetic fabrics and light garments will usually dry with little or no heat in about twice the time they would on a high-heat setting. Each load you dry at the lower settings will save hundreds of BTUs of heat energy, and the cost of those comes directly off your home energy bill.

If outside space is available, the greatest savings in clothes drying is to line dry most items, especially the heavy ones, and use the dryer only when weather conditions or time constraints require it. In most areas of the country, clothes will dry outdoors in no more than three times the time required to dry them in a dryer, less if there is wind.

Dishwashers

Like clothes washers, dishwashers run on preprogrammed cycles that cannot be adjusted much to save energy. The best that can usually be achieved in economy of operation is to run the dishwasher only with a full load and cold rinse dishes before loading them.

Most dishwashers have a heating element that increases water temperature during the washing cycle and warms the air inside the dishwasher after rinsing in order to dry the clean dishes. The heating feature is a convenience that helps dry dishes faster, but it is not a necessity. The dishwasher should be connected only to the hot water supply pipe so that it uses hot water exclusively during operation. In this configuration, no additional heating of the water is needed to adequately clean dishes.

The warm dishes will also dry without the aid of the heating element in about the same time as they would with its help, if the dishwasher door is simply opened immediately after the wash and rinse cycle is completed. This allows humid air to escape from the dishwasher and circulates dryer air around the cooling dishes to dry them.

Not using the heating element of a dishwasher will in most cases cut its power consumption almost in half on each operating cycle. If you feel that the heating feature is really needed for increased water temperature during the washing cycle, some energy savings can still be obtained by turning it off after that cycle is completed and allowing the dishes to dry in the open air.

Stoves

Even though electric ranges and other cooking appliances use a lot of energy when operating, they account for only a small fraction of total home energy use because they are used relatively infrequently. Likewise, there is little difference in the cost of operation of an electric range or a gas-fueled stove. Gas stoves use more energy than electric ranges, but because gas is less expensive than electric power, the cost of operating a gas stove is actually a bit less than the electric range. Table 6-2 shows the energy and cost of baking and cooking on various types of cooking units.

TABLE 6-2 Cooking Energy Use and Costs

APPLIANCE	TIME	TEMPERATURE	ENERGY CONSUMED	COST
Gas oven (pilot light)	1 hour	350°F	11,200 BTUs	$0.16
Gas oven (electric igniter)	1 hour	350°F	11,200 BTUs + .35 kWh	$0.21
Electric oven	1 hour	350°F	2.0 kWh	$0.24
Microwave oven	30 minutes	High	0.72 kWh	$0.09
Crockpot	7 hours	200°F	0.70 kWh	$0.09
Toaster oven	1 hour	350°F	0.33 kWh	$0.04

As Table 6-2 shows, the cost difference between gas and electric cooking is small and would never justify the cost of switching from one to the other if new gas lines or electric wiring had to be installed. Cooking in a toaster oven is the most economical, followed by using a crockpot or microwave oven. Full-size ovens are more costly because they are larger units that use more energy to attain and hold cooking temperatures.

It follows from this discussion that the best economy in cooking is achieved by using the smallest cooking appliance that will do the job and cooking only as much as you need. One cup of coffee water boiled in the microwave requires only a third of the energy needed to boil three cups in a coffee pot on the stove. To conserve energy when cooking, adjust your thinking to use the minimum equipment for the minimum time required. The following tips will help you pursue this minimalist philosophy when using electric and gas cooking appliances:

- Broil elements in an electric oven use much more power than the bake elements, usually between 3000 and 3600 watts compared with 2000 for the baking elements; a countertop broiler will use much less power and is adequate for all but very large broiling tasks.
- Self-cleaning ovens, gas and electric, use much more energy when cleaning than when cooking because they get much hotter than they do for cooking; use this feature sparingly and only in off-peak hours for electric units.
- Use crockpots or microwave ovens for as much baking and cooking as possible; they consume much less energy than an oven or a large range or stove burner.
- Open the oven door only when necessary; the inside temperature drops 25 to 30 degrees when you open the door. Get an oven with an oven light and a glass window in the door to avoid opening the door.

- Don't put aluminum foil on the bottom of a gas oven to catch drippings; it blocks some of the heat produced below the oven. Foil in an electric oven is all right as long as it is placed under the bottom heating element.
- Use glass and ceramic pans for baking; they retain heat better than metal pans.
- Don't use pilot lights on gas burners. If your existing stove already has pilot lights, replace it if you can with an electric ignition unit. If you can't replace the stove, turn the pilot lights out on the burners and use a clicker-lighter to light them. (Turning them off requires tightening the adjustment set screw fully. Don't just blow the flame out, because gas will still flow from the unlit pilot hole.)

Pumps

Pumps in homes are used to move water from place to place. Most home pumps are either sump pumps that remove ground water or surface leakage from basements and crawl spaces, or well pumps that lift water from a well for use in the home. Water is heavy and pumps use a lot of energy to move it. The most work for a pump is lifting water from a low elevation to a higher one. The more frequently a pump runs, the more energy it consumes. Similarly, the higher it must lift the water it moves, the more energy it uses. You can save energy by reducing the operation time of a pump and by reducing the elevation of its lift.

Sump Pumps

Three types of sump pumps are commonly used (Figure 6-12). The first is a pedestal pump. This is an upright electric pump with the motor located above the pump and

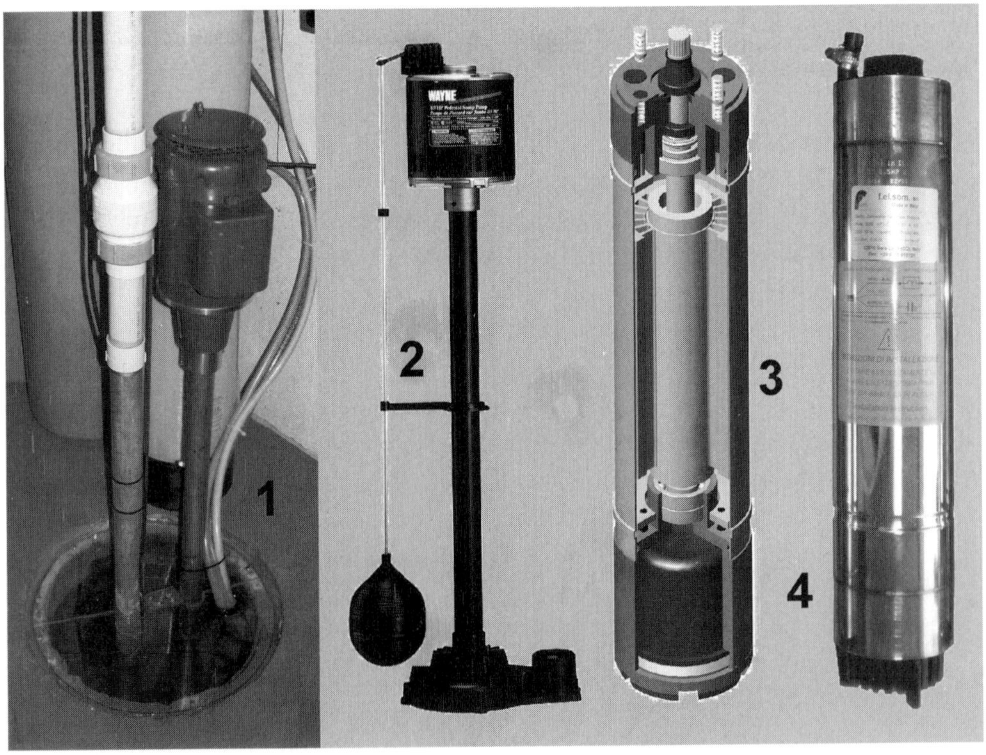

FIGURE 6-12

(1) A pedestal pump in a basement sump; (2) a pedestal pump with ball float actuator; (3) cutaway of a 4-inch submersible well pump; (4) sealed 4-inch submersible pump ready for installation in a well.

connected to it by a shaft. This arrangement allows the pump to be placed in the water while the motor, which is not meant to get wet, remains dry above it. The pump has a float-activated switch that turns the pump on when the water it is standing in reaches a certain level. This type of pump is ideal for basements needing frequent water drainage. It is the least expensive type but is usually noisier than the submersibles.

A submersible pump is installed in the ground and is specifically designed to function underwater. It has the same float-activated switch as a pedestal pump. Submersibles are more expensive, but are not subject to high water damage and have sealed, oil-cooled motors that run more quietly than a pedestal pump.

Floor sucker pumps are used mostly in basements, cellars, or crawl spaces that do not have an actual sump pit where water can collect. They are called floor sucker pumps because they remove water to within 1/8″ of the floor (or other surfaces such as a roof or the ground). They are available in various flow rates and can be controlled manually or by means of a water sensor or a float switch.

The more frequently a sump pump has to operate and the higher it has to lift the water it pumps, the more energy it uses. It follows that minimizing either of these factors will save energy and money for the homeowner.

All sump pump types should have a check valve on the water outlet pipe in order to prevent water from running back into the sump when the pump shuts off. Water flowing back and forth can cause the pump to turn on and off more than necessary, wasting energy and shortening the life of the pump.

Anything that can be done to reduce water flowing into a sump will reduce energy consumption and cost for pumping it out. Among the possible improvements the homeowner can make are the following:

- Caulk all cracks and openings in the cement foundation of the house. Pay special attention to any that show evidence of water leakage.
- Maintain rain gutters all around the home. Keep them free from leaves and debris. Add extensions to downspouts that will discharge rain water at least 4 feet away from the foundation.
- If surface water drains toward the home on any side, install a berm or drain to prevent it from reaching the foundation. In flat terrain, even 2 or 3 inches of fill may be enough to direct surface water away from the house.
- Install covers on basement window wells and be certain that rain gutters do not discharge toward them.
- Check nearby storm drains and water pipes to be sure they are not leaking water underground toward your home.
- Check the discharge hose on the sump pump to be sure that water coming out of it isn't flowing back toward the home.
- Extend the discharge line to the lowest point available, directly into a storm drain if possible. The lower the outlet is, the less energy the pump will consume.

Sump pumps are reliable units and don't usually require much maintenance, but if your pump is running frequently, it is a good idea to be sure that it is running efficiently. Among the things to check for:

- After a heavy rainfall, be sure no debris has washed into the sump. This is one of the major causes of sump pump failure. Check the inlet suction screen to make sure that there is no blockage.

- Check the pump motor for effective operation by pouring water into the sump area. If the pump is working well, the motor will turn on as the float rises with the water level.
- Ensure that the float of the pump moves freely. If it is obstructed, the pump may fail to start.
- If you see any oily film on the surface of the water in your sump, verify the condition of the pump's oil seal. The appearance of oil may indicate a faulty oil seal, which can cause the motor to burn out.
- While the pump is running, go outside to check that the outlet line is actually discharging water.

Well Pumps

There are two general types of well pumps: submersible pumps and jet pumps. Jet pumps are used for shallow wells less than 25 feet deep. None of the working parts of this type of pump is submerged in water. Deep well jet pumps that can be used for wells up to 75 feet deep have their ejectors installed in the well itself.

Submersible well pumps, as their name implies, are used underwater in wells. A small electric motor (called a driver) is installed in the well shaft, usually below the pump itself, and an electric cable is attached to the motor. Piping is fitted from the pump, through the length of the shaft, and into the home. Submersible well pumps may be set hundreds of feet beneath the water in a well.

Most well pumps are centrifugal pumps; they consist of a number of impellers and diffusers that spin on a common shaft (called stages) and push water up out of the well. When the pump is activated, the pump motor spins the impellers inside the pump casing. The impellers' rotation pushes water outward and then upward through the well shaft.

The diameter of wells is small and restricts the diameter of the pump. This is why submersible well pumps are very long. A standard 4-inch submersible pump measures from 24 to 48 inches in length, 3.9 inches in diameter, and spins at 3600 rpm. A 3-inch submersible pump measures 2.9 inches in diameter and spins at about 8000 rpm. In both sizes, the impellers have to be stacked on top of each other to exert enough pressure to force the water up through the pipe. The number of impellers and diffusers a pump has determines the head (or pressure) the pump exerts. This pressure determines the depth from which the pump can push water to the surface.

Pressure is usually stated in pounds per square inch (PSI), but it can also be stated as lift (as in vertical lift). One PSI is equal to 2.31 feet of lift. The taller the well pipe (that is, the deeper the pipe is set in the earth to reach the water in the well), the more pressure the pump must develop to force water up the pipe to the surface.

Most submersible well pumps require 230 volts of AC single-phase electricity. Pumps can be of the three-wire or two-wire variety. All three-wire units require a control box that includes a motor starting relay, thermal overload switch, and capacitors. It is surface mounted for easy access. Two-wire motors, however, have no external control box like the three-wire units; all of the components of the control box are inside the motor. The price is generally the same for both types. Because neither has a performance advantage over the other, choosing a model really boils down to aesthetics.

A standard electrically powered well water delivery system consists of a submersible pump in the well that delivers water into a pressure tank. A pressure tank extends the time between pumping cycles by storing some pressurized water for later use. This system prevents freezing problems by placing the pump deep inside the well and the pressure tank indoors.

This arrangement is energy intensive for several reasons. The pump must produce enough water volume to keep up with any potential demand, or the pressure tank will be depleted and the pressure will drop dramatically. The pump must also be powerful enough to push water out of the well and into the pressure tank against the pressure in the tank. This requires at least a 1/3-horsepower (hp) pump for a relatively shallow well and a ½-hp or larger pump for deeper wells.

The energy a well pump uses depends on the size of the pump, the amount of water used in the household, the pump's operating pressure, and the depth of the water in the well. The last of these factors can't be controlled by the homeowner, but the others can be optimized for maximum efficiency.

First, be sure the well pump is properly sized. A big pump may be advertised as a luxury feature for a home, but unless it is actually needed, it is really just wasting energy every time it activates. Many homes have ¾-hp pumps installed in their wells, when a1/2-hp pump would be adequate. Proper pump sizing can save a homeowner hundreds of kilowatt-hours (kWh) each year in energy consumption, as well as the cost of the larger pump.

Conserving water is another good way to reduce pumping costs. Leaky faucets, showers, and hoses waste water and increase demand on well pumps. The water from the well may be free, but you pay for every gallon of it that is pumped. Any measure you can take to reduce water usage correspondingly reduces energy cost for pumping.

Regular maintenance of a submersible well pump will also ensure that the pressure tank does not become waterlogged. If the tank becomes filled with too much water, the pressure sensor that controls the pump becomes overactive, causing the pump to activate too often and work too long. Because the startup phase of the pumping cycle requires the most energy, this hyperactivity causes pump energy consumption to greatly increase and also reduces the life span of the equipment.

SUMMARY

Maximizing the energy efficiency of an HVAC system means controlling the volume of air that the system must heat and cool, and the temperature level to which the system must heat or cool that air. Central heating systems are of two types, forced air and radiant. Forced-air systems draw air into a furnace where it is heated and then blown out again into the home. Radiant heating systems work by heating the air in the house through convection or by heating the house structure itself. Fuel for most furnaces is either heating oil or natural gas. Central air-conditioning (A/C) systems all move air the same way as in a forced-air heating system. Most A/C systems use the same ducts and vents as the heating system. The air conditioner cools air by refrigeration and then blows it back into the home through the duct system. Heating and cooling zones can be set up in homes to more evenly and efficiently distribute heat throughout the home. There are several additional ways to improve the energy efficiency of the home's HVAC system: seal and insulate ducts, maintain the HVAC system, and automate the HVAC system by adding zones or installing programmable thermostats. Residential space heaters heat a single room rather than an entire house. The main types of heaters are convective and radiant. Window or wall-mounted air conditioners are the cooling equivalent of space heaters.

Most residential water heaters in the United States are the tank type in which water is kept hot and ready for use. Their efficiency can be improved by adding a tempering tank or lowering heated water temperature. Tankless water heaters heat water instantly as it flows through the device. Clothes dryers, dishwashers, stoves, and pumps are all heavy users of electric power. The energy efficiency of each can be improved by how they are used.

Most home pumps are either sump pumps that remove ground water or surface leakage from basements and crawl spaces, or well pumps that lift water from a well for use in the home. The more frequently a pump has to operate and the higher it has to lift the water it pumps, the more energy it uses.

KEY TERMS

Air conditioning (A/C) The process of cooling the interior air in a building to a comfortable temperature.

Air duct Any metal, plastic, or other type of tube, round or rectangular, used to move air.

Blower A device that moves air like a fan but does so with a rotor rather than a bladed impeller.

Boiler An appliance for heating water to be circulated in a radiant heating system.

BTU (British thermal unit) A measure of heat equal to the amount of energy needed to heat 1 pound (0.454 kg) of water (0.1198 U.S. gallons) from 39° to 40°F (3.8° to 4.4°C).

Carbon dioxide (CO_2) A combustion product of burning fuels; toxic in high quantities.

Carbon monoxide (CO) A poisonous combustion product of burning fuels; produced in place of CO_2 when combustion is incomplete.

Central heating system Any heating system designed to warm an entire building rather than a single room or small part.

Clicker igniter An igniter for gas appliances that sparks to start the gas fuel burning; same as spark igniter.

Coal fired Any appliance using coal as is heat source.

Combustion chamber The part of a fuel-burning appliance where fuel is burned and heat produced.

Combustion products The elements and compounds created when a fuel burns; CO_2 and water are two usual combustion products but many others are possible.

Combustion residue Solid combustion products such as ash produced by burning fuel that must be removed from the combustion chamber periodically.

Compressor In A/C systems, a device for squeezing a gas into a smaller space at a higher pressure until it becomes a liquid; other types of compressors produce compressed air, liquefy natural gas, or compress other fluids.

Condenser A device that extracts heat from gas, causing it to become liquid; often used with a compressor in A/C systems.

Continuous-flow water heater A water heater that can produce a continuous stream of hot water at a set temperature.

Convective heater A heater that warms the air around it; the warmed air then warms other nearby objects.

Cooling coil The component of an A/C system in which heat is extracted from the air passing through it.

Damper A movable door in a duct that can be operated manually or automatically to permit or block the passage of air.

Duct tape Cloth-based tape, usually about 2 inches wide, used to seal ducts and for other uses.

Ductwork The system of ducts that moves air to and from an HVAC system.

Electronic ignition A method of igniting a gas appliance flame by means of an electric spark.

Evaporative cooler An air conditioner that works by the cooling effect of evaporating water rather than by refrigeration cooling; not effective in humid climates.

Exhaust vent A vent leading from an HVAC or fueled appliance to the outside air.

Filter A screen that removes pollution particles from conditioned air before it is distributed.

Flame arrester screen A device that inhibits an open-flame heating appliance from starting a fire.

Floor sucker pump A type of sump pump that draws water directly from a floor or other flat surface rather than from a sump pit.

Flow sensor A device that senses when water or another liquid is flowing in a pipe.

Forced-air heating system A heating system that uses a blower to distribute warmed air through ducts to all parts of a building.

Full-on/full-off tankless water heater A water heater that cannot adjust its rate of flow or its water temperature, but produces a set amount of hot water continuously.

Gas-fired Any appliance fueled by natural gas.

Ground water Water that flows beneath the earth's surface; source of well water and spring water; may also flow into basements or other below-surface openings.

Heat absorption The process by which heat passes from a warmer substance to a colder one; the cold object absorbs heat from the warmer one.

Heat exchanger Any device that transfers heat from one object or substance to another; a stove burner and an A/C cooling coil are two examples.

Heating element An electric component that heats when electricity passes through it because it has high resistance to the current.

Heating register The opening at the end of an HVAC system duct through which warmed or cooled air passes into a room.

Heating, ventilation, and air-conditioning (HVAC) system The integrated electromechanical system that supplies heating and cooling to a home or business. HVAC systems often use gas or oil in addition to electric power.

HVAC schedule The programmed operation of an automatic thermostat designed to get maximum efficiency from the HVAC system.

Impeller The paddle-wheel-like component in a pump that rotates to drive water out of the pump at high pressure.

Lift The vertical distance water must be raised by a pump from its intake to the point of discharge.

Mastic A viscous substance similar to caulk used to seal air ducts and air leaks.

Modulated tankless water heater A heater that can vary the temperature of the hot water it produces, making precise adjustment for comfort easier.

Multiple-unit HVAC A system consisting of several small HVAC units, each of which constitutes a separate zone in a building and can be independently controlled.

Oil fired Any device in which the fuel burned is heating oil.

On-demand water heater A water heater that produces hot water only when a hot water faucet is turned on; this type of heater does not store any hot water, but can produce it continuously.

Pedestal pump A type of sump pump that has its motor above the pump casing that sits in the sump pit; the motor is intended to always remain dry.

Pilot light A method of lighting gas- or oil-fired appliances that uses a small continuously burning flame to ignite the gas or oil.

Point-of-use tankless water heater A water heater placed as close as possible to the faucet from which its hot water is dispensed so that the delay in delivery will be minimal.

Polyurethane foam (PUF) insulation A type of high-R-value insulation now used in tank water heaters in place of the older fiberglass or similar type of insulation.

Pressure tank A tank that stores water pumped from a well under pressure (produced by compressed air in the top of the tank) so that a home water system will have continuous stable water pressure.

Programmable thermostat A thermostat that automatically raises and lowers an HVAC system's operating temperature according to a preset schedule.

PSI Pounds per square inch; the method of measuring water pressure in a pipe or the lifting power of a pump. One PSI is equal to a lifting power of 2.31 feet.

Radiant heater A heater that produces infrared-spectrum heat that passes through air without heating it much and heats solid objects as it strikes them.

Radiant heating system A heating system that pipes heated water from a boiler to standing radiators, baseboard radiators, or pipes set in a building's floor; the radiators or pipes warm the building structure or the air within it.

Radiator Any device from which infrared-spectrum heat is produced; also a type of heat exchanger that transfers heat from a fluid inside the radiator to or from the air passing by the radiator.

Refrigeration cooling Reducing air temperature by compressing a gas in a closed-loop system into a liquid and then allowing it to return to a gaseous state, extracting heat from the surrounding air as it does so.

Space heater Any heater designed to heat a room or similar small space rather than an entire building.

Spark igniter An igniter for gas appliances that sparks to start the gas fuel burning; same as clicker igniter.

Submersible pump A pump designed to be operated while submerged in the water it pumps.

Sump pump A pump located in a small pit in the basement or crawl space of a home that keeps the lower level free of water by pumping any leakage as needed.

Surface water Water flowing on the earth's surface from rain or other sources that can flow into the basement or crawl space of a home.

Tank water heater A water heater that heats and stores a quantity of water for later use; the hot water is constantly maintained at the set temperature, regardless of how little or how much is used at a time.

Tankless water heater A water heater that produces hot water only when a hot water faucet connected to it is turned on and water begins to flow; an on-demand water heater.

Tempering tank An auxiliary tank connected in series with a water heater to passively raise the temperature of water entering the water heater.

Therm (thm) A measure of heat equal to 100,000 BTUs; 10 therms is equal to 1 decatherm (Dth).

Thermal cutoff switch A switch on gas- and oil-fired appliances that cuts off the fuel if it detects no heat being produced; the gas type is also called a thermocouple.

Thermostat A device for controlling an HVAC system; it can be manually operated or programmed to operate on a schedule.

Vent A duct connecting a fuel-burning appliance to the outside; a vent may exhaust hot gases from combustion or warmed humid air, as from a dryer. An intake vent permits outside air to be drawn into a combustion chamber or HVAC system.

Water heater blanket An extra layer of insulation added to a tank water heater to increase the R-value of its insulation.

Water vapor (humidity) Water in a gaseous state always present in the atmosphere that creates humidity in the air; as air cools it can hold less water vapor, and this may produce liquid water condensation.

Well shaft The hole leading from the surface to the ground water that supplies a well.

Zone The section or part of an HVAC system that is separately operated and controlled; also a group of lights and other electrical devices controlled together to provide a specific scene or accomplish a specific task such as security lighting.

REVIEW QUESTIONS

1. What are the two factors that must be controlled to improve the efficiency of an HVAC system?
2. True or False: Forced-air heating systems heat the structure of a home to keep the interior warm.
3. What is the fluid that transfers heat in a radiant heating system?
4. What is the least expensive fuel to use for a heating system?
5. Why is the compressor/condenser unit of an A/C system located outside the home?
6. In most homes, what is the largest user of electric power?
7. How does dividing an HVAC system into zones increase its efficiency?
8. Why can't a radiant heating system also be used for air conditioning?
9. True or False: The further heated or cooled air travels from its source (the HVAC system), and the more heat loss or gain occurs in transit.
10. What is a damper, and what is its function?
11. How can the efficiency of an HVAC system be improved?
12. Does a convective space heater warm the air around it?
13. True or False: Programmable thermostats can maintain the temperature in one zone at 62° while also maintaining the temperature in two other zones at 75°.
14. Are large tank water heaters always more efficient than small ones because they store more hot water?
15. Which is more expensive to operate, an electric water heater or a gas-fired unit? Why?
16. True or False: A room A/C unit works best when mounted in the middle of the room so it can draw air in from the room and blow it out again to the entire room.
17. What is a tempering tank and what is its purpose?
18. What type of tankless water heater can maintain the precise temperature of the hot water it produces?
19. What type of cooking unit is the most cost-efficient for baking?
20. If the size of a pump is doubled, is its energy efficiency also doubled?

CHAPTER

7

ENERGY-EFFICIENT LIGHTING AND ENTERTAINMENT SYSTEMS

OBJECTIVES

After studying this chapter, you should be able to:

- Describe how incandescent lights, compact fluorescent lights (CFLs), and light-emitting diode (LED) lights work.
- Compare the energy efficiency and other advantages and disadvantages of the various types of lights.
- Increase home energy efficiency by adopting new lighting and automation technology according to a cost-effective plan.
- Achieve maximum energy efficiency from the entertainment and recreation systems and devices in your home.

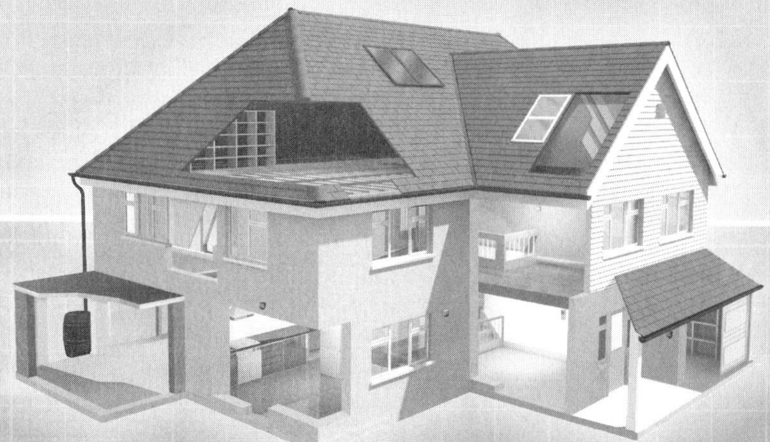

OUTLINE

I. Introduction
II. Incandescent Lights
  A. Incandescent Inefficiency—Light versus Heat
  B. Power Consumption
III. Energy-Efficient Lighting
  A. Compact Fluorescent Lights (CFLs)
  B. Light-Emitting Diodes (LEDs)
IV. Energy-Efficient Electronics
  A. TVs, Stereos, and Home Theaters
  B. Game Consoles
  C. Computer Systems
V. Other Power-Hungry Appliances
  A. Refrigerators and Freezers
  B. Exercise and Fitness Equipment
  C. Workshop Equipment
VI. SUMMARY

INTRODUCTION

In this chapter you will learn how important technology developments have made several new types of lighting cost competitive with incandescent lighting and made a large reduction in energy use possible simultaneously. You will learn how compact fluorescent lights (CFLs) and light-emitting diodes (LEDs) work, their cost and energy use, and their other advantages and disadvantages compared to incandescent lights. You will learn how to plan and implement a replacement program to adopt more energy-efficient lighting in your home, and, finally, you will become familiar with methods of improving the energy efficiency of the entertainment and recreation systems and appliances that are used in the home.

INCANDESCENT LIGHTS

THE INCANDESCENT LIGHT bulb, the type of electric light source most commonly used throughout the world over the last century, takes its name from the manner in which it produces light. An incandescent light uses an electric current to heat a metal filament wire to a high temperature until it glows or incandesces. The filament becomes so hot that it would quickly burn if it was not protected from contact with air by the glass bulb that surrounds it. The bulb is filled with inert gas or contains a vacuum, either of which prevents any oxygen from reacting with the filament.

The filament in a light bulb is connected to an electrical current by wires embedded in the glass that pass from the inside of the bulb out to a metal housing, called a base, to which the bulb is mounted. The base is shaped so that it can be screwed or plugged into a socket that keeps its electrical terminals in contact with the socket's current supply terminals.

Incandescent bulbs work well using AC or DC current and are produced in many sizes. Voltage ranges from 1.5 volts to about 300 volts, although home lighting almost always uses 120 volts AC or 24 volts DC. Light output is measured in lumens and depends on the amount of current flowing through the filaments. Typical light bulbs use from 25 to 150 watts of power at 120 volts and produce more light as their power consumption increases.

Since its introduction, many improvements have been made to the incandescent light bulb and literally thousands of different styles and types of bulb have been produced. The incandescent light illuminated the world for over 100 years and, because it represented such a vast improvement in illumination over its predecessors (candles and gas lights), its major shortcoming was overlooked or considered unimportant until recently.

Incandescent Inefficiency—Light versus Heat

All incandescent lights are inefficient in converting electricity into light. More than 90% of the current consumed by an incandescent light is transformed into heat that radiates from the filament as invisible infrared wavelengths. Less than 10% is emitted as visible light. Incandescent light bulbs are sold in sizes based on the amount of electrical power (watts) they consume. If two bulbs have the same voltage and are of the same type, color, and clarity, the higher-powered bulb will provide more light, as shown in Table 7-1.

TABLE 7-1 Incandescent Bulb Power and Light Output

POWER (WATTS)	LIGHT OUTPUT (LUMENS)	EFFICIENCY (LUMENS/WATT)
5	25	5
15	110	7.3
25	200	8.0
40	500	12.5
60	850	14.2
75	1200	16.0
100	1700	17.0
150	2850	19.0
200	3900	19.5
300	6200	20.7

The light output for different types of bulb varies slightly from the figures shown in the table. The lumen values for "soft white" bulbs will be slightly lower than those for standard "frosted" bulbs of the same power. Clear bulbs will emit a brighter light than equally powered coated bulbs.

Power Consumption

As can be seen in Table 7-1, higher-power bulbs are somewhat more efficient—that is, they produce more lumens of light for each watt of power consumed than those that use less power. But the improvement is small and even the largest bulbs still put out far more heat than light. This means that although incandescent bulbs produce a lot of light in the average home, they consume a lot of power to do it, as the following example illustrates.

A room-by-room survey of an average-sized home (about 3000 square feet) counted a total of 73 light bulbs (individual bulbs, not fixtures, many of which contain two or more bulbs) in the house, garage, and yard. The breakdown of the sizes of bulbs was:

1	300-watt bulb
2	200-watt bulbs
2	150-watt bulbs
49	100-watt bulbs
11	75-watt bulbs
6	60-watt bulbs
2	40-watt bulbs

If all these lights were turned on at once (which never happens), they would draw 7165 watts of current and probably trip the main breakers on the home's electric service. In reality, only a few lights at a time are generally used and most only for short periods of time. On average, each of the lights in this list is probably used about 1 hour per day, but that level of use still adds up to considerable power consumption.

If all the incandescent bulbs are used for just 1 hour per day, they will consume 7.165 kilowatt hours (kWh) of electricity per day, or 214.95 kWh in a 30-day month. The average cost of electric power in the United States in 2010 was $0.119 per kWh. At that rate, costs for incandescent lighting in this home amount to $25.58 per month and $306.95 per year. Doubling the efficiency of incandescent bulbs would cut that cost figure in half, and as the cost of electricity continues to rise, the savings would also increase. Unfortunately, this is not likely to happen.

Many attempts have been made to improve the efficiency of the incandescent light, but few have been successful (Figure 7-1). Lowering the voltage to the filament of a bulb, even by as little as 10%, will greatly increase the life of the filament, but it also reduces the light output by about 20%. Because the cost of light bulbs is low, due to mass-production techniques, improving their longevity is not as useful as upgrading their efficiency, but in this area little progress has been made. Incandescent bulbs have remained the default standard for residential and most commercial illumination because no viable alternatives were available until recently.

The increased use of electric power in recent years has at times threatened to exceed the generating capacity of the national grid. Use of more efficient electric lights is one method of reducing electric power demand or at least reducing the rate of increase in demand. To push consumers to accept these more efficient light bulbs in place of incandescent bulbs, in December 2007, the federal government enacted the Energy Independence and Security Act of 2007, which requires all general-purpose light bulbs that produce 310–2600 lumens of light to be 30% more energy efficient than current incandescent bulbs beginning in 2012. The new efficiency standards will apply to 100-watt bulbs as of January 2012 and expand to other sizes later, ending with 40-watt bulbs in January 2014.

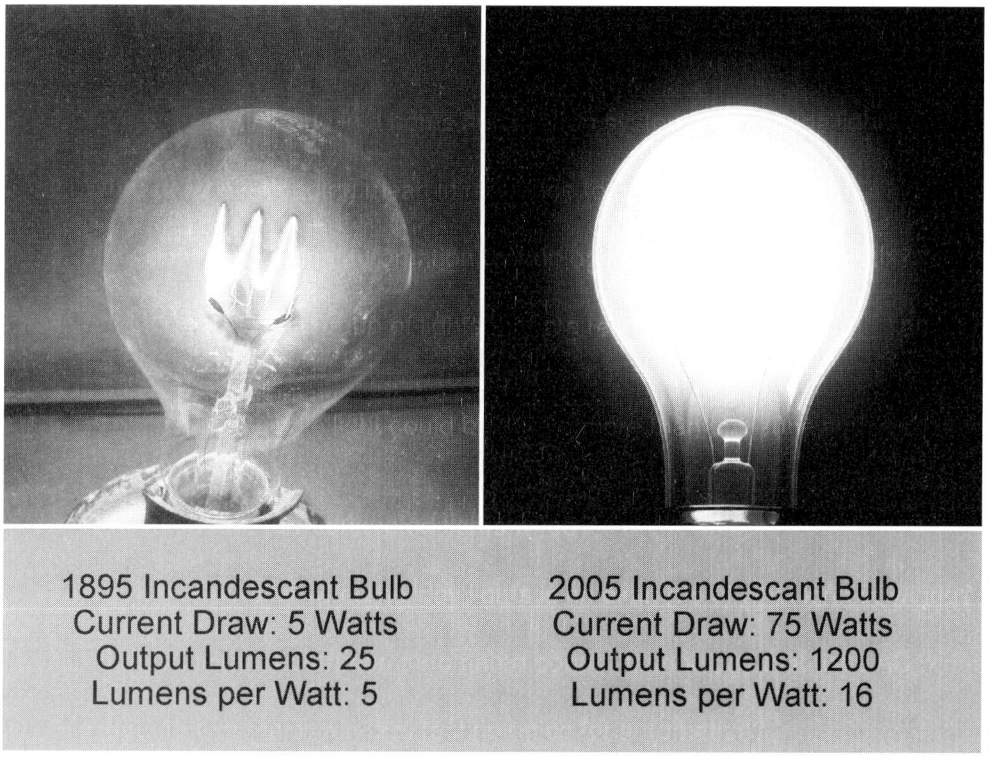

FIGURE 7-1

A comparison of the output of a century-old incandescent bulb and that of a modern one shows how little the efficiency of this type of bulb has improved over the years.

Light bulbs outside of the specified lumen range are exempt from the efficiency restrictions, so those consuming less than 40 watts or more than 150 watts will not be banned. Several types of specialty lights, including appliance lamps, rough service bulbs, three-way bulbs, colored lamps, and plant lights are also exempt from the efficiency requirement.

Under the 2007 law, by 2020 a second set of restrictions would become effective requiring all general-purpose bulbs to produce at least 45 lumens per watt. This is a standard that few, if any, incandescent lights can meet, and it will mandate the use of other types of lights, as discussed in following sections. Again, however, certain types of incandescent bulbs will be exempt from the standard, including reflector flood lights, three-way bulbs, candelabra bulbs, colored lights, and other specialty bulbs.

ENERGY-EFFICIENT LIGHTING

Although some incandescent light bulbs, such as halogen bulbs, may be efficient enough to meet the new standards set by the 2007 law, most will not. To meet the 2007 standards and the more restrictive 2020 standards, new types of electric lights have entered the market and are now vying to displace the dominance of incandescent lighting.

Compact Fluorescent Lights (CFLs)

Compact fluorescent lights (CFLs) have revolutionized energy-efficient lighting. CFLs are small versions of full-sized fluorescent tubes that have been used for decades in commercial lighting venues. The fluorescent light was invented in the 1890s and operates by passing a high-voltage current through a phosphor-filled glass tube. The current causes the phosphor to emit light, although of a somewhat different part of the light spectrum than that of an incandescent bulb.

How CFLs Work

A CFL is a miniature, high-output type of fluorescent lamp that is designed to replace an incandescent lamp of similar size and light output. CFLs are now made to fit into most existing light fixtures formerly used for incandescent bulbs. Compared to general-service incandescent light bulbs that produce an equal amount of visible light, CFLs use much less power and have a much longer rated life.

Although CFLs radiate light of a different color spectrum from that produced by incandescent lamps, new formulations of the light-emitting phosphors in these lights have improved the perceived color of their light by making it appear much closer to that produced by incandescent lights.

CFLs consist of two main components: a gas-filled tube (also called a bulb) and an electronic ballast. These can be separate parts that are connected by plugging the tube's connectors into the ballast. In all CFLs designed to replace similar sized incandescent bulbs, the ballast and tube are an integrated unit with a base that is compatible with standard sockets. The tube of a CFL can be a single-turn double-helix spiral; a double-turn, triple-turn, or quad-turn spiral; a circular bulb; or a butterfly shape. The electronic ballast contains a small circuit board that acts as a high-frequency AC-to-DC inverter. The high-frequency current produced by the ballast causes the phosphors inside the tube to illuminate.

Cost and Energy Efficiency

The efficiency of CFLs compared to incandescent lights can be stated very simply: CFLs require less power to produce a given amount of light and they last much longer than an incandescent light. The trade-off for this greater efficiency is that CFLs are more expensive to manufacture than incandescent bulbs, although continued research and development of new manufacturing techniques plus the economies of mass production have brought their cost down considerably.

The average rated life of a CFL is between 6000 and 15,000 hours, compared to the average rated life span of an incandescent lamp of about 1000 hours. The operating life of either type of bulb depends on several factors, including correct operating voltage, absence of manufacturing defects, ambient operating temperature, and mechanical shock. The life of a CFL will be reduced if it is turned on and off too frequently. To mitigate this problem, CFLs should be left on if the time period they will be off is 15 minutes or less.

As they age, CFLs produce less light than when they are new. Their light production will decline to between 70% and 80% of their original output by the end of their lives. The largest part of this light loss will occur during the time soon after the lamp is first used. This 20–30% reduction over many thousands of hours will not be apparent to the average user, especially if the original light level of the CFL was ample.

To produce a given amount of light, CFLs require only 20% to 33% of the power needed by incandescent lamps to produce the same illumination. Using a mid-range figure of 27% of an incandescent light's power use for a CFL of equal lumens output, we can compare the power savings that can be achieved by replacing the incandescent lights in the example given earlier in this chapter.

1	300-watt bulb replaced by 80-watt CFL equals 80 watts current
2	200-watt bulbs replaced by 55-watt CFLs equals 110 watts current
2	150-watt bulbs replaced by 40-watt CFLs equals 80 watts current
49	100-watt bulbs replaced by 27-watt CFLs equals 1323 watts current
11	75-watt bulbs replaced by 20-watt CFLs equals 220 watts current
6	60-watt bulbs replaced by 16-watt CFLs equals 96 watts current
2	40-watt bulbs replaced by 11-watt CFLs equals 22 watts current

If all the incandescent lights in this example home were replaced with CFLs and each were illuminated for 1 hour per day, they would consume 1.931 kWh per day and 57.93 kWh in a month. Using again the 2010 U.S. average cost of power of $0.119, the daily cost of CFL lighting is $0.23, the monthly cost is $6.89, and the yearly cost is $82.72. This compares with the $306.95 annual cost for incandescent lighting.

Replacing incandescent light bulbs with CFLs of equal lumen output is one of the easiest and surest ways of reducing home energy consumption. Before deciding exactly which lights in your home should be replaced with CFLs, you should be aware of some additional facts concerning CFLs. Figure 7-2 shows several styles and sizes of CFL bulbs that all have standard Edison-type bases. Note that the larger bulbs may be wider above the screw-in base than an incandescent bulb and may not fit into all sockets designed for these bulbs. The glass-enclosed CFL at the lower left is almost identical in size and shape to an incandescent bulb and will fit all sockets designed for these bulbs. Figure 7-3 shows a selection of CFL bulbs with small screw-in bases and pin connector bases that fit low-voltage track-light fixtures and other types of residential lighting.

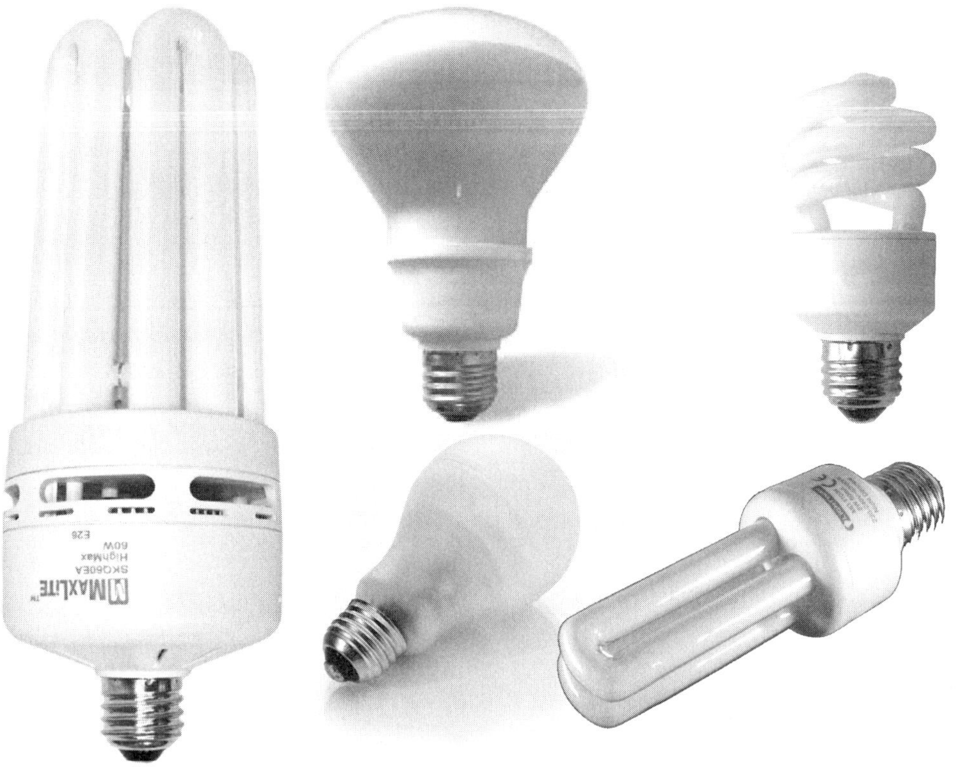

FIGURE 7-2

Several styles of CFL bulbs are available with Edison-type bases that will fit most residential fixtures.

FIGURE 7-3

CFL bulbs with various types of bases and connectors are available to fit many low-voltage light systems and display sets.

Advantages and Disadvantages

The most important advantage of CFLs over incandescent lights is that they use less energy and save money on electrical charges. The amount of savings will depend on the percentage of electricity used in a home that is used for lighting. It is estimated that in the average home about 25% of electrical use is for lighting. Using this average, if all the lighting in the home were converted to CFLs using about 27% of the current used by the incandescent lighting, the savings would amount to an 18% reduction in the home's total electrical charges.

This calculation probably exaggerates the actual savings somewhat because typically not all incandescent lights will be replaced, and those that are will not always be as efficient individually as rated. A more conservative estimate of savings is to reckon on a 12% reduction in the average home's electric bill realized by switching lighting to CFLs.

The second major advantage of CFLs is that they last longer than incandescent bulbs. Cost for CFLs has now dropped below $2.50 each for the common sizes. Those with more color-enhanced phosphors and other special features cost more, but the energy savings from using CFLs will more than repay the cost of replacing incandescent bulbs over the extended life of the new lights.

CFLs emit light from phosphors inside the bulb. Most CFLs contain a mixture of several phosphors, each of which emits one band of color. The group of phosphors in the mixture is designed to emit light in a combined color spectrum that closely resembles incandescent light or daylight. Each phosphor mixture is a compromise designed to produce the most appealing shade of emitted light while maintaining the energy efficiency of the CFL and keeping its cost to manufacture as low as possible. Every extra phosphor added to the coating mix reduces the light's efficiency and increases its cost.

High-quality CFLs use three or four phosphors to produce light that appears white and has a color rendering index (CRI) of around 80. CRI is a measure of how closely colors viewed under a light source match the same colors viewed in natural daylight, which has a CRI of 100.

CFLs can also be rated for light output based on the color temperature of the light they emit. This temperature is referred to as the correlated color temperature (CCT) and is measured in Kelvins (K), a unit of temperature measured on an absolute scale on which zero is equal to no heat at all. The higher a light source temperature is, measured in Kelvins, the more blue or blue-white it appears to the viewer. Natural daylight, produced by solar radiation and diffused by the earth's atmosphere, has a color temperature of about 5000 K. A soft white incandescent bulb produces light at about 3000 K. Bright white bulbs and cool white bulbs range between 3500 and 4000 K.

Names for CFLs that produce light at a specific color temperature and CRI are not yet standardized, and labeling variations and inconsistencies exist among different manufacturers. Some CFLs labeled "daylight" have a color temperature of 3500 K, whereas others lamps with a daylight label have color temperatures of at least 5000 K. Some CFL packages do not show any Kelvin value or CRI for the bulbs, but the federal government's Energy Star criteria for CFLs will probably be updated soon to require such labeling in the future.

At present, some CFL manufacturers label their products with a rather confusing three-digit code that is a combination of the CRI and the color temperature of the bulbs. The first digit in this code is the CRI of the lamp divided by 10 and the second and third digits represent the color temperature divided by 100. Thus, a CFL with a code of 840 would have a CRI of 80 and a color temperature of 4000 K.

With no standards firmly set for CFL names that match them to the accepted incandescent labels, each consumer should determine the specific CFL bulb that feels most comfortable in home use. Buying several different CFLs and trying each of them for a period of time is the most effective method of deciding which CRI and color temperature lamp should replace the incandescent bulbs in the home.

Incandescent lights turn on instantly and reach full brightness in a fraction of a second. CFLs turn on a bit slower, usually within a second, but may also not reach full brightness for as long as a minute. The colder the ambient temperature, the longer a CFL takes to warm up and reach full brightness. The turn-on time and the delay in reaching full brightness are being shortened as CFL technology improves, and some CFLs are now marketed as "instant on," with almost no noticeable warm-up period.

The light color of a CFL may be slightly different during its warm-up period, and this problem has been addressed by the introduction of a hybrid CFL. This lamp combines a halogen lamp that lights instantly with a CFL that has a warm-up period. Once the CFL reaches full brightness, the halogen lamp goes out. This type of lamp is more expensive and not quite as energy efficient as a standard CFL, but it does meet the need for instant full output illumination. As regular CFLs come close to this standard, the need for hybrids will probably decline.

All fluorescent lamps contain mercury as vapor inside the glass tubing. This includes CFLs, which contain about 3 to 5 mg in each light. The mercury is safely contained in the sealed tube, but because mercury is poisonous, it can become a problem if the CFL is broken or when it is discarded. Even small amounts of mercury contaminate the environment, and care must be taken to prevent the mercury in CFLs from being released. Spent lamps should be disposed of properly or recycled to keep the mercury in them contained. This is required by law in some states, but should be followed by all CFL consumers. Care should be taken not to break a CFL, but to keep its tube intact until it can be processed to recover the mercury it contains. Some retail outlets for CFLs now accept spent units for recycling.

Other potential drawbacks exist when CFLs are adopted to replace incandescent lights. These should be considered individually by the consumer to determine if they are significant.

- CFL lights may be somewhat larger than incandescent bulbs of similar lumen output, making them difficult to fit in some fixtures or locations.
- The electronic ballast of a CFL may fail on rare occasions; this can cause discoloration or partial melting of the ballast enclosure, noxious odors, smoke, or even fire.
- CFLs designed to fail safely at the end of their lives may not always do so; this can result in some burning and offensive smells. It can also make the spent bulb difficult to remove from its socket.
- Individual CFLs may not produce the rated lumens of light, making it necessary to replace the defective lamp with a new one.
- Regular CFLs do not work well with a dimmer; using them with a dimmer shortens bulb life and will void the warranty of certain manufacturers.
- Dimmable CFLs must be used with a dimmer switch matched to the lower-power consumption range of the CFL.
- Dimmed CFLs do not change their color temperature at low light levels as other light sources, such as the sun or incandescent lights do; dimmed CFLs tend to make people feel cold or even threatened rather than invoking the warm "firelight" feeling of other dimmed lights.
- Some CFLs are not suitable for pendant lamps or recessed lights fixtures because they build up heat when mounted and operated with their base up; CFLs for use in such applications are available but are more expensive (Figure 7-4).

FIGURE 7-4

Ballast failure caused both of these CFLs to overheat and partially melt or burn the outer casing; the CFL at the left was installed base up in a ceiling fixture, and this may have contributed to its failure due to heat buildup.

- The input stage of a CFL is a rectifier, which can cause some distortion in the home's AC power supply; this will only be significant in large CFL installations and should not affect most homes.
- Infrared remote controls and similar devices can sometimes read the infrared light emitted by a CFL as a signal; for this reason CFLs should not be placed near televisions, radios, remote controllers, or mobile phones.
- CFLs should be tested to be sure they work with and are not adversely affected by electronic timers and other control devices such as illuminated wall switches, motion sensors, and light sensors.
- Most CFLs are not designed for outdoor use and some will not start in cold weather; outdoor CFLs are available with cold-weather ballasts but their light output may drop at low temperatures.
- There can be large variations in the quality of light, cost, and turn-on time among different CFL manufacturers, even though their lamps appear identical and have the same CRI and color temperature ratings.
- Over time the ultraviolet light produced by CFLs can damage paintings and textiles made with light-sensitive dyes and pigments; ultraviolet light can also deteriorate polymer plastic, causing a loss of strength and yellowing.

Light-Emitting Diodes (LEDs)

Another type of high-efficiency light that is now being widely adopted for home use is the light-emitting diode (LED). LEDs are small, solid light bulbs that are extremely energy efficient. They were originally developed for use in instrument panels, electronics, pen lights, and similar applications. More recently, manufacturers have expanded the application of LEDs by producing cluster LEDs in which almost any number of individual LEDs lights are combined into a single unit that produces light equal in magnitude to large incandescent bulbs.

Current LED bulbs for home lighting are made using clusters of 30 to 180 individual LEDs that are encased in diffuser lenses. The lenses spread the light from the LEDs, which is emitted in a narrow intense beam, into a wider, more diffuse pattern similar to that of an incandescent or CFL bulb. Cluster LEDs are now available with standard bases that fit the sockets of common household light fixtures.

Early LEDs were expensive to produce using a sapphire-based technology and were therefore not cost effective to cluster for home lighting applications. New production processes using inexpensive silicon wafers have recently brought the cost of LEDs down sufficiently to make the clustered units' price competitive with CFLs and incandescent lights.

How LEDs Work

An LED is a special type of diode. A diode is a semiconductor device made of a material that has a varying ability to conduct electrical current. Most semiconductors are made of a base material that conducts electricity poorly. Impurities, small amounts of another material, are then added to areas of the base material, in a process called doping, that change the electrical conductivity of the material. The added impurities either add free electrons to the material or create holes where electrons can go. Either of these additions makes the material more conductive.

A semiconductor material with extra electrons is called an N-type material, because it has extra negatively charged particles. In N-type material, free electrons move from a negatively charged area to a positively charged area. A semiconductor with holes is called a P-type material, because it has extra positively charged particles that can attract electrons to the holes where they are missing. Electrons can jump from hole to hole, moving from a negatively charged area to a positively charged area.

A diode comprises a section of N-type material bonded to a section of P-type material, with electrodes on each end. The diode conducts electricity in only one direction. When the N-type side of the diode is connected to the negative end of a circuit and the P-type side to the positive end, current will flow from the N-type side of the semiconductor to the P-type side. Current will not flow the other way, however. If the P-type side is connected to the negative end of the circuit and the N-type side is connected to the positive end, no current flows because the electrons and holes of the semiconductor material are both moving in the opposite direction.

As current flows from the N-type side of the diode to the P-type side, some of the electrons in the material drop from a higher energy level to a lower one. When they drop, they release energy in the form of a photon of light. This happens in all diodes, but the light photons can only be seen if the diode is composed of certain materials. Visible light-emitting diodes (VLEDs) are made of materials that emit light photons at frequencies visible to the eye between infrared and ultraviolet wavelengths. Materials have now been developed to produce light at many wavelengths within the visible light spectrum, making it possible to manufacture LEDs of almost any color.

Any color, that is, except white, which is actually a combination of the primary light colors blue, green, and red. There are two ways of producing white-light LEDs suitable for home lighting use. One is to combine individual LEDs that emit the three

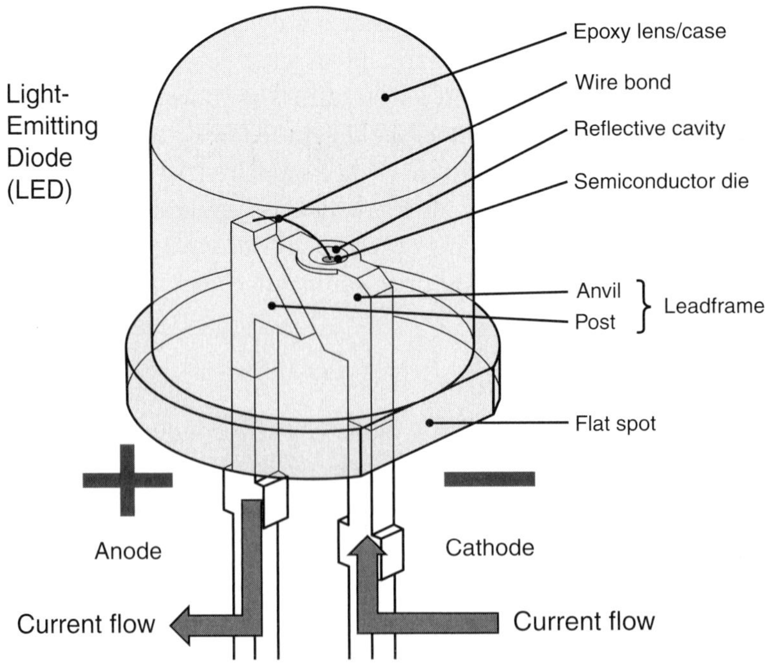

FIGURE 7-5

The structure of an LED is fairly standard; its color and brightness can be varied by the materials it is made of.

primary colors and then mix all the colors to form white light. The other is to use a phosphor material to convert monochromatic light from a blue or ultraviolet LED to broad-spectrum white light, in much the same manner that such phosphors work in a fluorescent light bulb. The latter method is the simpler and less expensive process and hence has become by far the most popular way of making white-light LED bulbs. These LEDs are called phosphor-based white LEDs.

The phosphor method involves coating an LED of one color, usually blue, with phosphors of different colors to form white light. When several phosphor layers of different colors are applied, the emitted light spectrum is broadened, thereby raising the CRI value of the resulting LED. Phosphor based LEDs have a somewhat lower energy efficiency than normal LEDs because they have more heat loss and some additional energy loss from phosphor-related degradation. This lower efficiency is compensated for by the simpler and cheaper design and production of a light source using a monochrome emitter with phosphor conversion, and most high-intensity white LEDs now on the market are manufactured using this process.

Figure 7-5 shows the components of an LED and the direction of current flow through it. This illustration shows only the LED itself; a phosphor-based LED is surrounded by an envelope containing the phosphors that transform its color to white light.

Cost and Energy Efficiency

LEDs are currently much more expensive than CFL or incandescent lights, but they are close to equal in cost efficiency with both over their working life spans. This is because they are more energy efficient and longer lasting by far than any other type of light bulb.

LED light bulbs for home lighting consume between 3 and 10 watts of electricity. An LED bulb equivalent in light output to a 60-watt incandescent bulb consumes only 6 watts of electricity, one-tenth the amount of the incandescent light and less than half the amount of an equivalent CFL. Table 7-2 shows a comparison between the three types of lights over a 25-year period of use.

TABLE 7-2 LED, Incandescent Bulb, and CFL Comparison

	LED	CFL	INCANDESCENT BULB
Projected life span	50,000	10,000 hours	1200 hours
Watts per bulb (equiv. 60 watts)	6	14	60
Cost per bulb	$35.95	$2.95	$1.25
KWh of electricity used over 50,000 hours	300	700	3000
Cost of electricity (@ 0.11 per kWh)	$33.00	$77.00	$330.00
Bulbs needed for 50-k hours of use	1	5	42
Equivalent 50-k hours bulb expense	$35.95	$14.75	$52.50
Total cost for 50-k hours	$68.95	$91.75	$382.50

Table 7-3 shows the cumulative cost of using each type of light bulb year by year over a 25-year period. From this table it is clear that CFLs are over twice as economical to buy and use after just 1 year as incandescent bulbs. LEDs are more economical than incandescent bulbs after 3 years, and more economical than CFLs after 16 years.

TABLE 7-3 LED, CFL, and Incandescent Bulbs Cumulative Costs

END OF YEAR	TOTAL COST FOR 60-WATT EQUIVALENT LED LIGHT FIXTURE	TOTAL COST FOR 60-WATT EQUIVALENT CFL LIGHT FIXTURE	TOTAL COST FOR 60-WATT INCANDESCENT LIGHT FIXTURE
1	$37.27	$6.03	$15.70
2	$38.59	$9.11	$31.40
3	$39.91	$12.19	$47.10
4	$41.23	$15.27	$62.80
5	$42.55	$18.35	$77.25
10	$49.15	$36.70	$153.25
15	$55.75	$55.05	$230.00
20	$62.35	$73.40	$305.25
25	$68.95	$91.75	$382.50

LED bulbs last up to 10 times as long as CFLs, and as much as 50 times longer than typical incandescent bulbs. Because they are solid pieces of electronic hardware and do not have filaments, they are not easily damaged by jarring, bumping, or vibration. They also produce very little heat, only about 4 BTUs per hour, compared to 85 for an incandescent bulb of equal brightness. They are smaller than other types of light bulbs and, because they contain no mercury, can be disposed of as ordinary trash, should the need ever arise.

LEDs are fairly new to the home lighting market and new styles and models are appearing regularly as the technology for making them improves and expands. This continuing development will probably result in lower-cost LEDs becoming available in the future, but for those who want to have the advantages of LED lighting now, selection should include several factors:

- Estimate the light output desired, not the wattage; a 3-W LED is equivalent in output to a 45-W incandescent bulb; a 6-W LED is equal to a 60-W incandescent.
- Choose a color for the warm or cool illumination you want; LED bulbs are available in several shades of white light.
- Standard Edison-base LEDs are available with several styles of bulb; most recessed or track-lighting types of pin sockets are also available now or will be soon.
- Select dimmable LED bulbs for circuits that include dimmer switches or controls; most standard LEDs will not work with dimmers.
- Buy high-quality bulbs from reputable manufacturers; inexpensive LEDs are available, but are made using low-quality chips that fail easily, negating long-life cost savings.
- Look for FCC and UL certifications on the LED package; they indicate the LED meets minimum safety and performance standards.

Among the LED bulbs currently available for home lighting are those shown in Figure 7-6. Each of these is described in following sections. LED technology is advancing rapidly and new styles and models of LED lights will continue to appear regularly.

FIGURE 7-6

LED bulbs with various types of bases and connectors are available to fit many low-voltage light systems and display sets.

Diffused LED bulbs have clusters of LEDs covered by a dimpled lens that spreads light from the individual LEDs over a wider arc. These lights are available in standard Edison bases and are suitable for use in area lighting for rooms, porches, reading lamps, accent lamps, hallways, and low-light applications where lights remain on for extended periods.

Diffused high-power LED bulbs are designed for standard household use. They produce light equivalent to that of a 100-watt incandescent bulb. The EvoLux and ZetaLux models are both UL Listed and FCC approved LED light bulbs for general household illumination.

Track-lighting LEDs are designed with pin connectors to fit in standard track-lighting setups. Unlike incandescent track lights, these LEDs do not get hot to the touch no matter how long they remain on. They are available in white-light versions or several colors.

Recessed, downlight, and spotlight LED bulbs are intended for use in standard recessed lighting pots and housings. These housings are referred to by the term *PAR,* which stands for parabolic aluminized reflector. The reflector can be part of the housing, but more commonly is part of the lamp that fits into the housing. Inside a sealed-beam incandescent PAR lamp, the light bulb is permanently fused

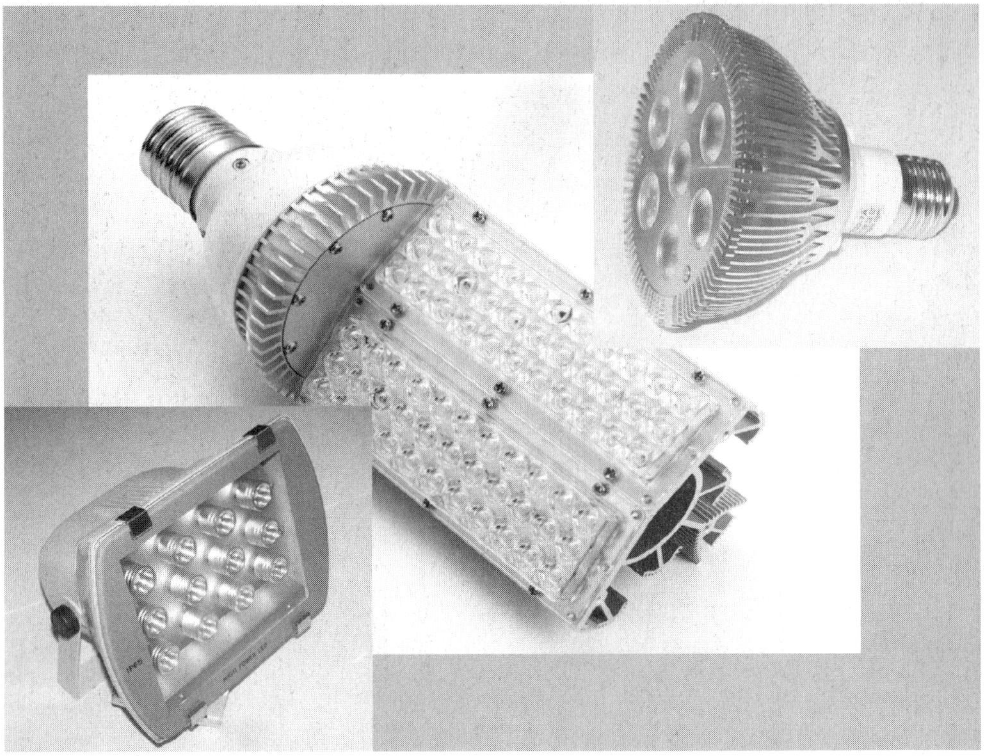

FIGURE 7-7

A 36-watt LED outdoor landscape floodlight (left), a 30-watt Moonraker LED bulb equivalent to two 300-watt incandescent bulbs (center), and a 16-watt outdoor LED floodlight (right).

to a reflector, which in turn is sealed tight with a lens. PAR lights are measured by size across the lens in eighths of an inch. Thus, a PAR 16 light is 2 inches in diameter, a PAR 20 light is 2.5 inches in diameter, and a PAR 30 light is 3.75 inches in diameter.

LED lights are available for PAR housings. When buying these, you should buy the correct diameter for the housing and also the correct lumen intensity of the light. PAR LEDs have light output and color quality similar to those of incandescent downlights, but draw only about one-fifth of the power. Heat buildup, often present in incandescent recessed lighting, is entirely eliminated by the use of LED bulbs.

Outdoor floodlight LEDs, such as those shown in Figure 7-7, are excellent replacements for incandescent bulbs, although they are not as bright as halogen bulbs. Their light output is directional, usually across about a 30-degree arc, and they are made weatherproof for continuous outdoor use. Unlike CFLs, their life span is not adversely affected by being turned on and off frequently and they function well in cold temperatures.

Advantages and Disadvantages

As you decide on whether to replace energy-inefficient incandescent bulbs, be aware of some other advantages and disadvantages that may affect your decision about using LEDs in specific applications.

All home lighting devices containing LEDs are safe under all conditions of normal use, and are classified as Class 1 LED products. LEDs do not contain mercury, although they may contain trace amounts of other hazardous metals such as lead and arsenic. According to federal standards, LEDs are not hazardous, but stricter California regulations still classify most of them as hazardous because of the small amounts of copper, lead, nickel, or silver they contain.

Other advantages of LEDs include:

- High efficiency, which means that LEDs emit more light per watt than any other type of home lighting including, incandescent and CFL bulbs. Unlike CFLs, their efficiency is not affected by the shape and size of the light.
- LEDs light up very quickly and reach full brightness immediately without any warm-up time required.
- Frequent on-off cycling does not affect either the performance or the life span of LEDs. They have no delay in restarting after being turned off and return to full brightness instantly when turned on again.
- LEDs can easily be dimmed, although not with the same type of dimmer used for incandescent lights in AC circuits. LEDs designed for dimming use pulse-width modulation or lowering of the forward current.
- LEDs radiate very little heat in the form of infrared energy, and the small amount they produce is dispersed through the base of the LED.
- LEDs usually fail by dimming over time after 20 or more years, rather than failing suddenly and completely after a year or so of service as incandescent bulbs do.
- LEDs produce directional light that can be tightly focused or spread over an arc. Their light can also be diffused to match that of incandescent or CFL bulbs. Focused or diffused light from an LED is still far more efficient than light from incandescent and fluorescent sources that is always diffused and can only be focused using an external reflector.

Disadvantages of LEDs for home lighting applications include the following:

- LEDs are more expensive to install, price per lumen, than incandescent or CFL lighting, although over time this differential is overcome by the longer life span of the LEDs.
- The light quality of LEDs differs significantly from that of sunlight or an incandescent light source. This means that the color of objects, particularly red ones, will be perceived differently under cool-white LED illumination than under sunlight or incandescent light. Continuing research is reducing this difference and new LEDs are now equal to CFLs in light quality and color perception.

ENERGY-EFFICIENT ELECTRONICS

A recent government Energy Star report found that use of consumer electronics products accounted for more than 25% of total household electricity consumption. The report noted that if two older-model TVs, a VCR, a DVD player, and three telephones found in the average home were all replaced with Energy Star compliant models, over 25 billion pounds of greenhouse gas emissions would be eliminated every year.

Many home electronics products use significant amounts of energy when they are in standby mode to power features such as clock displays and remote controls and to keep screens and other components warmed and ready for instant service. This constant use of power is referred to as the home's "phantom load" because it provides no obvious benefits but can be a significant part of total power use. The devices that contribute to the phantom load are sometimes referred to as "vampire appliances" because even when asleep they are constantly drawing power.

Electronics products that have earned the federal government's Energy Star rating meet the strict energy-efficiency guidelines set by the Environmental Protection

Agency and Department of Energy by using as much as 60% less energy to perform standby functions, while providing the same performance at the same price as less-efficient models.

Most electronics now being marketed meet the minimum standards required to carry the government's Energy Star label, but beyond the basic energy-efficient design this designation requires, other factors can affect the power consumption of electronic systems. The energy efficiency of the new HDTV you are planning to buy probably won't be the top consideration on its list of features, but like most electronic appliances, HDTVs with otherwise similar features do vary significantly in the amount of power they consume when operating and the amount they consume in standby mode.

TVs, Stereos, and Home Theaters

A recent test of four major brands of LCD HDTV sets with 50- to 55-inch screens showed that the most energy-efficient unit consumed 90 watts of power when operating. The least efficient unit consumed 170 watts and the others between 125 and 150 watts. Smaller screen sizes consume less power, but not proportionately less. Even an average 18-inch HDTV will use about 80 watts when operating, so product design rather than size is the most important factor for energy efficiency in electronics systems.

Energy efficiency should not be the first priority when buying a new TV, but as the figures just noted indicate, some TVs are definitely more efficient than others. If a set is operated about 8 hours per day (the current U.S. average), the savings from an energy-efficient set will be significant. The least efficient of the sets just noted will consume 496.4 kWh of electricity per year and cost $59.07 to operate. The most efficient will consume 262.8 kWh and cost $31.27 annually to operate.

Both the most and least efficient units consume about 30 watts when not operating but in standby mode, ready to start up instantly. For the 16 hours each day they are not in use, the TVs will each consume 175.2 kWh per year at a cost of $20.85, making a total annual cost for the most efficient unit $52.12 and the other $79.92. If other features of the two sets are substantially equal, the annual $27.80 savings on electric power for the more efficient one is worth having.

That savings can be increased further by not keeping the TV in standby mode when it isn't operating. To do this, the TV must be turned off, completely off, at the outlet where it is plugged in or by a switch on its AC plug. Some TVs have a manual power switch on the front of the unit, but this switch almost always puts it in standby mode with some power still being used to keep components warm and ready for instant use. An outlet or cord switch turns the TV completely off and, unlike the manual power switch on the TV, allows it to be turned completely on again by remote control, using an automated wall outlet or switch. For details on how either of these can be installed on a TV circuit, see the next chapter.

Audio systems of the past have also been bulky and energy inefficient. Older whole home audio systems that can play in multiple rooms are typically left on continually so that they will function when the volume is turned up in any area. They constantly burn energy, much of which is wasted as heat. New whole home audio systems generate less heat, take up less space, and, most important, many are zone specific. This means that each zone of the audio system can be used independently of the others and that energy consumption is confined almost completely to the component in use. These systems have a digital local amplifier built into each room's control unit, along with volume control and content selection via keypad or touch screen. This design results in better signal strength and much improved musical clarity with an expenditure of only 50 to 70 watts in the zone being used instead of powering a central distribution unit at 300 to 1200 watts continuously so that audio in every zone or room in the home can be accessed on demand. This zone system uses less energy, reduces wear on components, and improves sound quality.

To become Energy Star qualified, audio electronics must now consume less than 1 watt of power in standby mode. Some systems that meet this standard are now available, and most companies are evaluating their products as they are updated for their ability to meet the Energy Star standard. Millions of older systems remain in use, however, which consume more energy in standby mode. Like TVs and other electronic equipment that remains in standby mode when not in use, these systems can be shut off completely by means of a remote-controlled cutoff switch to conserve power.

Many home video/audio systems, especially those installed in home theaters, are rated at up to 1000 watts output power for the audio system alone, and may consume another 300 watts or more for the video components. The audio output power of these systems is not required continuously, but the input power necessary for the amplifier, tuner, video screen, and other components must be fed into the system constantly for it to be ready to deliver high-power output on demand. Very few home theaters ever need even 100 watts of output power to the audio system, and having more than that constantly available means wasting considerable power.

The power rating of an audio system should be selected for the size and acoustic properties of the room in which it will be used. The system needs no more power than necessary to provide the maximum safe volume of sound within the theater area. For all but the largest home theaters, this will mean a power rating of 200 watts or less.

Game Consoles

When the Electric Power Research Institute (EPRI) recently tested three top-selling video game systems to find out how they compared in terms of power consumption, results from the experiment showed that a Nintendo Wii system uses only one-sixth the power that a Sony PlayStation 3 or Microsoft Xbox 360 does in active mode. EPRI tested each system for 1 hour of active play using the EA Sports Madden 2011 football game, which is widely played on all three game consoles. EPRI found that the Nintendo Wii system used an average of 13.7 watts, the Sony PlayStation 3 used an average of 84.8 watts, and the Microsoft Xbox 360 used an average of 87.9 watts.

As with other electronic systems, you may not want to purchase a game console based mainly on its energy efficiency, but it may be one feature to consider, especially if the system is used heavily. Even if you don't have the most efficient system you can minimize wasted energy by turning the game console off when it is not in use.

Computer Systems

As noted in a previous chapter, desktop computer systems typically consume about 275 to 375 watts when operating and 50 to 60 watts in sleep mode. If used an average of 4 hours per day and left in sleep mode the rest of the time, a desktop system that includes a printer will consume about 2.2 kWh per day or 803 kWh per year. At a cost of $0.119 per kWh, the annual cost of operation will be $95.56. Half of this amount will be for operation of the computer and the other half for keeping it ready for operation in sleep mode.

If use of the computer can be confined to a single block of time, turning the system off for the remainder of each day should be considered. Even if the downtime is only 8 hours, from 11:00 p.m. to 7:00 a.m., savings of $16.25 per year can be realized. Doubling the downtime to 16 hours each day doubles the savings to $32.50 annually.

Even though these savings are significant, the cost of keeping a computer system on continuously may be justified, particularly if it is used for business as well as personal purposes. Because we have become accustomed to instant communication and feel that any delay is unacceptable, a continuous Internet link may be worth the price of keeping a home computer running all the time.

But if a cell phone can provide the same data link, it will be the most cost-effective option, at least as far as power consumption is concerned. Likewise, a notepad or laptop computer without attached peripherals and a smaller screen will require less power to operate than a full desktop system. Best practice for home computer systems and similar devices is to keep the most energy-efficient system operating continuously and shut the others down until actually needed. Even in a home with a LAN and several computer systems, there's no need to have all systems running all the time.

OTHER POWER-HUNGRY APPLIANCES

Several other appliances in the home consume large amounts of power because they contain relatively large electric motors. A 1-horsepower (hp) electric motor running on AC power typically consumes about 750 watts of power under a 60% to 75% continuous load. There is no way to reduce this power requirement other than by reducing the length of time that the appliance operates and its motor runs.

Refrigerators and Freezers

These appliances run in cycles that are timed by a temperature sensor inside the freezer or refrigerator. When temperature rises above the level the sensor is set for, it activates the refrigeration unit to cool the interior back down to the required temperature. The heat that the refrigeration unit of a freezer or refrigerator extracts from the interior of the unit is discharged into the room where the appliance is located. This is not energy efficient, but the only alternative is to vent the discharged heat to the outside. For most refrigerators in most homes, this is not practical or even possible because creating a vent to the outside would also create a leak for conditioned air to escape from the building.

The only ways to reduce power consumption by a refrigerator is to raise the inside temperature setting and to avoid opening the door frequently to access the interior. The best setting for food preservation in a refrigerator is 37°F. Most refrigerators don't show temperatures on their adjustment gauge. They usually have numbers (1 to 5 or 1 to 9) or letters (A to F). To get the correct setting, use an outdoor weather thermometer and place it on a middle shelf of the refrigerator. Set the adjustment gauge to a middle setting and close the refrigerator for half an hour or so. Check the thermometer's temperature reading and adjust the gauge up or down as needed until you get the 37-degree reading on the thermometer. Note where the refrigerator's adjustment gauge is set and leave the setting at that level.

Freezer temperature should be 0°F, and this should also be checked with a thermometer rather than adjusted by the unit's gauge alone. Once set, freezer and refrigerator temperature should remain constant without further adjustment. It's a good idea to check periodically with the thermometer to be sure there is no significant variation that could indicate a problem with the appliance's sensor or refrigeration system.

If temperature in a refrigerator or freezer does vary up or down a significant amount, the unit should be checked by a competent repairman. Among the causes of temperature variation in refrigeration systems are:

- Defective sensor and/or switch
- Defrost cycle problem (in refrigerators only)
- Compressor leak or other failure
- Interior cold air circulation obstructed
- Large temperature change in outside air

Any of these problems can cause the refrigerator or freezer to use more power than normal. Adjusting the temperature gauge on the unit may bring the interior temperature back to the desired level, but will have no effect on the cause of the problem or the excess power being consumed. For this reason, significant temperature variations in a refrigerator or freezer should always be checked by a repair person.

Unlike refrigerators, freezers do not need to be located in the kitchen or even inside the home's heated or air-conditioned living space. The most efficient way to operate a freezer is to locate it in an unheated basement, garage, or other space and provide a dampered vent to the outside. The vent can be closed in winter weather when discharge heat from the freezer is useful to partially warm the unheated room. It can be opened in warm weather to allow the discharge heat to escape. The temperature in the room where a freezer is located should not be lower than the freezer's interior temperature or the unit's sensor mechanism will not function properly.

The colder a freezer room is, down to near but not below the freezer's interior temperature, the more efficient the freezer's operation will be because the refrigeration unit has to overcome less heat transfer from the surrounding air. The few steps necessary to access a freezer placed in a cool room will be more than compensated for by the power saved in its operation and the elimination of the freezer's heat discharge into air-conditioned space.

Exercise and Fitness Equipment

Treadmills and similar exercise and fitness equipment have powerful motors and consequently consume large amounts of electricity in operation. Treadmill motors for home use are typically rated at peak-load and continuous-use levels. Peak load is the maximum power rating of the motor, whereas continuous use is an expression of the power required during normal use of the equipment.

Home-use treadmills are usually rated between 1.25 hp and 3.0 hp (continuous use). To choose a unit within this range you should consider the weight of the user(s) and the nature of the workouts planned. Heavier users require a larger continuous-use rating for the treadmill motor, but increasing the motor's size also increases its power consumption.

A 1.5-hp treadmill uses about 750 watts of electricity in continuous operation. A 3.0-hp motor uses about 1500 watts during continuous operation. Users weighing less than 150 pounds need no more than a 1.5-hp continuous-load treadmill. Users weighing over 275 pounds should have at least a 2.0-hp unit or larger. In between these weights, 1.75-hp or 2.0-hp continuous-load treadmill is adequate.

Do not buy a larger treadmill than you need. You'll pay more for it not only in the purchase price, but also in the amount of power it consumes every time you use it. A 1.5-hp unit will consume about 274 kWh annually if used for 1 hour per day. Operating cost will be about $32.58 per year. A 3.0-hp motor treadmill will use 548 kWh annually and cost $65.21 to operate. A unit more powerful than needed will not perform better or help the user lose weight faster than a correctly sized unit, but it will do more work and waste more power.

Other powered exercise or fitness equipment will probably not have a motor as powerful as a treadmill's, but it should still be sized and selected according to the weight of the user. Larger equipment always requires more power, and that power is wasted if it is not needed for the machine to perform at full capacity with the user.

Workshop Equipment

Workshop equipment, whether it consists of woodworking tools such as saws, drill presses, planers, and lathes, metalworking equipment, or other devices, should be sized and powered for the use it will be put to in the home. Very few home users require the speed or power of commercial-grade equipment built for continuous heavy use, but many buy such equipment in the belief that it will perform better than units designated for home use. As with other electrical and electronic equipment, bigger is not always better, and purchasing the appliances that adequately meet but don't exceed your real needs will save both on the purchase price and on your electric bill for years to come.

SUMMARY

All incandescent lights are inefficient in converting electricity into light. More than 90% of the current consumed by an incandescent light is transformed into heat. Less than 10% is emitted as visible light. Higher-power bulbs produce more lumens of light for each watt of power consumed than those that use less power. But even the largest bulbs still put out far more heat than light. At the rate of $0.119 per kWh, costs for incandescent lighting in the average home amount to about $25.58 per month and $306.95 per year.

Use of more-efficient electric lights is one method of reducing electric power demand. The federal government enacted the Energy Independence and Security Act of 2007, which requires all general-purpose light bulbs that produce 310–2600 lumens of light to be 30% more energy efficient than current incandescent bulbs beginning in 2012.

Compact fluorescent lights (CFLs) are more energy efficient than incandescent bulbs. A CFL is a miniature, high-output type of fluorescent lamp that is designed to replace an incandescent lamp of similar size and light output. Compared to general-service incandescent light bulbs that produce an equal amount of visible light, CFLs use much less power and have a much longer rated life. The average rated life of a CFL is between 6000 and 15,000 hours, compared to the average rated life span of an incandescent lamp of about 1000 hours. Using the 2010 U.S. average cost of power of $0.119, the monthly cost of CFL lighting in the average home is $6.89 and the yearly cost is $82.72. This compares with the $306.95 annual cost for incandescent lighting.

Another type of high-efficiency light that is now being widely adopted for home use is the light-emitting diode (LED). LEDs are small, solid light bulbs that are extremely energy efficient. An LED is a special type of diode. A diode is a semiconductor device made of a material that has a varying ability to conduct electrical current. The application of LEDs has been expanded by the production cluster LEDs, in which almost any number of individual LEDs lights can be combined into a single unit that produces light equal in magnitude to that of large incandescent bulbs.

LEDs are more expensive than CFL or incandescent lights, but they are more energy efficient and longer lasting by far than any other type of light bulb. An LED bulb equivalent in light output to a 60-watt incandescent bulb consumes only 6 watts of electricity, one-tenth the amount of the incandescent light and less than half the amount of an equivalent CFL.

Many home electronics products use significant amounts of energy when they are in standby mode and ready for instant service. This phantom load on use of power provides no obvious benefits but can be a significant part of total power use. Power use by TV sets, stereo systems, game consoles, computer systems, and other electronic devices can be reduced by disconnecting these appliances from electric power when they are not in use.

Refrigerators and freezers can be made more efficient by adjusting their inside temperatures to the correct level and by not opening them too frequently. Freezers can be further improved by operating them in an unheated area. Other motorized electric equipment for home use should not be larger or more powerful than necessary to adequately accomplish its purpose. Buying oversized equipment costs more, wastes power, and contributes little to the function of the equipment.

KEY TERMS

3-digit code A color code for lights that is a combination of the CRI and color temperature of the bulbs; the first digit is the CRI of the lamp divided by 10 and the second and third digits are the color temperature divided by 100.

Ballast A voltage transformer used to raise the voltage of current flowing to a fluorescent light.

Cluster LED A combination light in which numerous LEDs are combined into a single unit that produces light equal in magnitude to that of large incandescent bulbs.

Color-enhanced phosphors Light-emitting materials used in CFLs to change the color of their emitted light to be closer to that of incandescent light.

Color rendering index (CRI) A measure of how closely colors viewed under a light source match the same colors viewed in natural daylight, which has a CRI of 100.

Color spectrum The band of frequencies of electromagnetic radiation that produce light of all colors visible to the human eye.

Compact fluorescent light (CFL) A miniature fluorescent light made in several shapes that produces fluorescent light with a CRI close to that of daylight or incandescent light.

Continuous-load use A designation of the work that an electric motor can perform continuously without overheating or other failure; always less than peak-load use.

Correlated color temperature (CCT) A measure of the color of light emitted by a light source; measured in Kelvins (K), a unit of absolute temperature; the higher a light source's CCT, the more blue or blue-white it appears to the viewer. Natural daylight has a CCT of about 5000 K; bright white and cool white bulbs range between 3500 and 4000 K.

DC-to-AC inverter A device that changes AC current flow to DC flow; used in CFL lights; also called a rectifier.

Diffused LED bulb A cluster of LEDs covered by a dimpled lens that spreads light from the individual LEDs over a wider arc.

Dimmable CFL A CFL that can be dimmed using a dimmer switch matched to the low-power consumption range of the CFL.

Diode A semiconductor device made of a material that has a varying ability to conduct electrical current; diodes conduct electricity in only one direction.

Edison-type base The standard screw-type base on most incandescent light bulbs; CFLs and LEDs have been adapted to fit the sockets designed for Edison bases so they can replace incandescent bulbs.

Electronic ballast A type of ballast that uses a circuit board and electronic components to transform electric current voltage in an LED.

Energy Star rating A designation for electronics products that meet the strict energy-efficiency and price guidelines set by the Environmental Protection Agency and Department of Energy of the federal government.

Filament The fine coiled wire in an incandescent light that emits light (incandesces) when a current passes through it.

Fluorescent lamp Any type of tube, bulb, or CFL, that emits light from phosphors excited by an electric current.

Fluorescent light The light produced by phosphorescent materials that glow when excited by an electric current.

Frosted bulb A light bulb coated on the inside with translucent material that diffuses its light output.

Halogen bulb A type of incandescent light that uses halogen gas to increase the efficiency of its light output by up to 30% compared to that of ordinary incandescent bulbs.

Horsepower (hp) A measure of work used to define the power of electric motors and other engines; 1 hp is the amount of work (power) required to lift 565 pounds 1 foot in 1 second.

Hybrid CFL A lamp that combines an instantly lighting halogen lamp with a CFL that has a warm-up period; when the CFL reaches full brightness, the halogen lamp goes out.

Incandescent light Light produced by passing current through a high-resistance filament that produces both heat and light.

Infrared light Light with a wavelength longer than that of visible red light and cannot be seen by the human eye but only detected as radiant heat.

Infrared remote control A device for controlling electronic equipment by means of signals sent as infrared light pulses.

Light-emitting diode (LED) A light source consisting of a solid-state diode made of a material that emits light when a current is applied to one side of its circuit; more efficient than an incandescent light.

Light-emitting phosphor Any material that emits light when excited by an electric current.

Light spectrum The band of electromagnetic radiation with wavelengths from ultraviolet light (shortest) to infrared light (longest) and including all colors of visible light in between these extremes.

Lumen A measure of light output.

Mercury vapor The gaseous form of the heavy metallic element; it is extremely toxic but is a required component of fluorescent lights.

N-type material A semiconductor material with extra electrons; so called because it has extra negatively charged particles.

PAR (parabolic aluminized reflector) light A designation of enclosed light fixtures and light bulbs that refers to the size of the fixture and its bulb; PAR lights are measured across the lens in eighths of an inch; a PAR 16 light is 2 inches across, and a PAR 30 light is 3.75 inches across.

Peak-load use The maximum workload that an electric motor can perform for a short period using the most current it can draw; an electric motor will fail if run at peak load continuously.

Phantom load The load placed on a home's electric service from the electrical appliances and devices in the home that remain in standby mode when not in use.

Phosphor A substance that emits light when excited by an electric current.

Photon of light A unit of light that is emitted when an electron drops from a higher energy level to a lower one in a phosphorescent substance.

P-type material A semiconductor with holes for electrons; so called because it has extra positively charged particles that can attract electrons.

Sleep mode A state of an electric device in which its operation is suspended, but its circuits are still powered and ready for instant use on demand.

Soft white bulb A light bulb with a CRI much lower than that of daylight; it emits light with a warmer feeling that a bright or pure white bulb.

Standby mode Same as sleep mode.

Ultraviolet (UV) light High-frequency light not visible to the human eye but that causes some other materials to emit visible light (phosphoresce); UV light is highly destructive to human tissue and over time will destroy most other solid materials.

Vampire appliance or device Any electrical device that draws power to maintain its readiness for operation when not in use.

Visible light-emitting diode (VLED) An LED made of materials that emit light photons at frequencies between infrared and ultraviolet wavelengths.

Visible light spectrum The band of light wavelengths that can be seen by the human eye, ranging from violet to red and including all visible colors.

REVIEW QUESTIONS

1. Why does an incandescent light filament not burn up when it becomes hot enough to emit light?
2. True or False: Incandescent light bulbs function only using AC power of 120 volts or more.
3. Why is an incandescent light bulb energy inefficient?
4. Is an incandescent light bulb made more efficient by lowering the voltage to it by 10%? Why or why not?
5. What does the Energy Independence and Security Act of 2007 require for general-purpose light bulbs by 2010? By 2020?
6. What is a CFL and how does it produce light?
7. What is CRI and what does it measure?
8. What is CCT and what does it measure?
9. If a CFL's three-digit code is 842, what is its CRI and its CCT?
10. True or False: A CFL turns on more slowly and requires more time to reach full brightness than a comparable incandescent bulb.
11. What is an LED?
12. How is the color of an LED determined?
13. How is a colored LED's light output usually converted to broad-spectrum white light?
14. True or False: A diode, whether it is an LED or not, can pass current in either direction with equal efficiency.
15. Why are LEDs less hazardous than CFLs for everyday use?
16. Is the size of a TV screen the only factor that affects how much power it consumes?
17. True or False: An electronic device or system uses no electric power when it is in standby or sleep mode.
18. Which is more energy efficient for home use, a 200-watt audio amplifier with connected speakers in each of six rooms, or 20-watt amplifiers and speakers in each of the six rooms?
19. What is the best way to conserve the power used by a game console?
20. Which of these conditions is energy inefficient: a refrigerator not maintaining a set temperature low enough to preserve food adequately or a freezer maintaining a temperature below 0 degrees?

CHAPTER

8 HOME AUTOMATION AND LIGHTING CONTROL

OBJECTIVES

After studying this chapter, you should be able to:

- Describe the basics of home automation and lighting control design and planning.
- Define the major types of automation and control systems available.
- Describe the major components of an automation system and how they work.
- Describe how to install and set up an automation system in a home.
- Describe how to program an automation system.
- Describe troubleshooting techniques for eliminating problems in automation systems.

OUTLINE

I. Introduction
II. Home Automation and Lighting Control Design
  A. Load Requirements and Grounding
  B. Home Automation Plan
  C. Lighting Zones and Scenes
  D. Home Run-Connected Lights and Daisy Chains
III. Types of Automation Systems
  A. Power-Line Control Systems
  B. Wireless Control Systems
  C. Wire Runs and Wireless Zones
IV. Automation Control Components
V. Command Modules and Controllers
  A. Remote-Access Controllers
  B. Controlled Outlets
  C. Light Switches and Dimmers
  D. Fixtures (Luminaires)
  E. Automated Window Treatments
  F. Sensors
VI. Automation System Installation and Setup
  A. Safety Cautions for Wiring and Connecting Automation Components
  B. Installing Control Modules
  C. Installing Exterior Lighting Zones
  D. Installing Security Lighting Zones
  E. Programming the Automation System
VII. Troubleshooting an Automation System
VIII. Summary

INTRODUCTION

In this chapter you will learn how digital control systems can automate and control audiovisual systems, security systems, lighting, or any many other electric devices in a home through a system of command modules and control modules linked in a wired or wireless network. An automated control system can improve the energy efficiency of the home, enhance its décor, and increase the convenience of using many electronic and electrical systems. You will learn to define the major types of home automation and control systems and their characteristics and you will learn about the basic components that make up these systems. You will also learn how to install and set up an automation system and program it to function according to design parameters. Finally, you will learn how to troubleshoot the common problems that can occur in home automation systems.

HOME AUTOMATION AND LIGHTING CONTROL DESIGN

AUTOMATION DESIGN FOR a home can consist of only simple individual controls for a few lights and pieces of electronic equipment or it can be a whole-house plan that integrates all the electronic systems in the building and all the lighting both inside and outside into a single unified system controlled by a master program from a central location or from any one of several substations. The amount of automation provided should be at a level that the homeowner feels comfortable with and will use consistently. Automation for its own sake is a wasteful expense that probably will not be used enough to justify the cost of installing it. A good level of automation will make home life easier and more convenient for the homeowner.

Every home has lights that provide basic illumination. Some homes have elaborate lighting systems composed of light scenes (illuminated rooms or areas within a room) that not only provide light, but also beautify a room's décor or invoke a mood in those who see the scene or are within it. Whether a home's lighting system is minimal or extensive, it can be significantly enhanced by being automated. Automation enables lighting to respond quickly, and with relatively little effort, to a number of circumstances, including changes in natural lighting throughout the day, specialized illumination needs, social situations, and even the moods of the home's occupants. Although automation increases the utility of the lighting system, the electrical cost of operating it should not rise and may even decline due to greater efficiency in the timing and use of electricity.

Before the proceeding with an automation design, some homeowners may want to install new electronic systems, add additional lighting, or change some of the lighting fixtures already in the home. If electronic systems or lighting are to be upgraded or altered, this should be done before the automation design is created and provision should be made during installation for wiring, control consoles, and other fixtures that will be required as part of the automation design.

Electronic systems that can be brought into a unified automation design include the home's audio and video system (whole-house video distribution system, home theater, stereo system, and individual components, or all of these), security system, water management system, HVAC system, communications system, and home lighting. Each of these can be controlled as a separate system, but they can also be integrated in a unified automation design that allows them to work together to the maximum extent possible. This chapter does not cover the design or installation of any of these electronic systems or their individual controllers, but concentrates instead on the systems available for integrated control of all home electronic and electrical functions.

This chapter likewise does not discuss lighting design and installation except where it bears directly on the primary focus of home automation. Lighting design and installation is a large field and a subject on which books have been written. The automation design discussion assumes that the electronics and lighting system of the home are already installed and working and that the homeowner wants to automate them.

Caution: High-voltage wiring and fixture installation should be performed only by a licensed electrician or under the direct supervision of one. Wiring and fixture replacement descriptions in this chapter are informational only and should only be performed by persons trained and licensed to do so. Always shut off electric power in the area where wiring or other electrical work is being done.

Load Requirements and Grounding

A home automation system, whether it is wired or wireless, will not add any load to the home's AC electric wiring, but the adequacy of the in-house wiring should still be confirmed before any automation is installed. Both wired and wireless automation controls are subject to interference or "noise," which can be caused or increased by overloaded wiring.

Lights are usually not the cause of overloading in a home electric system because they are not the largest consumers of electricity in the system when they are used individually. When acting as a group, however, such as when an automation system turns all the controlled lights in a home on or off simultaneously, lights can have a significant impact on the power system. But individual lights being turned on or dimmed will not make much difference in the overall power consumption or load. A drop in line voltage won't cause a light to consume more power either. If line voltage drops, the light simply shines less brightly; that's how dimmers work.

Other electronic systems will also not require more power as a result of being automated. Most electronic systems consume only slightly more electricity when starting than they do when operating. Their operating power consumption is also seldom high enough to cause any fluctuation in the home power supply. Automation will simply make these systems use power more efficiently.

The one exception to this neutral effect of automation is in control of systems with large electric motors. When starting, motors momentarily draw three or four times the current that they require when running at speed. Before automating an HVAC system, or any other system that includes a large electric motor, the electrical wiring for the system should be checked to be sure it has adequate available power to prevent even temporary (startup) overload conditions. If necessary, a compensating device can be installed in the motor's circuit to reduce the startup surge of current drawn from the AC wiring so as not to adversely affect other equipment or the automation system.

A similar evaluation should be made before installing home lighting controls. If new circuits, wiring, or fixtures are needed, they should be installed before the automation or at the same time. New circuits, if necessary, can be used not only to reduce the load on older wiring, but also to connect additional new lighting that might not have been installed at all if the new circuits weren't available. Automation devices such as controlled outlets and switches can be installed in new circuits at the beginning, instead of replacing manual units as is necessary in existing circuits.

Home Automation Plan

All of these considerations point to the need for a home automation design plan for all electrical and electronic systems and for the automation program that will control it. The plan should start with a diagram of existing wiring drawn as completely and accurately as possible on a floor plan of the home. Add any new circuits that are required and plot how the wiring for these will be installed in the existing finished walls. Finally, design the control system, noting exactly which lights, outlet boxes, switches, and fixtures you plan to control; which lights will be controlled as zones (groups of lights controlled as single units); and where controllers will be located. Make the plan as detailed and complete as you can; every item planned before starting the hands-on work will enable you to avoid problems during the installation phase.

Label the plan with a unique label for each item. As you later install each device, it should be marked with the label you gave it in your plan. The plan can then serve as a schematic drawing of the completed lighting automation system that you can use in the future to troubleshoot and maintain it. Figure 8-1 shows a plan of the first floor of a home with all the electrical fixtures noted. Because the plan has been reduced in size for publication in this book, the fixture symbols have been enlarged and only those in the family room are shown labeled. Later, you'll see Figure 8-2, which shows the family room of this plan enlarged and with the zones and control devices that will be used in it noted.

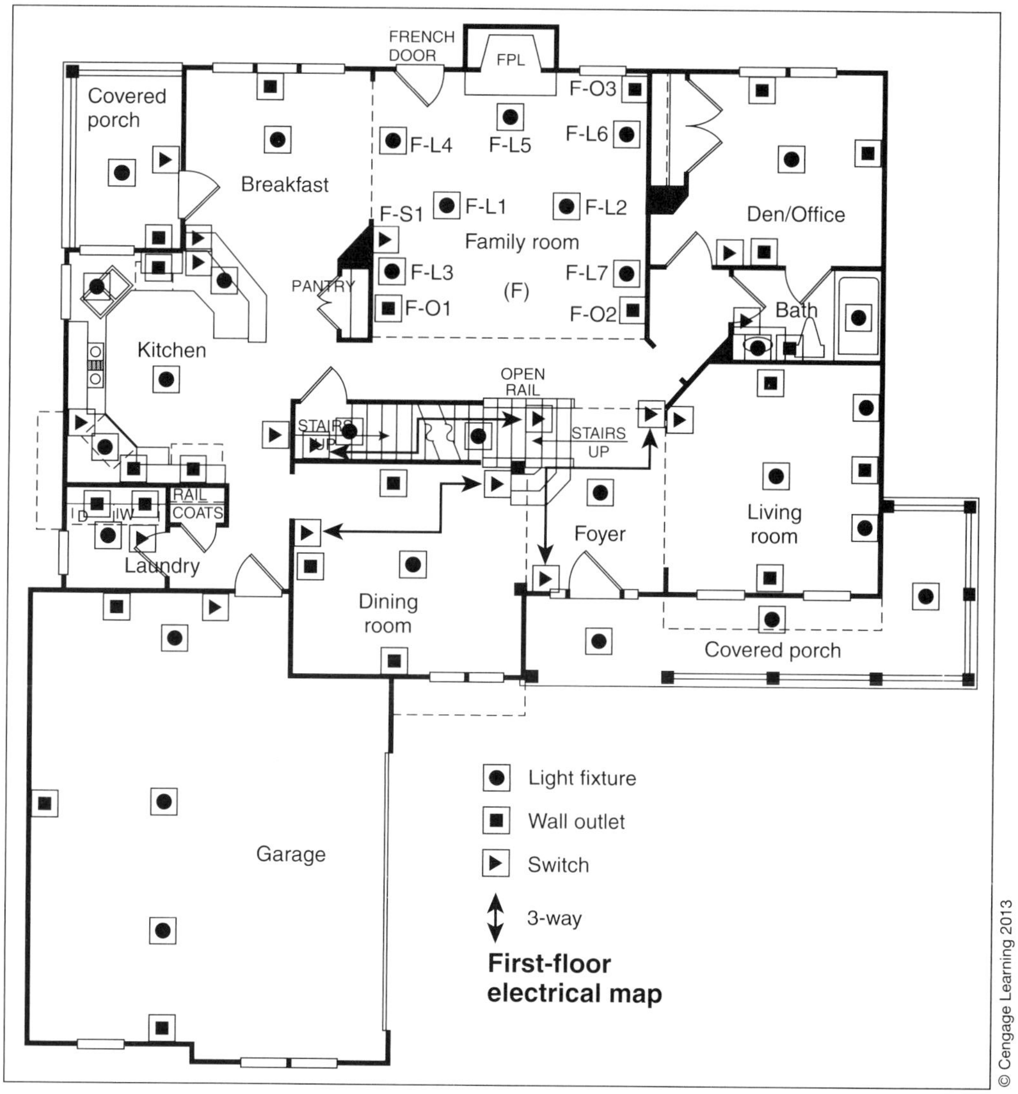

FIGURE 8-1

Electrical fixtures noted on floor plan, with those in one room labeled.

All of the AC fixtures in the home should be grounded, as should those installed as part of any new circuits added to upgrade the home's electrical system. Nongrounded fixtures are almost never found except in old homes constructed before grounding was a code requirement, but it is still wise to check that everything is in fact grounded as a safety measure, and also because any ungrounded circuit is likely to cause interference

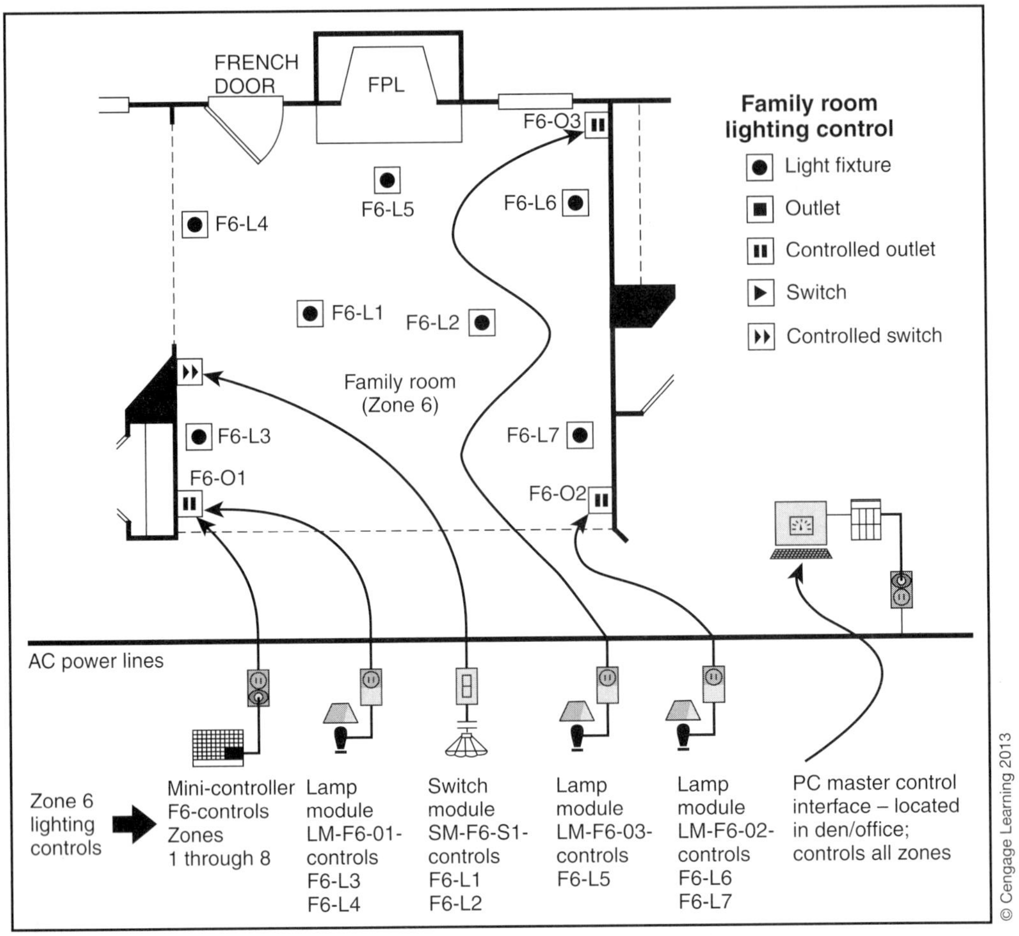

FIGURE 8-2

A lighting system zone and the controls in it.

in the automation system you install. This is true whether it is a wireless system or uses power-line technology. Poorly grounded or ungrounded fluorescent lighting is especially prone to generate interference in the automation system, as well as in other home technology installations.

Lighting Zones and Scenes

As you develop the lighting automation plan, you need to decide whether each of the lights will be controlled individually or be part of a zone. A lighting zone is simply a group of lights that are related by function (not necessarily by location, although they may be near one another as well). A zone, for example, could consist of all the lighting in the family entertainment room. These lights would nearly always go on or off together or be adjusted for light level as a group to fit the activity (watching a movie, reading, playing games, etc.) that is going on in the room. Each of the lights retains individual manual control, so the occupants can override the automatic controls in case of need, but most of the time, they function together as a zone.

A lighting scene might consist of the lights in a single zone, but can also be any set of lights designed to accomplish a specific lighting objective. A dining scene, for example, might dim the overhead dining room chandelier while turning on small spotlights to illuminate the paintings on the room's walls. Another lighting scene in a family room might dim the overhead lights to eliminate reflections on the glass while turning on backlighting behind the aquarium to make observation of its occupants easy. Still a third scene might turn off floodlights in the backyard garden while illuminating the flowers, fountain, and footpaths with indirect lighting for a romantic effect.

Other possible zone and scene configurations for home lighting include:

- Bedrooms
- Dining areas
- Indoor safety lighting (hallways, stairs, entries)
- Emergency lighting (battery-powered lights that activate in a power failure)
- Living room
- Interior decorative lighting areas
- Outside security lighting
- Outside safety lighting (sidewalks, steps, and outside entrances)
- Outside decorative lighting
- Outside special areas (swimming pool, hot tub, game area)

If any of these or other areas fit well with the automation you want for the home, plan the appropriate zone and the controls needed for it. Lights that always function together can be controlled by one device, if they are wired on the same circuit, or can be connected to outlets on a single circuit. This is especially true of lamps and other moveable lights, which can often be set up so that multiple lights connect to a single outlet. This allows one module to control all of them, up to the limit of the device's power capacity.

A light can belong to more than one zone, but this requires that it have an individual control, which can be directed for the lighting requirements of each zone. Lights that are part of a room lighting scene, for example, might also be part of a safety lighting zone for the home's interior core. Outside decorative lighting might also be part of safety lighting that turns on when someone leaves the home late at night, even though the rest of the decorative lighting remains off. Security lighting might override all other zones and turn every exterior light up to full power to give maximum illumination.

Note each zone on the lighting plan and add a zone code to the label of each light that is included in a zone. A light that illuminates a painting hanging in the living room might be labeled LVRM5-Z1, showing that it is the fifth light in the living room and it belongs to Zone 1. A light on a stairway descending to the basement family room might be designated STRS2-Z3-Z7, showing that it is a stairway light and belongs to the living room zone (Zone 3) and also to the safety zone (Zone 7).

Figure 8-2 shows the electrical fixtures and controls in a family room lighting zone. Each control module is labeled. In this plan, each light and fixture is identified with a unique label; all labels in the family room plan start with the letter F for family room followed by the number 6 to indicate they are part of Zone 6 in the overall automation plan. The rest of each label is a letter designating the type of device (O for outlet, L for light, etc.) and a number for that device.

Home Run-Connected Lights and Daisy Chains

When a light is connected directly to its power supply, it is said to have a home run connection. A lamp plugged into an outlet receptacle is an example of a home run-wired light, as is an overhead light connected to a switch on the wall. The light is wired directly to a power connection (a home run wire) inside the walls. Most, but not all, hard-wired lights (those that are wired as part of the home's electrical system) are home run lights. They are connected to the electrical system through a switch that allows them to be turned on and off.

Sometimes a switch controls two or more lights. In this case, not all the lights may be wired to the electrical system directly through the switch. One light may be wired to the switch and the second light connected to the first light, getting its power from the first light's connection to the electrical system (if there's a third light, it would be connected to the second light). This arrangement of connection is known as daisy-chain wiring (from the now little-known art of joining daisies by twisting the stem of one around the blossom of the next). Other examples of daisy-chain electrical connections are as follows: a series of lamps in which the first is plugged into a wall outlet, the second is plugged into a connection in the first, and perhaps a third into a connection in the second; or a television that is plugged into a wall outlet and a small backlight lamp that is connected to the television. Figure 8-3 shows two lamps with home run connections on the left and three lamps connected in a daisy chain on the right.

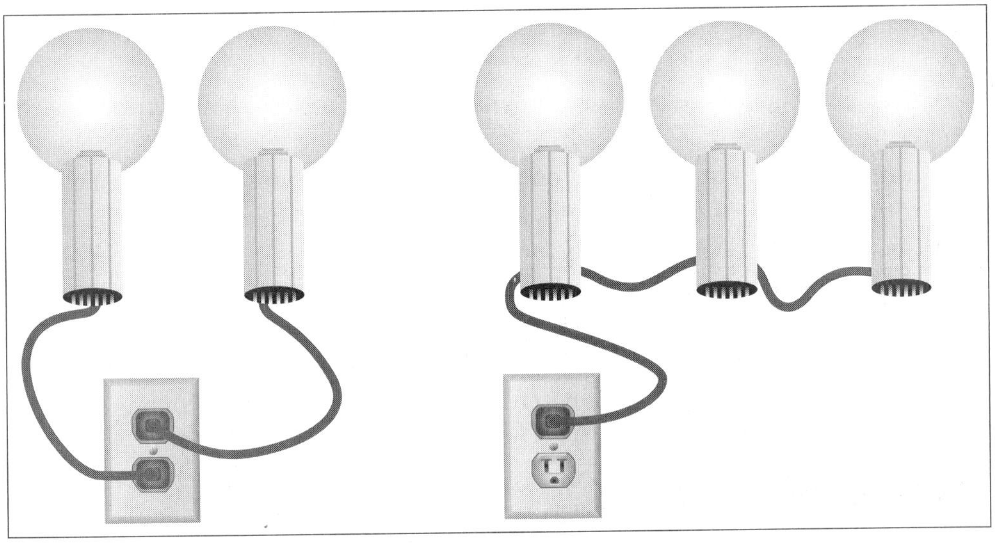

FIGURE 8-3

Lamps with home run connections (left) and daisy-chain connections (right).

Lights with daisy-chain connections always work together from a single control device because all the lights in the chain receive their power from one connection. They are wired in parallel, however, not in series, so that if any one light fails, the others in the chain will continue to work. Daisy-chained lights are useful because the arrangement permits several lights to be placed in one room without using all the wall outlets. Multiple track lights and indirect fluorescent lights are often connected as daisy chains. Decorative lighting, both indoor and outdoor, is usually in a daisy chain in multiple light segments. Parallel-wired holiday lights are daisy chains, and several strings of them can be connected one to another in an even larger chain. Only one automation device is needed to control an entire daisy chain, however, because it is all powered from only one controllable point.

Home run-connected lights that are part of the home's electrical system can be controlled individually from the switch that is always wired into their circuit. They can also be controlled from another switch wired into the circuit closer to the light than the first switch, if the first switch is left in the on position. Home run lights plugged into a wall outlet can also be controlled at the outlet or by another switch wired into the light's cord. Because each home run light has its own power connection, it must be controlled by a separate controller. Most plug-in controllers have two or more outlet connections, but all of these are usually controlled by the same commands and so all the lights plugged into the module work together as if they were one. Figure 8-4 shows two lights added in one corner of a room that has no built-in overhead lights (a matching pair was placed in the adjacent corner). These lights connect to a wall outlet and have no manual switch on their circuit. They are controlled, however, by a control module in the outlet and commanded by a controller.

FIGURE 8-4

Added lights controlled from an outlet module by an automated system.

TYPES OF AUTOMATION SYSTEMS

Automated control systems are specialized networks that use different protocols, hardware, and frequencies than computer networks or telecommunications networks. The major control systems were originally designed for lighting control, but they can also be used to control any other electrical device that operates on AC current or low-voltage DC current. Automated lighting networks have very narrowly defined functions. These functions consist of turning lights off and on and adjusting their brightness through a process called dimming. Automated systems can control many lights and electrical devices and can be programmed for complex lighting patterns and operation sequences, but the control commands to each light or device are simple and do not require large amounts of data transmission. Automated lighting systems consist of either power-line control networks, wireless control networks, or combination systems that use both technologies.

Power-Line Control Systems

Power-line control technology uses the AC power lines that supply current throughout the home to transmit control signals at frequencies higher than the 60-cycle AC electricity flowing in the same lines. This technology can thus send control signals over the same wires that provide electric power to the lights and other electrical appliances in the home. The first and still most common method of lighting control and automation uses X-10 power-line carrier (PLC) technology. Several other power-line technologies are also available, some of them more robust and sophisticated than X-10 products, but none to date as inexpensive. The most prominent of the newer power-line systems is Pulseworx, manufactured by Powerline Control Systems. It uses a signal standard known as Universal Powerline Bus (UPB) to provide control data over AC wiring. Some other manufacturers have adopted the UPB standard and produce products for lighting and electrical control using it, but Pulseworx is the leading control system employing UPB.

X-10 Lighting Control Systems

X-10 control devices were once manufactured by a single company, but the patent on the original technology expired several years ago and many firms now manufacture and market X-10 equipment. With over 30 years of sales, X-10 systems are now installed in an estimated 12 million homes in the United States and other countries, and the basic technology has been greatly expanded and improved. The original company, X-10 Limited, and its subsidiaries are still major manufacturers of X-10 equipment, but they have now been joined by many other firms who that produced a large array of control devices and controllers to command them.

There are many types of command modules for use with an X-10 system. All of them connect to an AC outlet and send out signals at a frequency of 120 KHz on the home's electric wiring. This frequency is way above the AC current frequency of 60 Hz and well below radio frequencies (RFs) so that neither interferes with the X-10 signals on the wiring. Other interference does occur (which is discussed later), but the X-10 signal frequency is as well placed in the electromagnetic spectrum as it could be to avoid both high- and low-frequency interference.

In an X-10 system, all the command signals are sent over all of the home's wiring and are received by all of the control modules. Each module responds only to the signals that have its unique address in the header of the data packet. X-10 control modules in a system are each assigned a letter (house) code and a number (device) code. There are 16 possible letter codes (A through P) and 16 number (unit) codes (1 through 16) for each letter, giving a total of 256 possible different control modules in an X-10 system (Even more control modules are possible if several are assigned the same codes and function together in a zone.)

Command modules have no code designation because they don't receive commands. They only send them. An X-10 system can have several command modules located in different areas of the home, and the control modules respond to commands from any of them so long as the commands are correctly addressed to the module. Table 8-1 shows the six possible commands that basic X-10 lighting control systems use. Not all control modules respond to all commands.

TABLE 8-1 X-10 System Commands

NUMBER	COMMAND	ACTION
1	On	Turns on any control module with the code specified
2	Off	Turns off any control module with the code specified
3	All on	Turns on all control modules in the system
4	All off	Turns off all control modules in the system
5	Dim	Dims any control module with the code specified
6	Brighten	Brightens any control module with the code specified

X-10 control modules come in many varieties. Table 8-2 shows the main ones that are intended for lighting control, the power ratings, the commands to which they respond (as noted in Table 8-1), and the function of each type of module. Note that lamp and outlet modules do not respond to the "all on" command because they may be used to control nonlight devices that should not be activated with this command.

TABLE 8-2 X-10 System Control Modules

CONTROL MODULE	POWER RATING	COMMAND RESPONSE	FUNCTION
110-V lamp	300 watts	1, 2, 4, 5, 6	Plugs into outlet, controls lights plugged into it
110-V lamp	600 watts	1, 2, 4, 5, 6	Plugs into outlet, controls lights plugged into it
110-V appliance	800 watts	1, 2, 3, 4, 5, 6	Plugs into outlet, controls appliances plugged into it
110-V outlet	500 watts	1, 2, 4, 5, 6	Replaces AC outlet, controls lights plugged into it
110-V switch	500 watts	1, 2, 3, 4, 5, 6	Replaces switch, controls lights connected to it
110-V three-way switch	500 watts	1, 2, 3, 4, 5, 6	Replaces switch, controls lights connected to it
Inductive switch	-	1, 2, 3, 4, 5, 6	Controls low-voltage lighting

In addition to lighting and appliance control, many other home systems are available with built-in X-10 power-line or wireless control. Wireless X-10 components can respond to additional commands beyond the basic ones described in Table 8-1. These expanded control functions allow X-10 systems to control security systems, sensors, cameras, audio systems, video devices, and many other automated mechanical functions. X-10 systems can open windows, raise and lower shades, control garage doors and gates, activate sprinkler systems and pool maintenance cycles, and manage all control functions through an interface with software on a PC or Mac workstation.

The following are types of command modules (controllers) that are available for X-10 systems. All of them send commands to X-10 control modules. The differences between them are mainly in how many devices the command module can control and how the command module itself is programmed to perform its work.

- Mini-controller—Usually controls up to eight control modules. Plugs into a wall outlet via a cord and sends commands on the AC wiring. Works like a handheld remote with buttons for each control module.
- Wireless mini-controller—Similar to a mini-controller, but wireless. Sends radio commands to an outlet-mounted receiver, which transmits them on the AC wiring.
- Tabletop controller—Larger than a mini. Manual remote control. Controls up to 16 control modules. Some can control multiple house codes, providing control for up to 256 devices.
- Programmable controller—Controls up to 16 modules and can use multiple house codes. Allows manual remote control and timed sequences of lighting and control actions.
- Telephone interface—Connects to telephone line and AC wiring. Controls up to 16 modules and can use multiple house codes. Allows manual remote control from any phone.
- Touch-panel interface—Controls up to 16 modules and can use multiple house codes. Allows manual remote control and timed sequences of lighting and other automated actions.
- Computer interface—Controls up to 16 modules and can use multiple house codes. Allows manual remote control, timed sequences of lighting actions, or sensor response actions.

Standard X-10 command modules and control modules use one-way communication only. Signals go from the command module to the control module, but nothing comes back from the control module to the command module. This means that there is no confirmation back to the command module that its instructions were received and carried out. Most of the time they are, but this lack of confirmation signals was the greatest weakness of past X-10 technology.

One of the best advances in recent X-10 products is the development of bidirectional (two-way) X-10 modules. These modules confirm receipt and execution of commands and are available (at higher cost) for most X-10 devices. They are worth the money, however, because they provide much more reliability in programmed sequences. If a command isn't confirmed, it can be resent and the program kept accurate through its timed sequence. Many of the touch-panel interfaces and nearly all of the computer X-10 interfaces use bidirectional modules.

UPB Lighting Control Systems

There are two principal differences between UPB power-line standard control devices and those using the X-10 standard: UPB devices send signals at a much lower frequency than X-10 devices (4 Khz to 40 Khz, as compared to X-10's 120-Khz frequency), and all UPB devices have two-way communication capability so that command signals can be confirmed and status reported on any device.

The UPB standard also uses spread spectrum signal hopping and sends its signals at a strength of 40 volts (X-10 signal strength is typically about 2 volts). These features make the UPB control signals very reliable and robust, even in power-line systems that have considerable noise and interference. UPB is also faster than X-10. It can send command signals at the rate of 5 per second as opposed to 1 for X-10 systems.

The Pulseworx lighting control product line includes all of the standard devices available in X-10 technology plus a few extras. In order to meet the demands of larger lighting scenes and multi-bulb fixtures, Pulseworx offers 1500- and 2000-watt wall-mounted dimmer modules and an 800-watt plug-in dimmer. These are larger than the X-10 maximum of 1000 watts for wall-mounted and 300 watts for plug-in dimmers.

Other Power-Line Systems

A hybrid power-line lighting control system, employing both wireless and power-line technology, is now on the market from Insteon. It uses proprietary technology and is not compatible with X-10, UPB, or other wireless systems. LonWorks is another power-line lighting control system developed by Echelon that has a large market base in Europe and Asia. LonWorks products are now available in the United States, but have not yet achieved a large market share. Last, the HomePlug consortium has announced the availability of a new power-line standard for home lighting control and related functions. It is known as HomePlug Command and Control (HPCC), but few products incorporating this technology have yet reached the market.

Wireless Control Systems

Nearly all the lighting control devices available in X-10 technology are also available using wireless technology. The largest wireless product supplier is Lutron, which offers a full range of home lighting control products that use its proprietary RadioRA technology. Lutron control modules do not offer the "all on" and "all off" commands that are available in X-10 systems, but other than that, just about every feature of X-10 control can be duplicated using RadioRA. Because its RF signals are much faster than X-10 signals, RadioRA controllers can approach the speed of an X-10 "all on" or "all off" command by sending individual signals in sequence to each module.

All RadioRA modules have two-way communication and automatic confirmation of commands from control modules. This wireless system can also be interfaced through an RS232 port to a computer-based control system. RadioRA components are somewhat more expensive than X-10 modules, but they offer high quality and reliability and, although still not as widely used as X-10 products, are making a definite place for themselves in the market.

In addition to the Lutron RadioRA wireless system, Vantage and Watt Stopper also offer wireless lighting control systems using proprietary radio technologies that are different from and not compatible with either RadioRA or X-10 wireless systems.

Leviton is the largest of several companies that use X-10 wireless technology as the basis for their lighting control systems. Recently Leviton has also started marketing lighting control products using the Z-Wave wireless technology originally developed by the Danish company Zensys. Z-Wave has become popular in Western Europe as a lighting control solution and is now moving into the U.S. market. In addition to Leviton, Z-Wave lighting control devices are made by Sylvania and Intermatic for the U.S. market.

Wire Runs and Wireless Zones

New wiring should be minimal for an automated lighting system, unless additional circuits are being installed. If that's the case, the new wiring should be integrated as part of a full home upgrade designed to bring the home's electrical system up to current electrical standards for all of the technology that will be operating in it. New circuit wiring should be done only under the supervision of a licensed electrician.

Because low-energy lights are not always as bright as incandescent bulbs and may dim somewhat more over time, the overall lighting level in a room may be lowered when the new lights are installed. This problem can be overcome by replacing existing light fixtures with new ones that contain sockets for more light bulbs or by adding additional light fixtures in the room.

Additional AC fixtures can often be added to a room without the need for more infrastructure wiring as long as the home's circuits are not overloaded by the increased power requirements. Such fixtures can usually connect to a wall outlet through a power cord run from the mounted fixture in a track or a raceway. This gives a permanent finished appearance to the lighting installation without requiring any new wiring within the walls.

Inadequate lighting in a room can be remedied with new multiple-bulb fixtures wired to the original junction box and controlled by the same switch(es). Figure 8-5 shows an example of a single overhead light connection that has been upgraded with multiple lights, providing improved appearance and better illumination.

Track lighting (lights mounted on metal tracks that contain their wiring) can be attached to the surface of walls or ceilings with screws and connected to a wall outlet.

FIGURE 8-5

Multiple lights wired from a single switched connection point.

Although it connects to the home's AC wiring, most track lighting is low voltage (usually 24 volts) and power is supplied to the lights through a step-down transformer. Track lighting can be controlled in the same manner as regular lighting using either power-line or wireless technology.

Figure 8-6 shows a single overhead connection that has been expanded into two chandeliers spaced 10 feet apart without any new structural wiring.

FIGURE 8-6

Dual chandeliers wired from a single connection point.

AUTOMATION CONTROL COMPONENTS

Each type of lighting control component is designed for a specific automation function. After completing the lighting automation design, you must determine the minimum number of controllers and control modules you need for the automation system. This section describes the function of each piece of equipment and any specific limitations that you need to be aware of when using it. Whether the automation system uses X-10 or UPB technology power-line components or wireless modules (RadioRA, Z-Wave, etc.), the various control devices function in much the same manner.

For lights in a single zone, try to connect as many as you can in daisy chains to minimize the number of control modules you need. If a light is in two zones, it must have its own control module (or share one only with other lights in the same zones that can respond independently to control commands for both zones). For the most professional appearance in an automation system, use modules that are concealed in standard electric fixtures or placed out of sight.

A combination of power-line and wireless components can be employed to create a home lighting system, but only one technology of each type can be used. X-10 or UPB devices are not compatible with one another so you must choose which type you want to use. Likewise, the wireless control technologies are all independent and noncompatible, so only one should be used in a system. Equipment that is nearly identical in function is available in each technology, so the decision of which to use comes down to finding the most cost effective and easiest to install for your particular application.

A few unique items in each technology address special needs. If your system has one of these needs, it may be the determining factor in deciding which solution to adopt. An X-10 socket module, for example, allows any standard light socket to be converted into a controlled light. The socket module screws into the regular socket and the light bulb screws into the module. The light can then be controlled by commands sent on the AC wiring. This solution is an alternative to using a controlled wall switch. It is useful for controlling lights that do not have remote switches but are turned on and off only by pull-string switches in the light fixture itself. None of the other technologies offer controlled sockets.

X-10 manufacturers have developed numerous wireless devices to augment the X-10 power-line products they sell. Not all the wireless components of the different companies will work together even though all can interface with X-10 power-line modules. For this reason, it is best to choose a single manufacturer for all your X-10 lighting system components, power-line and wireless.

The same is true if you are using a UPB hybrid lighting control system with power-line and wireless components. Get everything for your system from one manufacturer to be sure that the whole system will work together. If you must use a device from another company, be certain before you install it that it will function with the rest of your system.

COMMAND MODULES AND CONTROLLERS

The smallest command modules are similar in appearance to a television remote control. They can control up to eight control modules, but are not programmable. Commands for each control module must be entered manually by pressing the appropriate buttons on the command module. An X-10 command module must connect to an electrical outlet in order to transmit its signals on the AC wiring.

Wireless X-10 command modules are battery operated and send their signals to a separate receiver nearby, which is plugged into an AC outlet. The receiver regenerates the command signal and transmits it on the AC wiring. In homes with more than 2500 square feet of floor space, an X-10 signal regenerator may be needed both to strengthen the X-10 command signals and to eliminate noise from the signals so that they can be more easily understood by the receiving control modules.

Command modules and controllers capable of controlling more than eight devices or of being programmed for timed lighting sequences are usually tabletop or wall-mounted units. X-10 command modules range in size up to those that can control the full 256 devices that this technology is capable of handling. The larger command modules also operate manually so that individual commands are entered by pressing the appropriate buttons on the module to send the desired command to the control module.

The most elaborate X-10 command modules have touch screens and a graphic interface that allows them to be programmed by touching appropriate entry commands on the touch screen to store the commands in memory.

X-10 command modules also can interface with a computer, and command sequences can be programmed and stored on the computer. As with RadioRA technology, the X-10 computer software allows controls to be implemented based on time-of-day settings, on sunrise and sunset times, or on sensor input from lighting sensors or security system components.

RadioRA and other wireless handheld command modules are also battery operated and transmit their command signals directly to the various control modules or to a repeater unit located nearby that then repeats the commands in an amplified signal throughout the home. Homes with more than 3500 square feet of floor space may require a second repeater in order to cover the full home.

Many wireless command units are also programmable with timed sequences of instructions that are sent at preset times to adjust lighting, activate security systems, or perform other automation functions. RadioRA offers large command modules that can be preprogrammed with macros for specific lighting control sequences, but these do not have timed functions unless combined with an astronomic clock interface unit, which allows the command modules to control devices based on regular time-of-day settings or according to sunrise and sunset in the local area.

RadioRA command modules also can be interfaced to a computer using an RS-232 interface. Automation control programs on the computer can then control all RadioRA functions based on time-of-day settings, sunrise and sunset times, or on sensor input from light and motion sensors. The other wireless technologies can all interface with a computer for programmed control functions.

Remote-Access Controllers

RadioRA technology has wireless remote controllers that can operate the automated lighting system from a television universal infrared remote device, from a car visor remote unit, from a wired telephone, from a cell phone, or from a security system keyless entry device. Several of these controllers are more limited in function than a master control unit, but they are all adequate to command basic control sequences when away from the home. The keyless entry control and the car visor unit can also be programmed to control security gates and garage doors.

X-10 remote-access control devices are slightly more limited, but still include a telephone interface as well as battery-operated remote controls, keyless entry units, and car visor controls. The latter two, like their RadioRA counterparts, can also control gates and garage doors.

Controlled Outlets

Outlet control modules for X-10 installations can be obtained either as plug-in units, which are about the size of a small 9-volt power transformer used with many electronic devices, or as replacement outlet receptacles that have the X-10 control module built into them and must be wired into the AC wiring in the same manner as an ordinary receptacle. Control module outlet receptacles are identical in appearance to regular outlets and are available with one of the duplex outlets controlled, or both. The lights or other device to be controlled are simply plugged into the control module and receive their power through it. Plug-in outlet control modules accept all six X-10 commands (on, off, dim, brighten, all on, all off); the wired-in outlet control modules accept five of the six X-10 commands (on, off, dim, brighten, all off) in keeping with UL safety requirements for this type of device.

The receptacle devices are preferable both for appearance and durability, but are naturally more expensive. The newer X-10 two-way communication modules, which confirm receipt and implementation of signals, are presently available only as wired-in receptacles, but will probably soon appear as plug-ins as well.

RadioRA outlet control modules are tabletop units that plug into an outlet by means of a 6-foot cord. The module's plug has a "piggyback" outlet on its back side and the lamp that the module controls is plugged into this outlet in the module's plug. The module with its manual dimmer switch sits on the table next to the lamp. RadioRA replacement wall outlets are not available at present.

Light Switches and Dimmers

Both X-10 and RadioRA lamp control modules are capable of dimming lights as well as turning them on and off. Both technologies also have replacement wall switches available that, in addition to serving as manual dimmer switches, are controlled for automated dimming and brightening. As with X-10-controlled outlets, these switches must be wired directly into the AC circuits because they replace the standard manual switches. Both systems have single-pole switches and three-way switches. X-10 switches are also available for three-switch locations, and RadioRA can go as high as nine switches controlling a single lighting zone. Switch-type control modules accept all six X-10 commands (on, off, dim, brighten, all on, all off). RadioRA controlled switches accept on, off, dim, and brighten commands.

Fixtures (Luminaires)

Some X-10 light fixtures and lamps are available in which the X-10 control module is built into the device rather than attached outside at the wall. These lights must have continuous power supplied from an outlet or a switched circuit in which the power is always left on. The internal X-10 control module then turns the light on or off and dims it.

Built-in control modules are available with only a very limited selection of lamps and lights, but they do offer additional flexibility in locating controls. They are especially useful for lights that have inconvenient switches, such as basement or garage lights that are turned on by pull strings that are hard to find in the dark.

Built-in control modules are not available in RadioRA technology except as controls on low-voltage lighting sets. The control module can be attached to these and serve as an internal switch and dimmer. X-10 systems also have low-voltage control modules.

Automated Window Treatments

RadioRA systems do not provide any window shade control mechanisms, but these devices are available with X-10 technology. Although not really lighting controls, these modules can be very useful in controlling sunlight through windows, darkening entertainment centers in daylight hours, and other functions. X-10 control modules connected to motorized pulls can open shades, blinds, or curtains, close them, or partially open or close them.

Manual remote-operated window coverings are mostly used where the windows are high and are inaccessible for hand pulls. An automated shade or curtain system can be programmed like a lighting system to respond to time-of-day commands, sunrise and sunset macros, or other command sets.

Sensors

Light and motion sensors can be used for lighting control input on both RadioRA and X-10 systems that are programmed and controlled through a command module or a computer. Among the functions that can be programmed for both systems using sensor input are the following:

- Selected lights turn on when motion is detected by a sensor.
- Selected lights turn off when no motion is detected by a sensor for a specified time interval (unoccupied rooms).
- Selected lights turn on or brighten when ambient light level falls.
- Selected lights turn off or dim when ambient light level rises.
- Selected lights turn on when light beam is broken (more selective than motion detector).
- Selected lights turn off when light beam is broken (closed door or departing vehicle).

AUTOMATION SYSTEM INSTALLATION AND SETUP

Once you have an automation system planned and the necessary components on hand, you are ready to install the control modules and the command modules, set up the control program, and test the system. If you do these steps for one lighting zone at a time, even a large automation system can be installed and working in a short time.

Safety Cautions for Wiring and Connecting Automation Components

High-voltage wiring and fixture installation should be performed only by a licensed electrician or under the direct supervision of one. Wiring and fixture replacement descriptions in this section are informational only and should only be performed by persons trained and licensed to do so. If you are using X-10 or UPB plug-in modules, these can be installed by someone who is not an electrician, but the same safety precautions should always be observed when working around high-voltage electrical fixtures.

Before doing any installation, shut off the AC power in the area where you are working. This usually means shutting off some house circuit breakers while leaving others functioning in nonwork areas. Working in one power-off area at a time minimizes the inconvenience for others and allows you to run an extension cord from an outlet where the power is still on to the work area so you have light to work with while doing the installation. One final caution: If only some circuits in the house are turned off, do not assume every outlet or fixture you are working on is off. Test each one before doing any work to be sure the power is off. A small two-prong tester or a lamp will show you whether a fixture is safe to work on.

Installing Control Modules

Start with one zone defined on your plan and install all the control modules that are required for controlling lights and other devices. Wired-in outlets and switches should be installed just as if they were standard replacements for the devices already in the circuit. The various ways of wiring outlets and switches must be considered when installing control devices. Some points to remember include:

- Always-live duplex outlets should normally be replaced with matching X-10, UPB, or RadioRA control modules in which both receptacles are controlled. If one receptacle needs to remain always on, use a control module in which only one is controlled. The other remains always on by default.
- Switched duplex outlets should be replaced with control modules and the switch that controls them left in the on position. UL listing requirements for RadioRA, UPB, and X-10 wall-mounted control module switches and dimmers state that they must be connected only to permanently installed light fixtures and not to lamps or plug-in devices.
- Split duplex outlets, in which one receptacle is switched and the other always on, should be treated like switched outlets. If only the switched receptacle is to be controlled, the control module should be placed on the outlet, not on the switch, and the switch left on. If both receptacles need to be controlled, then the outlet should be replaced with a control module and the replacement control module should be wired so that both receptacles are always live. This disables the switch that formerly controlled the receptacle and leaves the control module able to control both.
- Single-pole switches should be replaced with single-pole control switches wired the same as the original switch.
- Three-way switches should be replaced with three-way switch combinations. For X-10, UPB, and RadioRA, three-way switch combinations consist of one control module switch and one companion switch that does not have a control module in it, but still allows manual operation of the switch. Both switches must be replaced in a three-way circuit, not just the one that has the control module in it.
- Multiple switch sets above three-way should be replaced with one control module and the other switches as companion switches. X-10 and UPB can only handle up to four switch combinations, but RadioRA can go to nine.

As you install control modules, remember to check loads on each one. RadioRA and UPB devices are built for heavier duty than most X-10 devices. They can accommodate 500 watts of power on switch modules and lamp modules. Standard X-10 devices have 300-watt power-load maximums, although units are available with 500-watt capacity and more. Be sure you install control modules with enough current capacity to handle the load you are putting on them.

Also note that control modules should not be placed where a high-current appliance is likely to be plugged into them. Heating appliances, such as hair dryers, waffle irons, space heaters, and toasters, typically use from 800 to 1500 watts of power when operating. This exceeds the load capacity of RadioRA, UPB, and X-10 control modules and will burn them out in a short time. Plug-in modules can be removed when a high-current appliance is used and then put back into the outlet with no loss of control function. Wired-in control modules cannot be taken out, but will definitely burn out if overloaded, so be cautious of where you install them.

Low-voltage light sets require the special low-voltage control modules available in X-10, UPB, and RadioRA systems. These control modules will normally be wired into the light circuit directly rather than being installed as an outlet or switch replacement.

When the control modules are in place and the lights and other devices are connected to them, set up the master command module or controller in its assigned location. Program the zone and test it to be sure everything works as you want it to, then proceed to the next zone and continue the installation process.

If the automation system has multiple command modules or controllers, each should be tested with at least one zone's control modules to be sure it functions correctly in its assigned location. For portable remote controllers, this means testing from several locations to be sure the radio transceiver unit can pick up commands from anywhere in the house. For car visor units and keyless units, it means testing from outside the home to be sure that the wireless transceiver is still within range.

Installing Exterior Lighting Zones

Once you have set up all the interior control zones and programmed the light scenes and other functions you want for each of them, install the exterior lighting zones. These usually consist of a decorative lighting zone and a security zone. Set up the security lighting zone first, as it probably is the larger zone and requires more power. You want to be sure you have enough capacity for controlling all the security lighting before going on to decorative lighting. In the latter, some compromise can be made with bulb sizes and numbers if the load on the electrical system appears to be getting high. In the security system, you want to have all the illumination necessary.

If any new circuits or lights are being added outside, check your state and local code requirements to find out whether they need to be wired with armored cable (AC) or metal-clad cable (MC). Be sure that you follow the NEC requirements as amended by any local or state ordinances when installing outside cable. Also be careful about grounding outside circuits. Don't assume that because the cable is buried in the ground, it does not need to be grounded. That is not true. The same rules apply for grounding outside circuits as inside circuits.

Installing Security Lighting Zones

Security lighting is usually programmed with time-of-day settings or to come on and go off at sundown and sunrise. If security lighting control is augmented by sensor input, then these parameters need to be programmed into the system in addition to, rather than in place of, the time settings. Security lighting should not come on during daylight hours, even if a motion sensor detects movement of someone approaching the house. Other security features, such as video surveillance cameras, may activate in daylight from the same sensor input, but the lights should probably stay off during non-nighttime hours.

Programming the Automation System

Automated lighting programs are based on time-of-day schedules, event schedules (sunrise, sundown, etc.), sensor input (motion detected, light level, etc.), or combinations of all three. Every program is a unique sequence developed for the individual home and its occupants. A typical program using time schedules, events, and sensor input is noted in Table 8-3. The programs you develop may include some of these features and many others.

TABLE 8-3 Typical Automation System Schedule

TIME/EVENT	LIGHTING ACTION
Sunrise	Outside security lights turn off for daylight hours; all window shades open.
Sunrise + 30 minutes	East-side shades close; south-side shades activate close-cycle shutting 20% hourly until closed.
Sunrise + 240 minutes	West-side shades activate close-cycle shutting 20% hourly until closed.
Local noon	East-side shades activate open-cycle opening 20% hourly until fully open.
Sundown	Outside safety/security light zone turns on.
Sundown + 15 minutes	All shades close. All inside light zones turn on, if room is occupied. Family room set to activity scene lighting level. Living room set to entertainment scene lighting level. Decorative interior and exterior lights turn on. Pool lights turn on.
8:00 p.m.	Family room zone set to entertainment scene lighting.
11:00 p.m.	Pool lights turn off.
11:30 p.m.	Outside safety light zone turns off (turns on again if motion detected in zone). Outside and inside decorative light zones turn off.
11:45 p.m.	All inside light zones turn off unless individual rooms are still occupied.

Note that lighting control systems timed to events such as sunrise and sundown vary daily by as much as 3 hours from winter solstice to summer solstice for areas located in the middle latitudes of the continental United States. Northern latitudes vary even more and southern ones somewhat less. These events, as well as local noon, are not affected by time changes (Daylight Savings Time) or by a location's position within a time zone. The data tables from which they are calculated are based on Greenwich Mean Time (GMT) and the geographical location specified.

Time schedule events follow the clock, including changes into and out of Daylight Savings Time. In cases where event programming and time schedule programming are used together, some lighting changes may need to be adjusted during summer or winter time. In Table 8-3, for example, if the home using this program were located in Seattle or Boston, the pool lights would only be on for an hour or so in midsummer because sundown occurs late in these northern cities. For situations like these, a program might include a few time-of-year adjustments to make allowance for longer or shorter daylight hours.

In addition to a timed program, such as the example in Table 8-3, command controls in each room of the home can be used to control modules. These controllers can override the timed program's instructions to the modules, but the program will resume control when the time of the next programmed instruction to the modules arrives.

Sensor input to a control program and event scheduling of commands both require a computer-based master control system. Time schedule commands can be entered manually on many of the larger command modules in X-10, UPB, and RadioRA systems. Complex time schedules are much more easily entered on a graphic interface program such as the X-10 touch screens or computer programs.

One strategy that can be programmed into an automated system is a load-reduction strategy. This strategy programs the control system to minimize electrical use during peak-demand hours (usually during warm afternoons and early evenings) in order to reduce the home's peak use of power. This strategy is one to consider seriously in areas where utility companies use demand pricing to charge for electric service.

TROUBLESHOOTING AN AUTOMATION SYSTEM

Problems with automation systems, whether X-10, UPB, or RadioRA, come mainly from interference or attenuated signals. For X-10 and UPB, the separation of the two legs of the home's electrical system can also be a problem.

AC electrical service to homes arrives as a 220-volt current that is divided once it enters the home into two "legs" of 110 volts each. One of these legs powers each of the home's 110-volt circuits and both power the 220-volt circuits to the stove, clothes dryer, and other 220-volt devices. Both legs of the AC power work together in most homes, but they may not be actually joined in a manner that allows the higher-frequency X-10 or UPB signals to travel easily on both legs. If the signals cannot move freely on both legs, then they likely will not reach some control modules on one leg or the other.

The solution to this problem is a bridge installed on the in-house side of the service line, which connects the hot wires of both legs of the house AC current to one another. A bridge can be installed at the electric service box by an electrician and a new model is now available that can be installed by a homeowner or technician on any 220-volt outlet (such as for a range or clothes dryer). The bridge, which is available as an X-10 accessory from several manufacturers, provides a path for X-10 or UPB signals to reach control modules on both legs and should eliminate the problem of one leg transmitting no signals or signals too weak to be read by the control modules.

Weak X-10 signals can also occur in a home with a large AC electrical system. X-10 signals are only about 2 volts in strength and attenuate with distance from the source. A regenerator, placed in an outlet some distance from the command modules but close enough to read their signals clearly, can solve this problem. It reads all command signals on the AC wiring and regenerates them at full strength without any interference or noise they may have picked up in transit. In very large homes, two regenerators, placed some distance apart from one another, may be needed to keep signals strong throughout the house, but in most a single unit is enough.

Even with a bridge and a signal regenerator in place on the AC wiring, some X-10 control modules may have trouble receiving commands from controllers. Among the causes of specific problems with individual control modules are the following:

- Reverse-wired outlets with the hot wire connected to the neutral side of the outlet and the neutral wire to the hot side. Correcting the wiring often solves the problem.
- Switched circuits in which the switch is on the neutral wire instead of the hot wire. Again, correcting the wiring often solves the problem.
- AC suppressor or filter that blocks signals. Removing the blocking device is the only solution. Lighting circuits usually don't need suppressors or current filters, but if other equipment that does need one or both is on the same circuit, some rearrangement of connections will be necessary.
- Noise from other electrical equipment that interferes with the X-10 signals. You can locate noise-producing electrical equipment using a radio tuned to a blank (no station) area of the broadcast band. A filter on the offending equipment often cures this problem, but if not, moving it to another circuit may be necessary.
- General noise that comes from an outside source that cannot be eliminated. This type of interference is tough to combat. A filter on the incoming AC service line may help. Sometimes, an extra regenerator to keep the command signals well above the noise level works. If neither solution solves the problem, the only other option is to locate the source of the noise and somehow shield the home's wiring from it.

UPB signals are also subject to interference and noise, but much less so than X-10 signals. UPB signals are 40 volts strong and are set at a lower frequency than X-10 signals. Both of these features make them less subject to interference than the weaker X-10 signals. If interference is a problem in a UPB control system, the same possible causes and cures as described for X-10 systems may apply.

Wireless RadioRA and X-10 systems can also be troubled by noise. RF frequency noise is difficult to filter out. Filtering each wireless receiver is simply too expensive and complex. The only workable answer is to find the source of the noise and filter that device.

Minor noise levels in wireless systems can usually be overcome by simply boosting the wireless signals so that they can be clearly read over the noise level. Both RadioRA and X-10 signal regenerators are routinely used in large homes where signals attenuate, but these can also be used in smaller installations where their signal boost may be enough to overcome background noise.

SUMMARY

Home automation systems can be used to control any number of electronic systems and lighting fixtures. Automation systems should be planned before installation to include control of as many electric devices, systems, and lighting fixtures as desired by the homeowner. These may include automation of lighting in zones set up according to function or location, audiovisual systems, security systems, water management systems, door and window automation systems, and others. Lights and other devices connected with home run wiring require individual control modules in order to be automated. Lights and devices that are wired in daisy-chain fashion can all be controlled by a single control device at the head of the chain. Control of multiple devices in a zone is more economical using daisy chains.

Automated control systems use X-10 or UPB power-line technology or wireless X-10 or RadioRA technologies to communicate on the automation network between command modules and control modules. Control modules come in several varieties and respond to signals from command modules to turn lights and electrical devices on or off, or to dim or brighten them. These modules plug into existing outlets or can replace existing outlets and switches in the AC wiring system. Modules hard-wired into the AC system should be installed by an electrician. Those connected to existing outlets can be done by the homeowner or a technician.

X-10, UPB, and RadioRA controls are installed by setting up control modules in separate zones, which can be individually controlled from the command modules in the system. Command modules can be small handheld devices or large computer-based modules that can be programmed to control hundreds of lights and sensors. In addition to in-home command modules, both power-line and wireless technologies have wireless remote units and telephone interface units for communicating with the lighting system at a distance. Control modules must interface correctly with the devices they are designed to control. There are several types of outlet control modules and switch control modules, as well as low-voltage modules and relay activators.

Automated control systems can be programmed according to a time-of-day schedule, an event schedule, or a combination of both. They can also be used to reduce peak load on the home's electrical service. Attenuation of signals, noise, and RF interference are the main problems encountered in automated lighting systems. All of these can be overcome in most cases.

KEY TERMS

Always-live duplex outlet An AC electrical outlet that is not controlled by a switch and always has current available.

Automation The process of controlling a lighting or other electrical system remotely, either with manual commands from a controller or by a programmed set of instructions from a computer.

Command module An X-10 technology controller that sends commands (manually generated or programmed) to control modules that control lights.

Controllers Devices for managing the operation of another device remotely, either through a wired connection or wirelessly; devices that send commands to control modules in an automation system.

Control module In X-10 systems, any of the devices that receive lighting control instructions from the controller and implement them by adjusting the lights.

Daisy-chain wiring Wiring light fixtures in parallel one to another with only the first light being connected directly to a power supply and each of the others receiving power in succession down the chain.

Fluorescent lighting Lighting tubes that contain phosphorescent material and glow when a high-voltage current is passed through them. The tubes are made in various lengths up to 8 feet.

Home run connection A wired connection that links a light directly to its power supply without the current to it passing through any other fixture except a switch.

Light scenes Illuminated rooms or areas within a room that not only provide light but beautify a setting or invoke a mood.

Load In electrical systems, the amount of current (amps) flowing in the wiring at a given time.

Macros Sequences of commands in a specified order that a programmable device stores in memory and executes at preset intervals.

Overload Current flow in an electrical circuit above its capacity to carry safely.

Power-line technology (PLT) A communication and control system in which data signals are transmitted on AC wiring using a different frequency and voltage than the regular AC current flowing in the circuits.

Programmable Capable of storing instruction sets for later execution in sequence.

RadioRA A proprietary wireless lighting control technology developed by Lutron and used for home lighting control.

Single-pole switch An AC basic switch that opens or closes a circuit to control power to a fixture.

Split duplex outlet An AC electrical outlet in which one receptacle is always live and the other is controlled by a switch.

Switched duplex outlet An AC electrical outlet controlled by a switch that turns current to it on or off.

Three-way switch A type of AC switch that is used in pairs to turn a circuit on or off from two locations.

Zone The section or part of an HVAC system that is separately operated and controlled; also a group of lights and other electrical devices controlled together to provide a specific scene or accomplish a specific task such as security lighting.

Z-Wave A wireless technology developed by the Danish company Zensys that is popular in Western Europe and is now available in the United States.

REVIEW QUESTIONS

1. True or False: A zone in a home automation system can include lights and other electrical systems.
2. What does overloading mean in reference to an electric circuit?
3. How does power-line automation technology transmit its data signals?
4. A lighting zone is a group of lights that are related by ____________ and may also be in or near the same ____________.
5. Explain how a single light could belong to more than one zone and what kind of control it would need to have.
6. How are home run-connected lights wired?
7. A string of parallel-wired holiday lights is an example of what kind of wiring?
8. What is the most common automated control technology that uses AC wiring to transmit control signals?
9. What are two major wireless technologies for automation?
10. X-10 systems can use a maximum of ________ house (letter) codes, each of which can have a maximum of ________ unit (number) codes for individual modules.
11. What does a telephone interface for an automation system do?
12. How do RadioRA signals differ from standard X-10 signals?
13. How does a programmable control module differ from a manually operated controller?
14. An outlet controlled by a switch in an AC circuit should be automated by what type of control module?
15. In addition to controlling lights, what other systems can an automation system control?
16. What two functions could a motion detector serve in controlling the lights in a room?
17. How is a split duplex outlet different from a standard duplex outlet?
18. Attenuated (weakened) signals in an automation system can be strengthened by what added component?
19. In a RadioRA system, RF interference might be reduced by what means?
20. On July 4, does the sun set later in Houston or Minneapolis? How about on New Year's Day? (Hint: Houston is in southern Texas; Minneapolis is in northern Minnesota.)

CHAPTER

9

SOLAR POWER FOR YOUR HOME

After studying this chapter, you should be able to:

- Describe the passive and active ways in which sunlight can be turned into a useful energy source for home use.
- Describe passive methods of using and blocking solar energy for heating and cooling.
- Describe how passive and active solar water-heating systems work and the components in them.
- Describe how photovoltaic (PV) solar generating systems work and the components in them.
- Describe how to install a solar water-heating system for home use.
- Describe how to install a grid-connected or stand-alone solar electric generating system.

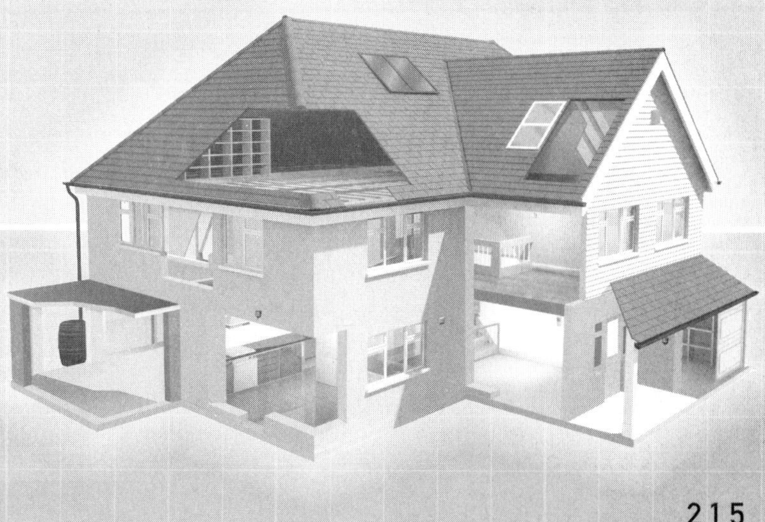

OUTLINE

I. Introduction
II. Solar Radiation
   A. What Is Solar Power?
   B. Using the Sun's Free Energy
III. Passive Solar Energy Use
IV. Solar Water Heating
   A. Active and Passive Heating Systems
   B. Installing a Solar Water-Heating System
V. Solar Electricity Generation
   A. Types of Photovoltaic (PV) Modules
   B. Installing a Grid-Connected Solar Power System
   C. Installing a Stand-Alone Solar Power System
VI. Summary

INTRODUCTION

In this chapter you will learn how energy from the sun can be used in the home to provide heat and power electric appliances. You will understand how the passive heating effects of the sun can be enhanced or blocked to warm or cool the interior of the home. You will learn how the sun's heat energy can be concentrated and transferred to a liquid so that it can be moved into the home and utilized to provide hot water or interior heating. You will also learn how sunlight can be transformed into electrical power through the photovoltaic effect, and you will be able to describe the equipment required to construct a photovoltaic solar generating system for your home. Finally, you will learn how to install a solar electric generating system that is connected to the electric grid or a system that provides electric power independent of the grid.

SOLAR RADIATION

SOLAR RADIATION IS the general term used to describe the electromagnetic radiation emitted by the sun. Every location on the earth receives sunlight for at least part of the year. The amount of solar radiation that reaches any one place on the earth's surface varies according to five factors:

- Geographic location
- Time of day
- Time of year
- Local landscape
- Local weather

Geographic location, the local landscape, and local weather should each be evaluated when considering whether to install solar power systems in a residence. Time of day and time of year are factors that equally affect the amount of solar radiation at all locations on the earth and so are not relevant when choosing a favorable site for solar power installations.

Location is critical because the sun strikes the spherical surface of the earth at different angles, ranging from perpendicular directly overhead down to horizontal at the horizon. When the sun's rays are vertical, they travel straight down through the atmosphere and the surface receives the most energy possible. The more slanted the sun's rays are, the more atmosphere they travel through and the more scattered and diffuse they become, with less energy remaining when they finally strike the surface (Figure 9-1). The further north or south of the equator a location is, the more slanted are the sun's rays that reach it. The polar regions never get a high sun, and because of the earth's tilted axis of rotation, these areas receive no sun at all during part of the year.

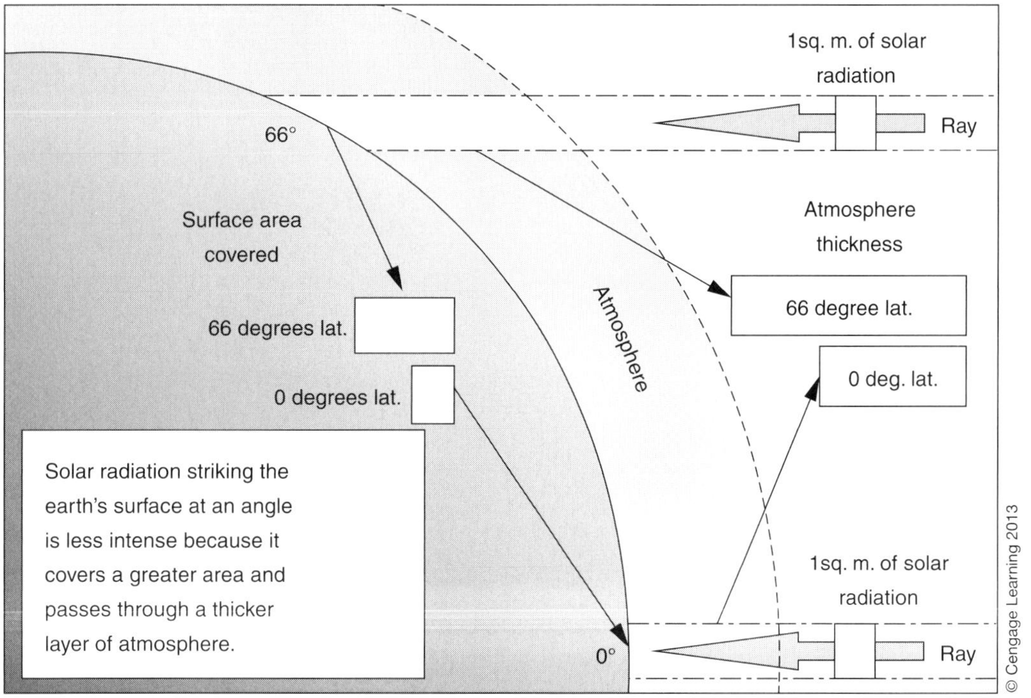

FIGURE 9-1

The intensity of solar radiation striking the earth's surface is reduced when it passes through a thicker layer of atmosphere and when it strikes at an angle rather than vertically.

The 23.5-degree tilt in the earth's axis of rotation is important in determining the amount of sunlight striking its surface at a particular location. The tilt results in longer hours of daylight in the northern hemisphere from the spring (vernal) equinox through summer to the fall (autumnal) equinox and shorter hours of daylight from the fall equinox through winter to the spring equinox. In the southern hemisphere, the seasons and the length of daylight hours are reversed. Days and nights are both exactly 12 hours long on the equinoxes, which occur each year around March 23 and September 22.

The United States, which lies in the middle northern latitudes, receives more solar energy in summer days not only because they have more hours of daylight, but also because the sun is nearly overhead. The sun's rays are far more slanted during the shorter daylight hours of the winter months. Cities like San Francisco, Denver, St. Louis, and Philadelphia, all situated about 40 degrees north latitude, receive nearly three times more solar energy in June than they do in December.

Local weather also affects the amount of solar radiation available. When sunlight passes through the atmosphere, as shown in Figure 9-2, some of it is absorbed, scattered, and reflected by the following:

- Air molecules (atmosphere)
- Water vapor (humidity)
- Clouds
- Dust
- Smoke and other pollutants
- Rain and snow

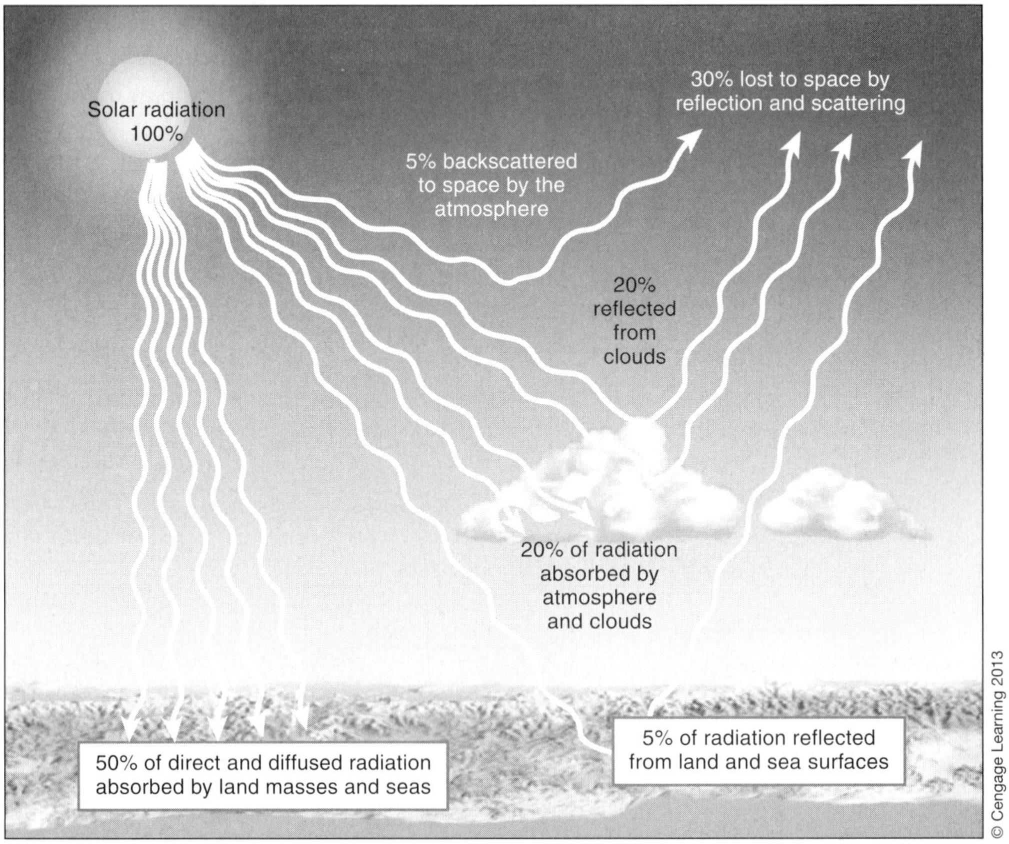

FIGURE 9-2

Atmosphere, clouds, reflection from land and water surfaces, absorption, and diffusion all affect how much of the sun's energy reaches the earth's surface.

Radiation affected by these factors is called diffuse solar radiation. Solar radiation that reaches the surface without being diffused is called direct-beam solar radiation. The sum of the diffuse and direct solar radiation is called global solar radiation. Atmospheric conditions can reduce direct-beam radiation by 10% on clear, dry days and by 100% during cloudy days. Diffusion also varies with the time of day. In the early morning and late afternoon, the sun's rays travel through more atmosphere than at noon when the sun is at its highest point. On a clear day, the greatest amount of solar energy reaches the surface around solar noon.

What Is Solar Power?

Simply put, solar power is energy from the sun transformed into an energy source that can do useful work or provide useful heat. Saying that, however, is much easier than doing it. Energy from the sun that reaches the earth comes in the form of radiant energy, mostly heat (infrared radiation), light, and ultraviolet radiation. A small portion comes as radio waves, x-rays, and other wavelengths of radiation that are even more difficult to convert into a useable source of power than are the sun's heat or light radiation.

The amount of energy that reaches the earth from the sun is huge. The energy that strikes the earth's surface in 1 hour is more than sufficient, if it could all be converted into useable forms, to supply all the planet's current energy needs for a full year. The solar generation potential of the U.S. is shown in Figure 9-3. And all of it is free, delivered daily without charge to every part of the earth. Why then, it can reasonably be asked, are we paying $100 per barrel for oil and similar amounts for other forms of energy? There is a simple answer to that question too: We are paying the cost of converting stored sun energy into a useable energy form.

Most of the energy sources on earth originally came from the sun. Fossil fuels (oil, natural gas, coal) are all forms of sun energy that was originally absorbed by living things and then transformed over millions of years into fuels that can be burned (oxidized) to release their stored sun energy as heat. Hydropower is also sun energy that causes evaporation of surface water and generates weather systems in the atmosphere. Evaporation and weather systems both use sun energy to lift water from the oceans and deposit it as rain or snow in higher elevations on land. As this water flows down again to the sea, its movement provides energy to drive turbines that produce much of our electric power.

Nearly all the energy we use in our daily lives can be traced back to sun energy. Only geothermal energy (heat from within the earth that rises to the surface) is not a form of sun energy. It is produced by the force of gravity compressing and heating the materials of the earth's interior. Geothermal energy is also plentiful beneath the earth's surface, and free for the taking. But extracting this energy from the ground and transforming it into useable electric or steam energy is expensive. Once again we must pay for the cost of conversion.

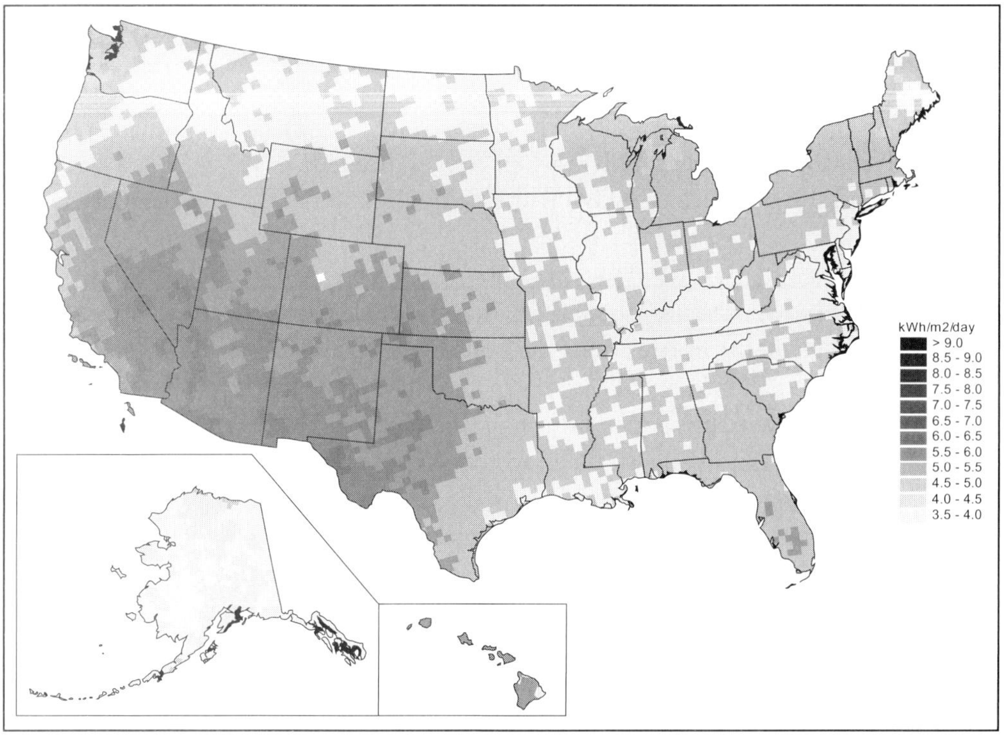

FIGURE 9-3

The estimated average kilowatt hours per day that can be generated by a 1-square-meter area solar panel in the United States. This model considers factors of latitude, climate, altitude, and pollution.

Using the Sun's Free Energy

People have long sought for ways to utilize the sun's free and abundant energy directly, without having to pay the conversion costs associated with fossil fuels or hydropower energy. There are at least three ways of using sun energy directly to provide heat or produce electricity. Each of these methods can be used in the home to reduce or eliminate the cost of other sources of energy.

The first way of using the sun's energy is called passive use, and every home uses this method to some extent each day. Passive use is simply allowing natural sunlight to warm the home and its contents by shining through windows and heating the exterior walls so that heat transfer will bring some of the heat inside. Conversely, passive use may consist of blocking sunlight to minimize its heating effect when it is not wanted.

The second method of using sun energy is by direct heat transfer. This means developing mechanical devices that concentrate the sun's energy and transfer some of its heat directly into a substance from which it can later be extracted and used. The most common solar device for direct heat transfer is a water heater that collects solar heat and transfers it to water or another liquid. The heated liquid is then moved inside to heat the home or to heat more water.

The third way of using solar energy is by transforming it directly into electricity by means of the photovoltaic effect. This effect causes some substances, when properly constructed and connected, to generate an electric current when their exposed surfaces are struck by sunlight. The photovoltaic effect is not a heat transfer process. It is produced by a change in the molecules of the photovoltaic substance so that they release a moving stream of electrons (electricity) when sunlight strikes the substance.

PASSIVE SOLAR ENERGY USE

Passive use of solar energy consists of using the sun's rays to provide heat when it is beneficial to the home's environment and, conversely, preventing the sun's rays from adding heat to the home environment when it is not wanted.

Several passive methods of utilizing or blocking the sun's rays have been discussed in previous chapters. These include:

- Placement of most windows and door openings on the side of the house where they are the most beneficial. In hot climates, large windows and doors should be on the north side of the house, where little direct sunlight will strike them. Homes in cold climates should have most windows on the south side, where sunlight passing through them can help warm the interior.
- Awnings or shades placed on all windows to block sunlight when it is not wanted for interior warming.
- Deciduous trees and plants placed strategically, as shown in Figure 9-5, to shade the home during the hot summer months, but permit sunlight to pass through their leafless branches to warm the interior in winter.
- Insulation in ceilings, walls, floors, and around doors and windows to prevent heat transfer. Although insulation is usually intended to prevent heat

loss from the interior of the home in cold weather, it is equally as effective in blocking heat transfer into the home from the hot outside in summer.

- Storm windows, storm doors, shutters, or other insulated window coverings to help keep heat from sunlight out of the home as well as prevent heat loss from inside during cold weather.

Other methods that can maximize the benefit of using or blocking passive solar heating are architectural design features that are expensive to retrofit in an existing home but should always be considered in new construction.

For energy efficiency, homes in cold climates should avoid high ceilings, open lofts, and multi-story entryways and atriums. These are all attractive design features, but they consume more heat energy than low-ceiling rooms. Warm air rises, and high ceilings, lofts, and similar vertical spaces contribute to poor heat distribution. For this reason, a multistory home is very difficult to heat or cool evenly. If the upper level is warm the lower part will be too cool. If the heating is increased to warm the lower level, much of the warmed air will rise into the upper level and overheat it or simply be wasted.

Low ceilings and heating/cooling zones that are separated from one another by closed doors are most efficient for heating a home in a cold climate. If high ceilings are already in place or considered a must-have design feature, ceiling fans can help circulate the interior air more evenly, but any cost benefit to heating or cooling will probably be more than offset by the power required to operate them.

In warm climates high ceilings are a good design feature for the same reason that they are not in cold climates. Warm air rises and can be vented from the top of a high ceiling. This allows circulation of cooler air below and can reduce demand for air conditioning.

FIGURE 9-4

Light colors on exterior walls and roof combined with dense foliage of south-side shade trees help keep a home cooler, especially in areas with high solar radiation, such as the southwestern United States.

Anything that can be done to bring the sun's energy inside the home in cold weather and prevent it from entering during warm weather is a passive technique for using solar energy. Direct sunlight can easily heat a flat metal panel above 160 degrees. Keeping that energy off every exterior surface of the home during hot weather is a very effective way of reducing energy consumption for air conditioning. Conversely, getting as much sun energy inside the home during cold weather is equally effective in reducing heating bills.

If sunlight cannot be kept off some parts of the building, make those surfaces as reflective as possible so that the least amount of heat is absorbed. Light colors on exterior walls and roofs reflect heat better than dark ones (Figure 9-4). Dark blue, green, and purple tones absorb more heat than yellow, orange, or red ones. White is the most reflective color of all. Metalized film on the outside of windows blocks 85% or more of the sun's heat from passing through the glass. Use of passive solar techniques can be as simple as the color you choose to paint your house, but it can pay significant benefits in reduced energy costs.

SOLAR WATER HEATING

The second method of using solar energy directly is for heating water. A large amount of energy is required to heat water, and sun energy is particularly well suited for this task. Most sun energy is in the form of heat when it arrives at the earth's surface, and so no transformation is required for it to be used in heating water. All that is required is a means of transferring the sun's heat energy into water and a storage tank to contain the heated water until it is used.

To heat water using solar energy, a collector is used. The collector is a large, flat container through which water or another liquid that is called the *working fluid* flows. It is placed on a roof or a wall facing the sun, which heats the working fluid passing through it. The collector can be a simple frame with a glass top on its upper face and a flat solar absorber on its bottom side. The absorber is usually made of sheet metal attached to copper pipes and painted black. The working fluid is either pumped or driven by natural convection through the copper pipes, picking up heat from the absorber as it does so.

The heated water is stored in a hot water storage tank. The heat transfer fluid (HTF) that flows through the absorber may be the hot water from the tank, which flows directly from the tank, through the absorber, and back to the tank. More often the absorber uses a separate loop of piping containing antifreeze and a corrosion inhibitor that delivers heat to the tank through a heat exchanger similar to that of an air conditioner. The exchanger coil is positioned inside the storage tank and the HTF continually transfers heat that it has gained from the absorber to the water in the tank. Heat-exchange water heating systems require a small pump to keep the HTF circulating constantly and are called active systems.

A somewhat more efficient, but also more expensive type of solar collector uses evacuated heat pipe tubes (EHPTs) to transfer heat to the working fluid (Figure 9-5). EHPTs are composed of double glass tubes, one within the other, with the space between the tubes evacuated to reduce heat transfer. The inner tube of each EFPT contains an absorber plate fused to a heat pipe. The heat from the hot end of the heat pipes is transferred to the working fluid in a heat exchanger called a manifold. The manifold is insulated and encased in sheet metal or plastic to protect it from the elements.

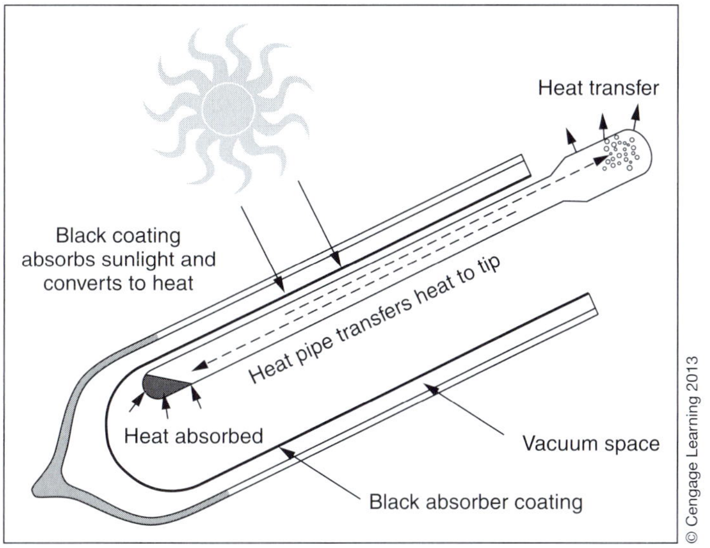

FIGURE 9-5

An EFPT converts sunlight to heat inside a vacuum-insulated tube; the absorbed heat is then transferred to a working liquid for use in heating water.

The vacuum that surrounds the inner tube of an EFPT greatly reduces convection and conduction heat loss to the outside, thereby achieving greater heat-collection efficiency than flat-plate collectors, especially in colder conditions. Their advantage is reduced in warmer climates, because the warmer ambient air temperature reduces heat transfer.

A third type of solar collector, almost always used for space heating, uses air as the working fluid. A fan circulates the air over the collector plate within a sealed glazed case. The warmed air is then forced back into the heating system or directly into the building's interior.

Active and Passive Heating Systems

Residential solar water-heating installations can be either active or passive. Active systems use a pump to circulate the HTF, but passive systems rely on municipal water pressure or convection to circulate the HTF. Both types include an auxiliary energy source for heating water to a preset minimum temperature when solar power alone is insufficient to do so. The auxiliary heat source can be an electric heating element or a gas or fuel oil heater that is built into the storage tank and activated when needed so that adequate hot water is always available. Figure 9-6 shows the simplest type of passive solar water-heating system connected to a gas or electric tank-type water heater. This type of solar system must be drained during periods when outside temperatures drop below freezing to avoid damage to the collector.

Water pressure in the incoming cold water line is the only force moving water through the system and so cold water is only fed to the collector when hot water is drawn out of the storage tank. This type of system uses no outside energy, but is limited in the amount of water it can heat and when it will be heated. This passive system can be made much more efficient by the addition of a closed loop of pipe and a pump that continually circulates water from the tank through the collector. This enables the collector to heat water all day, regardless of whether or not any is being used from the tank. An electronic controller shuts off the pump when the collector is not receiving any solar heat energy.

The volume of the storage tank in a solar heated system that does not have an auxiliary heating method needs to be larger than with other heating methods in order to

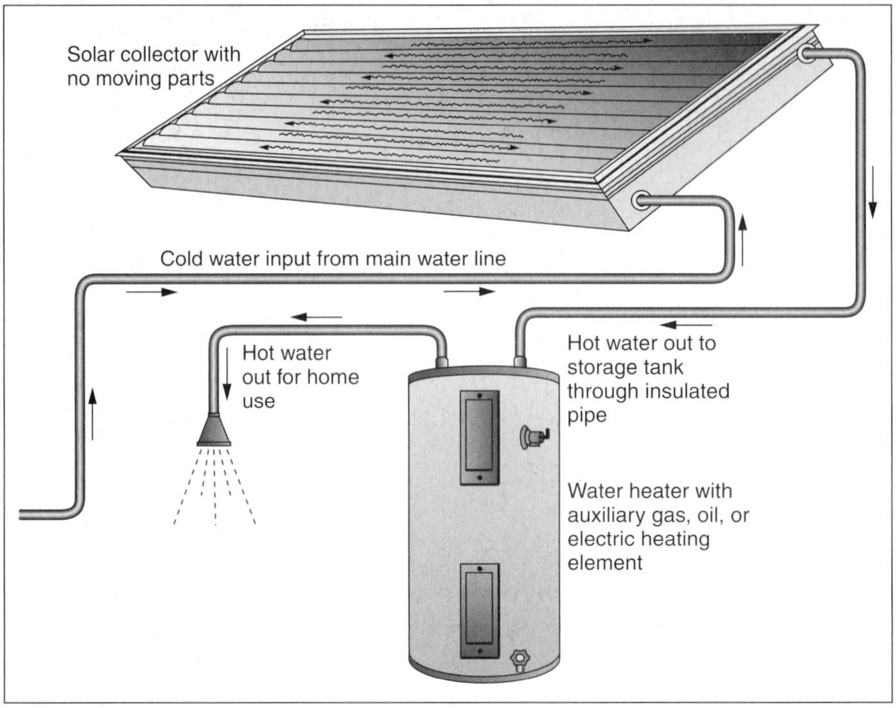

FIGURE 9-6

A simple passive solar water-heating system raises the temperature of the water in the water heater tank and reduces the amount of additional fuel needed for heating water.

provide enough stored hot water at night and during bad weather. Also, solar heaters do not usually raise the water temperature as high as gas or electric units. This means that a greater volume of hot water may be needed because it is mixed with less cold water to obtain a comfortable temperature level.

Solar water heaters can also be used in conjunction with a hot water central heating system, but because the need for central heating is greatest at night and in winter when solar heat potential is low, hot water heating systems do not pair well with solar heaters in most cases. Solar water heating for household use such as washing and bathing is a better match than central heating because the supply of energy for heating water and demand for the hot water are more closely timed to one another. A solar hot water system can provide at least half of domestic hot water energy for nearly all homes in the United States and up to 85% of heating energy for homes in warmer climates. Figure 9-7 shows an active split solar system in which the HTF is glycol antifreeze that circulates through a closed-loop heat exchange system, gaining heat from the collector and transferring it to the water in the first storage tank.

The pump operates whenever solar energy provides it with electric power, which is also the times when solar energy will heat the fluid passing through the collector. This arrangement enables the solar unit to heat water throughout the day and store it for later use. It is the most efficient solar water-heating method for home use.

The water-heating system in Figure 9-7 can also be connected to a tankless water heater. In this configuration the auxiliary heating element in the storage tank is eliminated and the tank is used only to store water heated by the solar heating unit. This heated water is then sent on demand through the outlet pipe to the tankless water heater, which brings its temperature up to a higher level, if necessary, and dispenses it through the faucet.

Another use for solar heated water is in swimming pools. This application is perhaps the best matched of any for solar heating, and more than 300,000 pools in the United

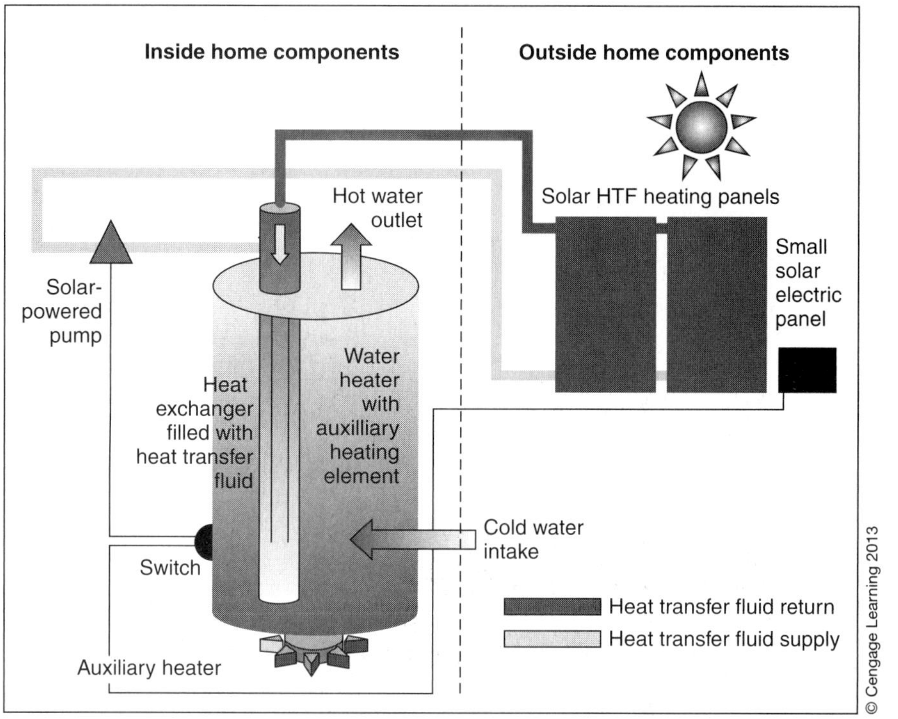

FIGURE 9-7

A solar-powered pump circulates the HTF through the collector, where it gains heat, and back through the exchanger, where it heats the water in the tank.

States are now heated by solar systems. Like inside solar water-heating systems, pool systems can be simple stand-alone units or supplements to fuel-fired pool heaters. In mild climates a stand-alone system can usually supply all the heating needed for a pool, completely eliminating the cost of fuel-fired heating. In colder climates, large solar heating panels can dramatically lower the cost of keeping a pool at a comfortable temperature.

All swimming pool solar heating systems use a pump to circulate the pool water through the collector and back to the pool. The pump is powered by solar energy and should be large enough to pump the contents of the pool through the collector once every 24 hours. Figure 9-8 shows a hybrid solar heating system for a swimming pool.

Swimming pool heating systems do not require any storage tank and do not operate with water pressures as high as those found in culinary water systems. To minimize the energy needed for the pump in a swimming pool system, the collector should be located as close to the level of the pool as possible. The less vertical distance the pool water must be lifted to the collector, the less energy the pump will use.

Installing a Solar Water-Heating System

The main component of a solar water-heating system is the solar collector panel, which can be either glazed or unglazed. The first step in installing a solar water-heating system is to determine what type of collector to use. A glazed panel looks more like a photovoltaic solar panel. It is a flat, rectangular frame with a metal back and a glass-covered face. Glazed panels are more durable than unglazed and can endure below-freezing temperatures without damage. Sunlight passes through the glass faceplate of the panel and collects on a dark absorber plate behind the glass. Tubing looped back and forth under the glass next to the absorber plate circulates an antifreeze solution. As the fluid in the tubes is warmed, it is pumped to a heat exchanger.

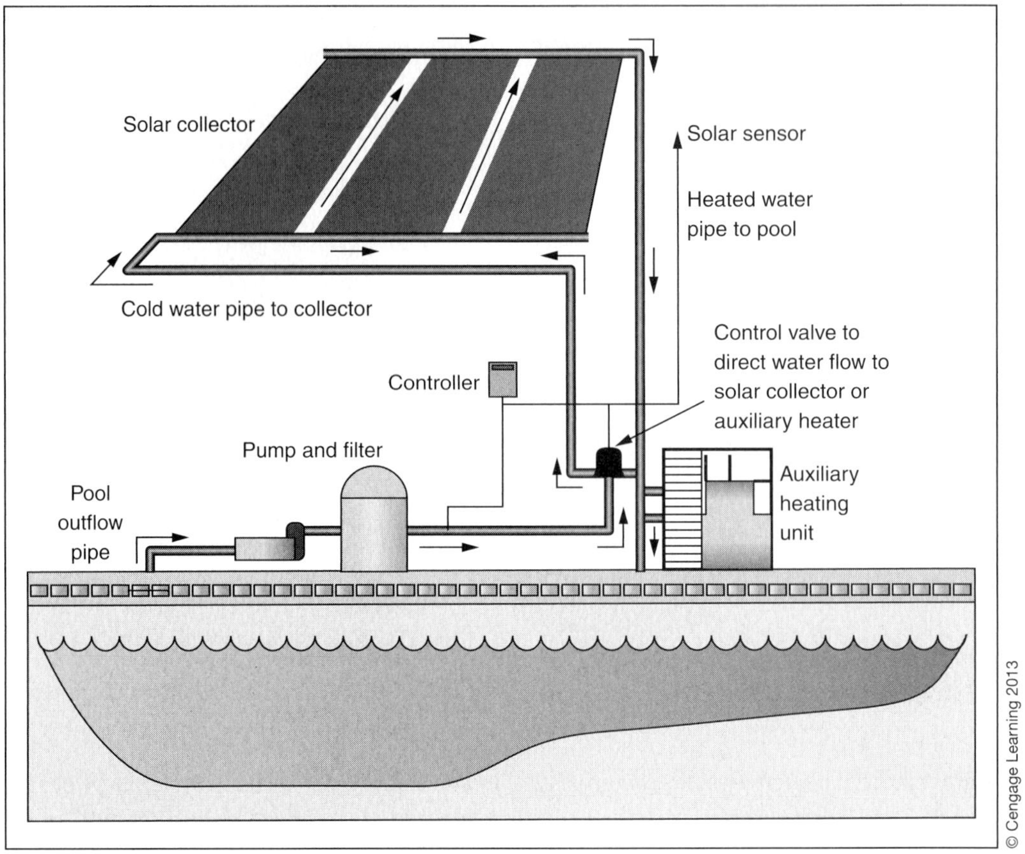

FIGURE 9-8

A hybrid solar heating system for a swimming pool uses a solar-powered pump to circulate the pool water through the collector and includes an auxiliary fuel-fired heater for additional heating capacity.

Unglazed collectors are made of black hose or pipe set in an open frame that is directly exposed to the sunlight (no cover). The pipe is coiled or looped into a compact unit in the frame, which has a black back surface to help absorb heat. Water flowing through the pipe is warmed directly and stored in a holding tank until used. Unglazed collector panels must be drained during periods of freezing temperature and so cannot function in cold climates as well as glazed collector panels. Unglazed collectors are less expensive than the glazed type and are nearly always used for heating swimming pools, which only require heating in moderate weather.

Whichever type of collector you select, it is installed outside, usually on the roof of the home, but sometimes at ground level, especially if it is for heating a swimming pool. When the collector is in place, piping must be run from its entrance connection to the pump that will circulate fluid through it. Another pipe is run to the storage tank (if the collector is heating water directly) or to the heat exchanger (if the collector is heating antifreeze solution). A third pipe connects the storage tank or heat exchanger to the pump, forming a loop that allows the fluid in the collector to be continuously circulated by the pump.

If a heat exchanger is used, another loop of pipe runs from the exchanger to the storage tank and back to the exchanger. Water flows through this pipe and picks up heat from the exchanger as it passes through. The water can be circulated by a second pump or, alternately, the exchanger can be located inside the storage tank, where it is surrounded by water. In this configuration the water in the storage tank absorbs heat directly from the exchanger coil and its antifreeze solution.

An intake pipe brings cold water to the storage tank and an outlet pipe connects from the storage tank to the home's delivery plumbing. The pump's operation is controlled by a sensor that keeps it running whenever there is sufficient sunlight to raise the temperature of the circulating fluid. When the solar system is not able to heat water sufficiently for all home uses, the auxiliary heater in the storage tank makes up the difference between supply and demand.

SOLAR ELECTRICITY GENERATION

A third method of using solar energy is by generating electricity directly from the sun's radiation. This process is known as photovoltaic (PV) power generation. A small solar electric or photovoltaic system can be a reliable and pollution-free producer of electricity for the home. Because PV technologies use both direct and scattered (diffuse) sunlight to create electricity, the solar radiation available across the United States is sufficient to operate small solar electric systems. The amount of power that can be generated by a solar system at a particular site depends on how much of the sun's energy reaches that location. PV systems, like all solar technologies, function most efficiently in the southwestern part of the country, which receives the most solar energy. PV systems in other locations that receive less energy must be sized up in order to produce power equal to that of smaller systems located in the Southwest.

Solar electric systems are also known as photovoltaic (PV) systems because they use arrays of solar cells to convert sunlight into electricity. Solar cells are the basic units of a PV system and are made up of semiconductor materials. When a semiconductor material absorbs sunlight, the solar energy knocks electrons loose from the atoms that make up the material. This phenomenon is called the *photoelectric effect*. These free electrons flow into a circuit built into the solar cell and form an electric current. Only sunlight of certain wavelengths works efficiently to create electricity through the photoelectric effect. For this reason, PV systems can still produce electricity on cloudy days, but not as much as on a sunny day.

Types of Photovoltaic (PV) Modules

Silicon is the most popular semiconductor material for solar cells, and four varieties of this material are now being used, each of which has different properties that affect how efficient an electric generator it is:

- Monocrystalline or single crystal silicon
- Multicrystalline silicon
- Polycrystalline silicon
- Amorphous silicon

The photovoltaic efficiency of these materials is determined by how far specific wavelengths of sunlight can penetrate into them before being absorbed by the material. A small absorption coefficient means that light is not readily absorbed by the material. A large coefficient indicates that the material easily absorbs light of a specific wavelength.

The wavelengths that must be absorbed by the material are those with sufficient energy to move electrons from their bound state within the atoms of the material to a free state. This free state allows the electrons to be involved in conduction and form part of an electric current flowing out of the cell. The lower energy level of a semiconductor is called the *valence band*. The higher energy level created by sunlight energy being absorbed so that its electrons are free to move is called the *conduction band*. The bandgap (symbolized by Eg) is the energy difference between the conduction band and valence band. Light with energy below the material's bandgap cannot free an electron, and so the material will not absorb it or generate an electric current from it.

In addition to being manufactured as individual units that are combined into modules and arrays, solar cells are now being made in other forms. Thin-film solar cells are made of layers of semiconductor materials only a few micrometers (millionths of a meter) thick. This thin-film technology has made it possible for solar cells to be combined with various noncrystalline materials to form dual-function building components, including:

- Rooftop or solar shingles
- Roof tiles
- Building facades
- Glazing for skylights or atria

These materials, which form part of a home's structure as well as being capable of electric power generation, are now available in the residential market. In addition to their dual function, they are attractive and durable, but at present are lower in photovoltaic efficiency and more expensive than an equal-capacity small solar electric system.

An individual PV or solar cell produces only a small amount of power. To generate enough current to produce useable power, 40 or more solar cells are interconnected in series to form panels or modules. In addition to its solar cells, a PV module or solar panel consists of these components:

- A transparent top surface, usually glass, that protects the solar cells from weather
- An encapsulant, usually thin sheets of ethyl vinyl acetate, that binds the top surface, solar cells, and rear layer of the module together
- A rear layer, a thin plastic sheet that seals the back of the module and prevents water and other contaminants from entering it
- A frame around the module's outer edge, usually made of aluminum

PV modules range in output from 10 to 300 watts. To get still more power, several modules can be installed and wired in parallel to produce more current at a given voltage. The multiple modules are usually mounted in a rack to form a PV array. PV arrays can be mounted at a fixed angle facing south, or they can be mounted on a powered tracking device that turns at a constant rate to follow the sun, allowing them to capture the most sunlight over the course of a day. Figure 9-9 shows the structure of a photovoltaic module.

Because they are modular, PV systems can be sized to meet any electrical generation requirement. They can be connected to the home's AC wiring to provide some or all of its power during the daylight hours when the PV system can function. PV systems that provide power only to the home where they are located are referred to as standalone or off-grid systems. If the home is connected to the local electric utility through a smart meter, the PV system can also connect to the local electric distribution system and feed any surplus power it generates into that grid. A PV system that is connected to the local electric grid is referred to as a grid-connected system.

Installing a Grid-Connected Solar Power System

Solar power systems should be installed only by a qualified electrician or electrical contractor. The information that follows is to help you select the size of system you want

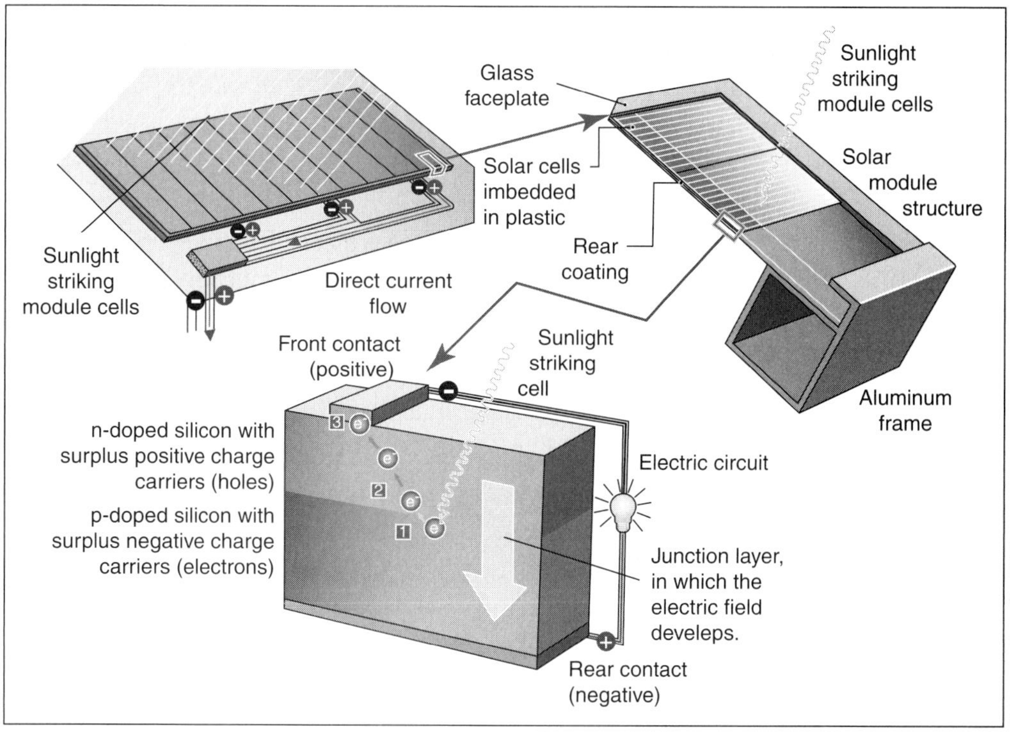

FIGURE 9-9

Dozens of individual photovoltaic cells imbedded in plastic make up a PV module that produces a continuous DC current when struck by sunlight.

and determine the cost of the solar generation panels and other equipment you will need to integrate the generated solar power into your home's AC system. Unless you are licensed electrician, do not attempt to do the installation yourself. Professional help is absolutely essential to be sure your system is safe for people in your home to use and also safe for others outside your home who may come in contact with it. Electric power can be deadly if equipment is not properly installed and operated. Be safe. Get the professional help you need to do the job right and follow the electrical code requirements for your area when installing a solar system.

Most homeowners prefer a grid-connected system to a stand-alone system because the grid system allows the solar panels to generate some or all of the home's electricity when sunlight is adequate to do so, but also permits electricity to be fed to the home from the municipal electrical grid whenever solar power alone is not sufficient to meet current needs. Any excess electricity produced by the solar system is fed back into the grid.

Power utilities in most states offer net metering, an arrangement where electricity generated by a grid-connected solar energy system is measured by the home's electric meter as it is fed into the grid. You pay the power provider only for the net amount of electricity (the difference between what the home received from the grid and what it produced and fed into the grid) during each month. If net metering is not offered, the value of solar power fed into the grid is lost to the homeowner (although it is still used by the utility), but the convenience of a grid-connected system still makes it the preferred choice for most homeowners.

A grid-connected solar power system requires some additional equipment beyond the basic solar generation components. This equipment is called *balance-of-system* equipment and is required in order to safely condition the electricity the solar system produces and connect it to the home's AC power lines. The electricity produced must also comply with the power provider's grid-connection requirements, for which the following items will be necessary:

- Power conditioning equipment
- Safety equipment
- Meters and instrumentation

Specific grid-connection requirements vary depending on the utility that provides the grid power. This is another reason to secure the services of a professional electrician who is knowledgeable about the grid-connection requirements in your area before purchasing components for your solar energy system. Figure 9-10 shows the components of a two-meter grid-connected solar generating system.

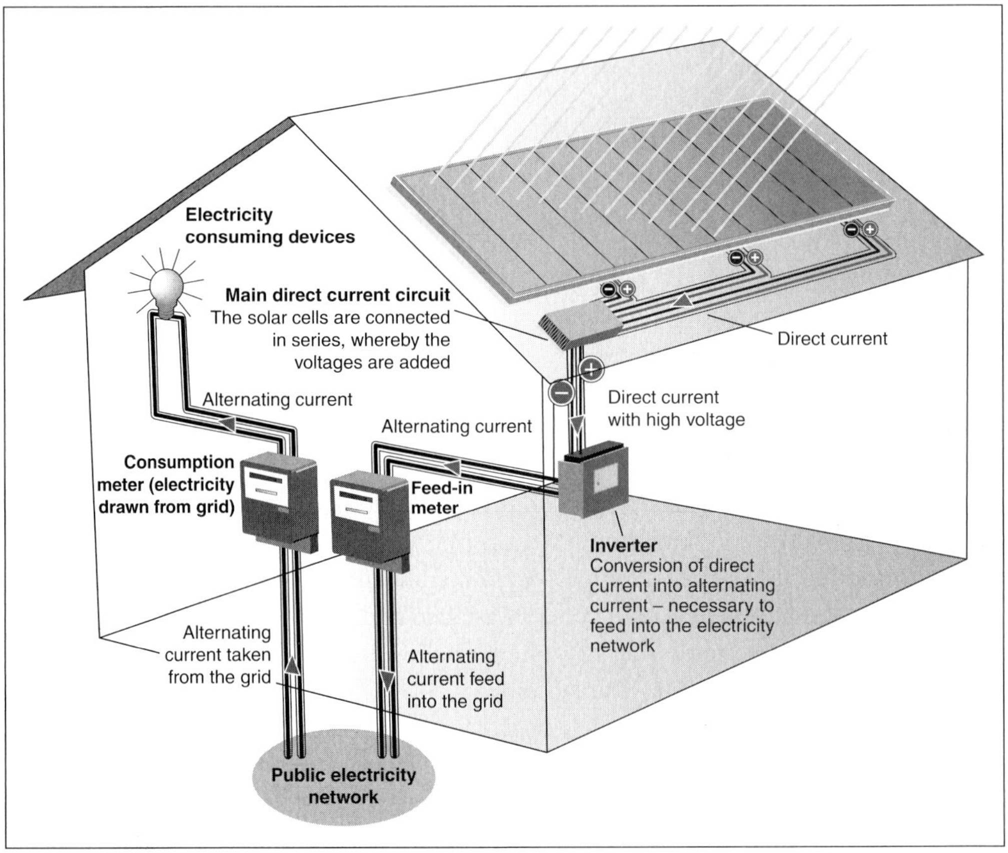

FIGURE 9-10

This solar generating system feeds power into the grid through one meter while the home's electric current is brought into the home from the grid through another meter.

Power Conditioning Equipment

Most home electrical appliances and equipment operate on 120- or 240-volt, 60-cycle alternating current (AC) electricity. AC current oscillates (varies) between −170 volts and +170 volts at the rate of 60 complete cycles per second. If you could see the current flow, it would look like a sine wave flowing up to plus 170 volts and down to negative 170 volts 60 times each second.

Because the value of AC voltage is continually changing from zero up to the positive peak, through zero down to the negative peak, and back to zero again, the value of the voltage is less than the peak voltage. Peak voltage (170 volts) is therefore not a good measure of actual voltage value. To describe the effective value of a varying voltage or current, we use the root mean square voltage (VRMS), which is 0.707 (about 71%) of the peak voltage (Vpeak). The Root Mean Square (RMS) value of AC current is the equivalent of steady direct current (DC) at the same constant voltage.

Photovoltaic (PV) solar energy technologies produce DC electricity. To run standard AC appliances, the DC electricity must be converted to AC electricity. This is accomplished by feeding the DC solar power through an inverter that converts the constant DC power into oscillating AC power at a frequency of 60 cycles per second and a constant voltage of 120 volts. A good-quality inverter will produce AC current with a smooth sine wave that is free of bumps and distortion that would make its output unsuitable for powering most electronic equipment or for feeding into the larger power grid.

Your home power conditioning equipment must match the voltage, phase, frequency, and sine wave profile of the electricity produced by your solar generating system to the power flowing through the local grid. High-quality current with consistent voltage and a smooth sine wave is required for sensitive electronic equipment, such as computers or microwave ovens, which cannot tolerate much power distortion.

Inverters must also operate safely, which in practice means that they must meet the requirements of a grid-interactive inverter listed by an organization such as Underwriters Laboratories. Underwriters Laboratories is a safety-testing and certification organization that has developed safety requirements for inverters in cooperation with the National Fire Protection Association (NFPA), which publishes and updates the National Electric Code. These requirements, referred to as UL 1741, apply to power-producing stand-alone and grid-connected renewable energy systems. The inverter for your solar generating system, as well as all of the other balance-of-system components needed to connect to the grid, should comply with the UL standard.

An inverter that delivers high-quality AC power will probably cost more than one with a lower power standard. Other specifications for the inverter can also affect its cost, but should be matched to the requirements of your solar generating system. These include the voltage of the incoming DC current from the PV panels, the AC output voltage and wattage required by the loads, the peak power required for the starting surge of some equipment, and monitoring features such as meters and indicator lights.

When deciding on the size of the inverter, you should plan for any future expansion of your solar generating system. A grid-connected system cannot be enlarged by just replacing an inverter with a larger unit. Wiring, disconnects, fuses, and other aspects of the system will also need to be replaced, which can greatly increase the expense. It is usually more cost effective to purchase an inverter and associated equipment with the largest input and output rating that you will need in the future and can afford now rather than to replace it with a larger one later. Figure 9-11 shows a completed solar energy system for generating electricity and heating water installed on the roof of a home.

Safety Equipment

Safety equipment for a solar energy system is similar to what any outdoor electrical installation should have and is designed to protect the system from being damaged or harming anyone who comes in contact with it.

Automatic safety disconnects protect the wiring and components of a renewable energy system from power surges caused by lightning, short circuits, static electric discharges, and equipment malfunctions. Manual disconnects enable the system to be shut down safely for maintenance and repair. Safety disconnects ensure that a grid-connected system can be isolated from the grid when necessary to prevent it from feeding power into the grid when people are working on the grid transmission and distribution systems.

FIGURE 9-11

The solar energy system on the roof of this home includes an array of PV panels on the right side of the roof and a solar water-heating panel on the left side.

Grounding equipment provides a well-defined, low-resistance path from the solar generating system to the ground to protect it against current surges from lightning strikes or equipment malfunctions. Each PV module and each component of the balance-of-system equipment should be separately grounded as well as any exposed metal such as mounting frames and equipment boxes that might be touched by anyone.

Surge protectors installed in the solar circuit can also give some additional protection to the system in the event that it or nearby power lines are struck by lightning. Nothing can prevent damage from a direct hit by lightning, but a surge protector can minimize the effect of high-voltage current induced in the wiring of the system by a nearby strike.

Meters and Monitoring Equipment

Meters and other instruments allow you to monitor a small solar energy system's output voltage and wattage, the amount of power being consumed in the home, and the amount that is being fed into the grid or used to charge storage batteries. These instruments are not required for the continuous monitoring of the system, but are useful to have for checking its performance periodically.

For a grid-connected system, the only required meter is a smart meter that will keep track of the electricity the home system produces and the electricity used from the grid. Many power providers use a single meter called a net meter to record the excess electricity a home energy system feeds back into the grid and the amount the home draws from the grid. The meter spins forward when drawing electricity from the grid, and backward when the home system is producing excess electricity and feeding it into the grid. This type of meter enables the utility company to charge based on the net power usage. Power providers that do not allow net meters usually require the homeowner to install a separate meter to measure the electricity the home energy system feeds into the grid.

Connecting to the Grid

Whether your local power company permits the use of net meters or requires a separate meter for the home energy system, it will have other definite requirements that any home energy system must meet before being allowed to connect to the grid. These requirements vary, but all power providers must address common issues in connecting small energy systems to their grid, such as regulations regarding safety, power quality, metering standards, power rates, contracts for purchase of power, equipment maintenance agreements, and liability insurance. Before installing a solar energy system, always contact the local power provider to find out about its requirements.

Several organizations are developing national guidelines for home energy system equipment manufacture, operation, and installation. Your power provider will have specific information about the standards it has adopted and how to implement them. The three most widely recognized organizations that have developed home energy system guidelines are as follows:

- The Institute of Electrical and Electronics Engineers (IEEE) has written a standard designated 1547–2003 that provides technical requirements and tests for grid-connected operation. It covers all grid-connected distributed generation, including renewable energy systems.
- Underwriters Laboratories (UL) has developed the UL 1741 standard to certify inverters, converters, charge controllers, and output controllers for power-producing stand-alone and grid-connected renewable energy systems. UL 1741 verifies that inverters comply with IEEE 1547 for grid-connected applications.
- The National Electrical Code (NEC), produced by the National Fire Protection Association, deals with electrical equipment and wiring safety, including installation of home energy systems.

Most states and power providers have adopted the IEEE, UL, and NEC standards and codes, although frequently with local variations and additions. For this reason, it is best to get your installation information from the local power provider rather than relying on the national published documents.

To connect your home energy system to the grid, you will probably need to sign an interconnection agreement with your power provider. In addition to specifying the technical requirements that the home energy system must meet, the agreement will probably require you to carry liability insurance, and pay some connection fees and charges.

Homeowner-utility agreements usually take the form of a feed-in tariff (FIT). They are also known as a standard offer contract or an advanced renewable tariff. A FIT is an agreement designed to promote investment in renewable energy technologies by offering a long-term contract to purchase the power produced by a renewable energy producer. In an FIT, the price paid by the utility for the renewable power produced is based on the cost of generation of each different renewable technology. Wind power, for instance, would be offered a lower per-kWh price, whereas solar PV power would currently be offered a higher price, reflecting its higher production costs. To view a typical FIT program offered by a public utility, visit the Ontario Power Authority website at http://www.microfit.powerauthority.on.ca.

Liability insurance protects you from having to pay the power provider or anyone else who might be injured in the event of accidents resulting from the operation of the home energy system. Most homeowner insurance policies provide at least $100,000 of liability coverage, but you should verify that the policy covers events connected to your home energy system. Some power providers may require a higher limit of coverage and some may also require the homeowner to indemnify them for any potential damage, loss, or injury caused by a home energy system. Rather than accept that additional risk, however small, you might want to consider making your home energy system a stand-alone unit.

Identify any additional costs that an agreement requires you to pay and decide whether a home energy system is still a good investment with these charges included. Permit fees, engineering/inspection fees, charges for a second meter (or a smart meter), maintenance charges, and insurance fees can add up to a heavy additional sum. The utility company will naturally try to pass as many of these costs to the homeowner as possible, but they are all negotiable and you should try to get the best deal for the front costs of your home energy system installation before you commit to install it.

The Public Utility Regulatory Policy Act of 1978 (PURPA) requires power providers to purchase excess power from grid-connected small renewable energy systems at a rate equal to what it costs the power provider to produce the power itself. To implement this requirement, FIT agreements have been developed to compensate producers for the cost of producing power.

Utility companies must either install a separate meter on your home energy system to measure the power it feeds into the grid, or install a single, two-way net meter that measures the difference in the power your home consumes and what its home energy system produces. The separate meter is not as cost beneficial for the homeowner as the net meter arrangement. This is because you pay the retail rate for the all the electricity you use, but the power provider purchases the excess generation from your home energy system at its wholesale (cost-of-generation) rate, which will probably be considerably lower. With net metering, the same meter measures both the power the home uses and the amount its energy system generates and subtracts the generated power from the total amount used. The meter spins forward as the home draws electricity, and it spins backward as excess electricity is fed into the grid. The net meter credits all the excess electricity your home system generates to your electricity account at the retail rate. If you use more electricity than your system produces, you pay retail price only for that extra electricity. If your system produces more than you have used, the power provider usually only pays for the extra electricity at wholesale cost, but you get credit at retail rate for all the power your system produces up to the full amount your home uses.

Some power providers will carry over the balance of any extra electricity your system generates from month to month, which can be a large advantage for a solar energy system that has a seasonal high in its generating potential. If, at the end of the year, your system has produced more power than you have used, you forfeit the excess generation to the power provider.

Legal Considerations

Each state and community has its own set of codes and regulations that you will need to follow in order to add a solar energy system to your home. These regulations may determine the size and type of system you can install, who can install it, and whether it must be a grid-connected or a stand-alone system.

Electrical systems must always be installed according to local building code requirements and be inspected to ensure that they comply with standards. Your solar energy system will likely be required to pass one or more electrical inspections at the beginning and/or end of the installation process. To pass inspection, a system should be composed of certified equipment that complies with local requirements and appropriate technical standards, and should be installed according to NEC requirements as adapted for the local area. Article 690 of the NEC provides guidance on equipment and wiring safety for small renewable energy system installations, and specifically discusses photovoltaic systems. For more information on Article 690, see Sandia National Laboratory's 1996 guide to installing NEC-compliant photovoltaic systems, which can be downloaded from the Sandia Internet site.

You may also have to obtain a conditional-use permit or a variance from the existing building code before the local zoning authority will issue a building permit for your solar installation. Check what is required with the local building code office before you buy a renewable energy system.

Before installing a solar energy system, which must have sunlight to operate, you should check to be sure no structure can be built on adjacent property that will block sunlight from your system. If this is a possibility, you might seek an easement from the property owner specifying that no structure that blocks the sunlight from your system will be built. Such an agreement is binding regardless of changing land ownership.

You might also want to check your own property deed to determine if any language in it could prevent you from adding a renewable energy system to your own property. Some communities have covenants or other regulations specifying permitted or forbidden uses of property. Most regulations support renewable energy systems but some might prohibit them for aesthetic or noise-control reasons. In addition, it is good public relations to discuss your intentions with your neighbors to avoid any future objections.

Installing a Stand-Alone Solar Power System

For some homes, generating power using a small renewable energy system that is not connected to the electricity grid makes economic sense and may be a method of dramatically reducing the home's environmental impact. A stand-alone electric system does not require some of the equipment that a grid connected system uses, but it does require a means of storing power for use during times when the system is not able to generate electricity. Stand-alone systems are most often chosen for homes located in remote areas where a connection to the grid is either not available or so costly that a home power-generating system is much more economical.

Stand-alone systems are most successful if they employ a combination of power-generating technologies and also adopt as many as possible of the other techniques covered in this book to reduce their electric energy requirements. These strategies include installing hybrid generating systems (solar and wind turbine), having a fuel-fired emergency backup system, and shifting some electric energy needs, such as water heating and home heating, to other sources of energy.

Setting up a stand-alone solar generating system is not as complex as creating a grid-connected system because there is no need to deal with a public utility or to build the system to meet its technical requirements. A stand-alone system should still comply with local government requirements and should also follow the NEC requirements for safety and quality installation, but these regulations are usually easier to comply with in a system that is entirely contained on a single parcel of property.

In addition to its photovoltaic panels, a stand-alone system also requires some additional balance-of-system equipment to safely transmit the electricity it generates to the load that will use it. In the simplest stand-alone configuration, the current generated by the PV panels can be connected directly to a load such as a pump or an electric appliance. Owners of most whole home systems, however, will want some power to be available for use around the clock, including the nighttime hours when the PV panels are not producing electricity. To have power always available, the stand-alone system must have batteries to store it and a charge controller to direct current flow.

If the system will be used to power electronic equipment or other devices that require clean and consistent electricity, it will need the same kind of inverter and power-conditioning equipment that a grid-connected system employs to produce high-quality electricity. The same type of monitoring equipment will also be required to control and maintain the system while it is operating. No meter is necessary because the electricity generated is all being consumed within the home, but having a meter to measure the solar system's total output over time provides useful data for determining the value of the system.

The wiring of a stand-alone system is different than that for a grid-connected system because the PV panels are wired directly to the battery bank through a charge controller. The controller regulates the amount of current going to the batteries to prevent overcharging. Current from the battery bank goes to the inverter, where it is transformed to AC current at the correct voltage and then is fed into the home's AC wiring. Figure 9-12 shows a schematic wiring diagram for a stand-alone solar generating system.

Batteries for stand-alone systems store electricity for use during times that the solar system is not producing electricity at night or not producing enough during cloudy or stormy weather to meet home needs. Storage batteries should be the "deep-cycle" type that last 5 to 10 years and will feed back about 80% of the energy channeled into

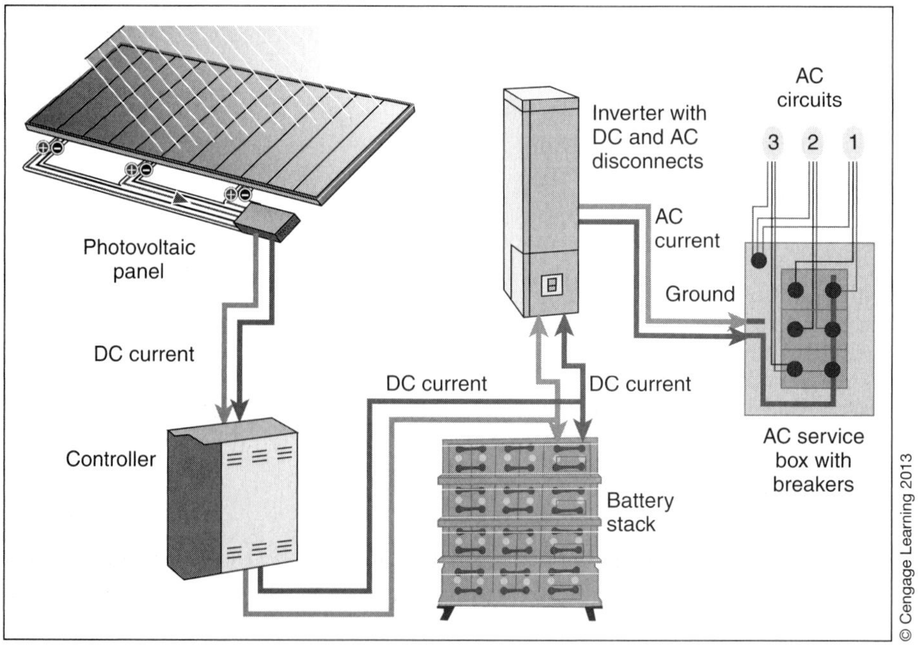

FIGURE 9-12

A schematic diagram shows a stand-alone solar generating system with PV panels connected through a controller to a battery stack and an inverter.

them. These batteries are usually of lead-acid composition and are designed to provide electricity continuously over long time periods. They are not adversely affected by repeatedly charging and discharging up to 80% of their capacity, in contrast to automotive batteries, which are shallow-cycle and can be damaged if they repeatedly discharge more than 20% of their capacity. Figure 9-13 shows a battery stack for a solar generating system located in a storage room.

FIGURE 9-13

A battery stack of 24 batteries on a metal rack with an attached plastic shield over the connection posts. Battery arrays like this can also be arranged in a horizontal array if space permits.

A stand-alone system's battery stack needs to be of sufficient size to store power for the home's needs during nighttime hours and periods of cloudy weather. The hours of darkness are constant, and power use can be calculated closely for these periods, but the amount of cloudy weather varies widely depending on location. The number of batteries installed in a home system should be determined by the home's power requirements during nongenerating periods, but the cost of additional batteries for storage beyond average needs should not exceed the cost of adding a fuel-fired backup generator to cover unanticipated reduced generating capacity or lengthy downtimes in the solar generating system.

Batteries contain dangerous chemicals and emit both hydrogen and oxygen gas when being charged. For safety, they must be located in a ventilated space and isolated from living areas. No device with an open flame or one that can produce an electric spark should be in the same room with the batteries. They should also be protected from freezing temperatures or high heat, both of which will reduce their capacity and shorten their life spans. The battery stack should have easy access for maintenance, repair, and replacement, and you should plan to recycle worn out batteries.

SUMMARY

Solar radiation is the general term used to describe the electromagnetic radiation emitted by the sun. Solar power is radiant energy from the sun transformed into an energy source that can do work on the earth. The energy that strikes the earth's surface in 1 hour is more than sufficient, if it could all be converted into useable forms, to supply all of the planet's current energy needs for a full year. One way of using the sun's energy is called passive use, allowing natural sunlight to warm the home and its contents or, conversely, blocking sunlight to minimize its heating effect when it is not wanted. A second method of using sun energy is by direct heat transfer, using mechanical devices to concentrate the sun's energy and transfer its heat into a substance from which it can later be extracted and used. A third way of using solar energy is by transforming it directly into electricity by means of the photovoltaic effect. Solar water heating is accomplished using a solar collector through which water or another liquid, called the working fluid, flows. The heat transfer fluid (HTF) that flows through the collector may be water that flows directly from a storage tank, through the collector, and back to the tank. More often the collector uses a separate loop of piping containing antifreeze and delivers heat to the tank through a heat exchanger similar to that of an air conditioner. Solar heating systems can be active (pumped) or passive. An active system uses electricity to power the pump; a passive system moves the HTF by convection or water pressure in the line. Solar electric systems are known as photovoltaic (PV) systems because they use arrays of solar cells to convert sunlight into electricity. Solar cells are the basic units of a PV system and are made up of semiconductor materials that generate electricity through the photoelectric effect. Silicon is the primary semiconductor material used in solar cells. PV modules, made of many solar cells, range in output from 10 to 300 watts. Several modules can be installed in an array and wired in parallel to produce more current at a given voltage. PV modules produce DC power that must be changed to AC house current by an inverter and conditioned by additional equipment for use in electronic appliances. A grid-connected solar power system allows the solar panels to generate some or all of the home's electricity and also permits electricity to be fed into the grid when not needed by the home. Electric utility companies are required by law to pay homeowners for power that they generate and feed back into the grid. Solar power generating systems can also be stand-alone systems that feed their electric power directly into the home's AC wiring and are not connected to the larger electric grid.

KEY TERMS

Absorber plate The black-painted metal sheet placed behind the tubing at the bottom of a collector panel to absorb heat for transfer to the tubing and fluid inside it.

Absorption coefficient The measure of how readily a material absorbs light. A small coefficient means poor light absorption; a large coefficient indicates the material easily absorbs light.

Active system A type of solar water-heating system that uses a pump to circulate the heat transfer fluid (HTF).

Article 690 of the NEC A section of the National Electric Code that provides guidance on equipment and wiring safety for small renewable energy system installations, including photovoltaic systems.

Auxiliary energy source In a solar heating system, a second source of heat energy to provide additional heat when solar energy alone is insufficient.

Balance-of-system equipment Additional equipment beyond the basic solar generation components or wind turbine components that is required to make a renewable energy DC generating system compatible with the home's AC electric circuits and the electric grid.

Bandgap (symbolized by Eg) The energy difference between the conduction band and valence band in molecules of a semiconductor material.

Battery stack A connected bank of batteries used to store electricity generated by a renewable energy system for later use.

Charge controller An electronic device that regulates the amount of current going to the batteries in a battery stack to prevent overcharging.

Conduction band The higher energy level of atoms in the molecules of a semiconductor material created by sunlight energy being absorbed so that the electrons are free to move.

Convection The movement of heat through a liquid or gas by internal circulation within the material.

Disconnect A safety switch used to cut off power from a generating circuit or storage battery.

Electric grid The system of electric generating and distribution lines that provides electricity to all areas of the country.

Evacuated heat pipe tube (EHPT) A type of solar collector composed of double glass tubes, one within the other, with the space between the tubes evacuated to reduce heat transfer; the inner tube contains an absorber plate fused to a heat pipe.

Fall (autumnal) equinox The day in September of each year on which the amount of daylight and darkness are equal.

Feed-in tariff (FIT) Also known as a standard offer contract or an advanced renewable tariff; an agreement designed to promote investment in renewable energy technologies by offering a long-term contract to purchase the power produced at a rate based on the cost of generation.

Geothermal energy Energy produced in the heated interior core of the earth that can be used to provide heat and generate electricity.

Glazed panel A heat collector panel that has a glass faceplate over the interior absorber plate and tubing.

Grid-connected system A PV system that is connected to the local electric grid.

Heat transfer fluid (HTF) The fluid that flows through the absorber tubing in a solar collector. It may be hot water from the storage tank or another fluid that goes to a heat exchanger.

IEEE 1547–2003 An IEEE standard that provides technical requirements and tests for grid-connected equipment and operation.

Infrared radiation Invisible solar heat energy with a wavelength just longer than that of red visible light; most solar heat energy is infrared.

Institute of Electrical and Electronics Engineers (IEEE) A professional electrical industry group that develops standards for electrical equipment, safety, and operations.

Inverter An electronic device that converts direct current (DC) into alternating current (AC) and adjusts current voltage for use by AC appliances and the electric grid.

Latitude The distance of a location north or south from the equator, measured in degrees, minutes, and seconds.

National Electric Code (NEC) A detailed set of specifications for electric wiring and construction developed by the NFPA and used nationwide with some local variations.

National Fire Protection Association (NFPA) The professional organization that publishes and updates the National Electric Code and other fire protection codes.

Net metering An arrangement where electricity generated by a grid-connected solar energy system is measured by the home's electric meter as it is fed into the grid.

Passive solar energy Solar energy striking the earth's surface that can be enhanced or blocked to increase its usefulness for heating or cooling a building.

Passive systems A solar water-heating system that relies on municipal water pressure or convection to circulate the HTF.

Peak AC voltage (170 volts) The highest level of electric voltage in an AC current since wave; peak voltage for AC house current is + or − 170 volts.

Photoelectric effect The generation of an electric current in a semiconductor material when it absorbs sunlight; the solar energy knocks electrons loose from the atoms of the material which then flow out of the material as electricity.

Photovoltaic (PV) module or panel A large group of solar cells mounted in a frame and interconnected in series to produce a larger current at higher voltage than a single cell.

Power conditioning equipment Electronic devices required to regulate and monitor electric current to maintain its voltage, amperage, and waveform to precise standards.

Public Utility Regulatory Policy Act of 1978 (PURPA) A federal law that requires power providers to purchase excess power from grid-connected small renewable energy systems.

Root Mean Square (RMS) The value of AC current that is the equivalent of steady direct current (DC) at the same constant voltage.

Root mean square voltage (VRMS) The AC voltage equivalent of a given DC voltage; it is equal to 0.707 (about 71%) of the peak AC voltage (Vpeak).

Semiconductor material A material that conducts electricity in only one direction and can be made to generate electricity or light by means of energy being input to it.

Silicon The most popular semiconductor material for solar cells; available in at least four different types.

Solar array A grouping of solar panels connected in parallel to generate an increased current flow over that of a single panel.

Solar cell A small electronic device made of a semiconductor material that generates an electric current when struck by sunlight.

Solar collector A panel made up of an absorber plate and tightly coiled tubing that collects solar heat and transfers it to an HTF in the tubing.

Solar power Energy from the sun transformed into electricity using photovoltaic technology or concentrated into a heat energy source to generate electricity and/or provide other heating functions.

Solar radiation The general term used to describe the electromagnetic radiation emitted by the sun.

Split solar water-heating system A system that heats HTF in a closed loop and sends it to a heat exchanger for transfer to the water in a storage tank.

Spring (vernal) equinox The day in March of each year when daylight and darkness are of equal length.

Stand-alone system A solar generating system that is not connected to the grid, but provides power independently to a home electric panel.

Thin-film solar cells Solar cells made of layers of semiconductor materials only a few micrometers (millionths of a meter) thick that can be integrated into building materials to make them solar panels.

UL 1741 Standard developed by Underwriters Laboratories for certifying the utility and safety of power-producing stand-alone and grid-connected renewable energy systems.

Ultraviolet radiation The invisible wavelengths of solar radiation just shorter than the violet light in the visible spectrum.

Underwriters Laboratories A safety-testing and certification organization that has developed safety requirements for inverters and other renewable energy equipment.

Valence band The lower energy level of atoms in the molecules of a semiconductor material.

Water vapor (humidity) Water in a gaseous state always present in the atmosphere that creates humidity in the air; as air cools it can hold less water vapor, and this may produce liquid water condensation.

Working fluid Water or an antifreeze fluid that is circulated through a solar collector to pick up heat for transfer directly to the water or to a heat exchanger for heating water.

REVIEW QUESTIONS

1. True or False: *Solar radiation* is the general term for the electromagnetic waves of energy emitted by the sun.
2. What factors affect the amount of radiation that reaches any one place on the earth's surface?
3. What are three ways of using the sun's energy to provide heat or produce electricity?
4. Awnings and shades placed on windows to block or admit sunlight for cooling or warming is what kind of use of solar energy?
5. A large flat device containing an absorber that is used to transfer solar heat into a working fluid is called a _____________.
6. What does the term *HTF* mean, and where is it used?
7. What is the difference between an active and a passive solar water-heating system?
8. In an active solar heating system, is it more energy efficient to install the collector on a rooftop or at ground level?
9. __________ __________ are the basic units of a PV system; many of them are combined in series to make a ________ __________; several of these can be combined into a solar __________.
10. When a semiconductor material absorbs sunlight, and the solar energy knocks electrons loose from its atoms so that they flow into a circuit and form an electric current, the phenomenon is called the ________________ __________.
11. True or False: A small absorption coefficient means that light is not readily absorbed by a material; a large coefficient indicates that the material easily absorbs light of a specific wavelength.
12. How do thin-film solar cells differ from regular solar cells?
13. True or False: A stand-alone solar generating system can use a net meter to deduct the amount of electricity generated from the total amount used by a home, but a grid-connected solar system cannot.
14. What are three additional components that a grid-connected solar generating system requires in addition to PV modules?
15. What is the function of an inverter in a solar generating system?
16. If a net meter is not permitted to be connected to a solar electric generation system, how can the homeowner be compensated for the electricity a grid-connected system feeds into the grid?
17. What is the IEEE 1547–2003 standard?
18. What is the PURPA law?
19. True or False: A stand-alone solar generating system can be connected directly to an incandescent lamp, a toaster, or another appliance that can function on DC power.
20. What is VRMS?

CHAPTER

10 WIND TURBINE POWER FOR YOUR HOME

OBJECTIVES

After studying this chapter, you should be able to:

- Describe how horizontal- and vertical-axis wind turbines work and the differences between them.
- Identify the characteristics of a suitable location for installing a wind turbine.
- Describe how to install a wind turbine in a residential setting.
- Identify the balance-of-system equipment necessary for installing a stand-alone wind turbine and a wind turbine connected to the electric grid.
- Describe the safety equipment needed to ensure that a wind turbine can be safely operated in all weather conditions.
- Describe the components and connections necessary for installing a hybrid wind turbine and photovoltaic generating system.

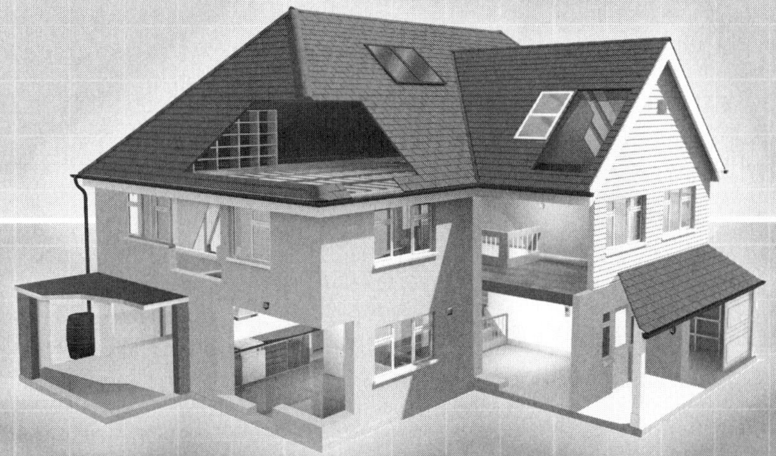

OUTLINE

I. Introduction
II. Wind Power from Solar Radiation
  A. How Wind Turbines Work
  B. How Much Wind Is Enough?
  C. How Large a Wind Turbine?
  D. Vertical-Axis Wind Turbines
III. Wind Turbine Towers
IV. Balance-of-System Equipment
  A. Safety Equipment
  B. Meters and Monitoring Equipment
  C. Connecting to the Grid
  D. Legal Considerations
V. Hybrid Solar Panel and Wind Turbine Generating Systems
VI. Summary

INTRODUCTION

In this chapter you will learn how another form of energy from the sun can be used to generate electricity for use in the home. You will understand how solar heating of the earth's atmosphere develops weather systems that produce winds with enough energy to drive wind turbines and generate electricity. You will be able to describe the different types of wind turbines, the styles of towers used for turbines, and the other components that make up the balance-of-system equipment in a complete wind turbine generating system. You will learn how to install a wind turbine generating system that is connected to the electric grid or a system that provides electric power independent of the grid. You will also learn how wind turbines can be combined with solar photovoltaic (PV) arrays to form hybrid generating systems that can provide more electric power more consistently than either system can provide alone.

WIND POWER FROM SOLAR RADIATION

IN THE PREVIOUS chapter you learned how the sun's radiation energy can be utilized to heat water and generate electricity. Wind is another manifestation of solar radiation, which heats the earth's surface unevenly and causes air currents to form in the atmosphere above the surface. These air currents are the winds that power weather systems continuously across the entire surface of the planet.

How Wind Turbines Work

Wind turbines convert the kinetic energy in wind into mechanical power that runs a generator to produce clean electricity. Wind turbines, which are still often mistakenly called windmills, are vastly more advanced technologically than their ancient, water-pumping antecedents. Their rotors have three or even two blades in place of the windmill rotor's traditional four, and those blades are aerodynamically designed to capture the maximum energy from the wind. Wind pressure against the blades turns the turbine's rotor and the shaft on which it is mounted. The shaft is connected through an expansion gearbox to a generator that spins more rapidly to produce electricity. Figure 10-1 shows the internal structure of a large wind turbine.

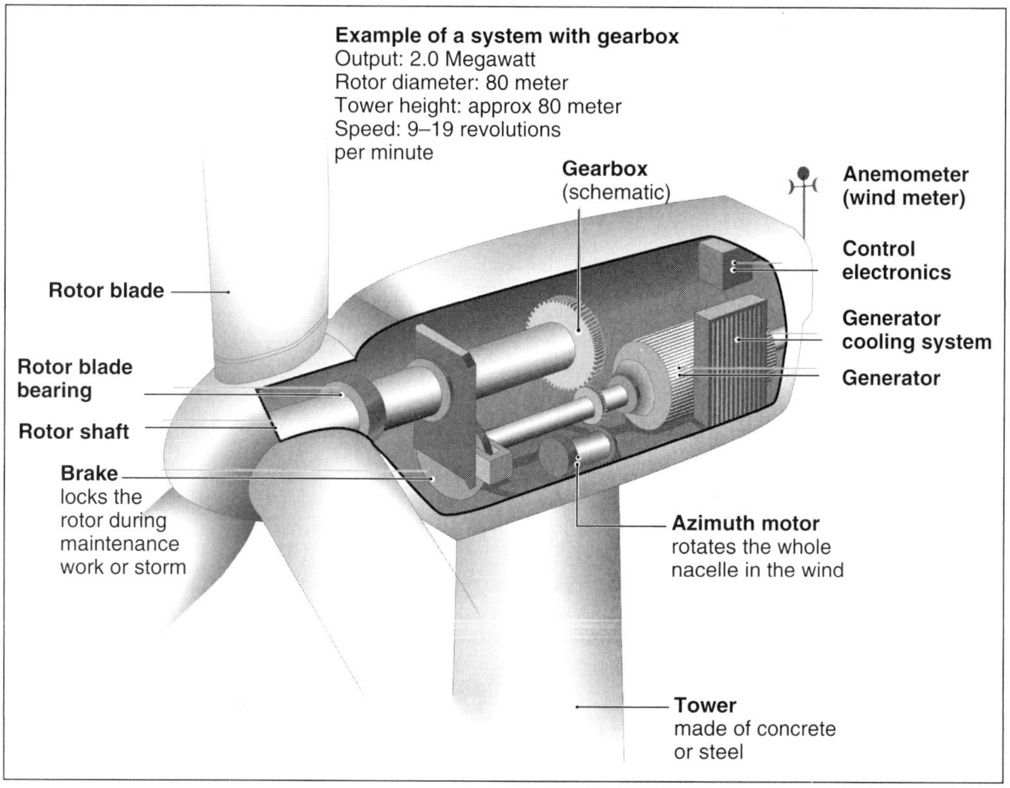

FIGURE 10-1

A cutaway drawing shows the working parts of a 2-megawatt wind turbine.

Wind turbines increase in efficiency as they increase in size. The unit pictured in Figure 10-1 is capable of generating 2 megawatts of electricity. It stands on a tower that is 80 meters high (267 feet) and has a three-bladed rotor that measures 80 meters (267 feet) in diameter. It reaches maximum generating capacity while rotating at a rate of 19 revolutions per minute. Still larger turbines are now being constructed that can generate up to 5 megawatts of electricity. These turbines are mounted on towers that stand 124 meters high (413 feet) and have three-bladed rotors measuring 114 meters (380 feet) in diameter. Though it has more than double the generating capability of the 2-megawatt turbine, the 5-megawatt unit rotates only 8 to 13 times per minute to reach its full generating capacity.

Residential-scale wind turbines are far smaller and slightly less efficient than the commercial behemoths, but they are otherwise similar in most characteristics.

FIGURE 10-2

This schematic shows the patented "in the hub" design of Hummer Wind Power's 2-kilowatt wind turbine.

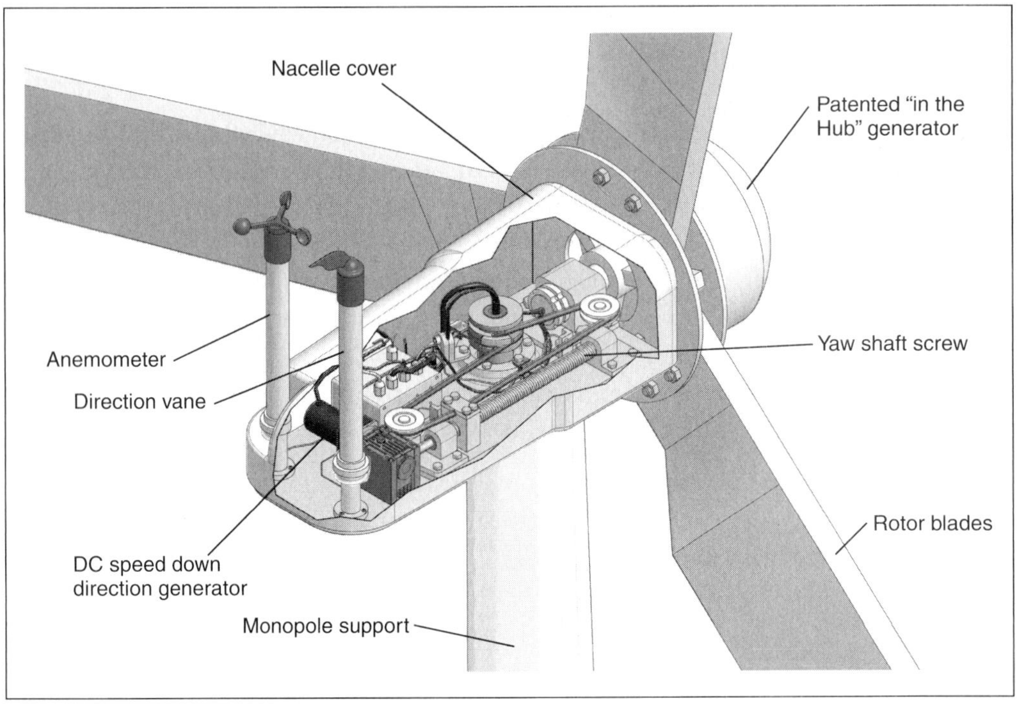

Figure 10-2 shows a 2-kilowatt wind turbine manufactured by Hummer Wind Power. This turbine is mounted on an 18-meter (60-foot) tower and has a rotor diameter of 3.8 meters (12.5 feet). Its generator is mounted in the hub of the rotor, which gives the unit a startup wind speed of less than 7 miles per hour.

How Much Wind Is Enough?

Wind has tremendous energy that can be used to produce electricity, but wind force is not distributed evenly on the earth's surface. Almost all areas experience at least some wind a few days each year, but wind force sufficient to power a wind turbine requires more than an occasional gentle breeze.

To decide whether a wind turbine generator is practical for your home or other property, you must first determine if the location has enough wind to justify the expense of installing the system and operating and maintaining the equipment. Generally, any location with an annual average wind speed of 12 miles per hour or greater has sufficient wind energy to drive a turbine for enough hours each year to make the system cost effective. A location with an average wind speed of 10 miles per hour or less does not have enough wind energy for a cost-effective wind turbine installation. A location that has an average wind speed between 10 and 12 miles per hour might be a practical site for a wind turbine, but will need to have a wind resource assessment performed to determine if it is actually suitable. Such an assessment adds significantly to the cost of a small wind turbine system and will result in a negative finding for least half of the sites studied. Unless you're really determined to have a wind turbine generator at a marginal site, it is best to regard the 12-mile-per-hour average wind speed as the minimum for installing a small system.

Figure 10-3 shows the wind power density across the United States. Darker shades on the map indicate stronger average wind and a higher kinetic energy level in the

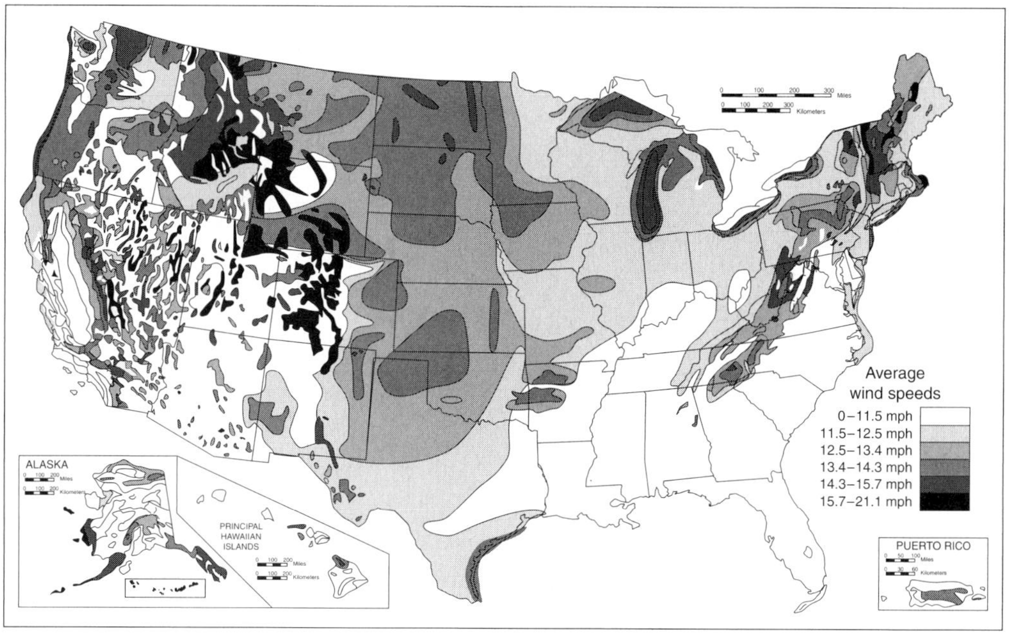

FIGURE 10-3

Average annual wind density and power in the United States.

wind, but this map alone is not a good guide for locating wind turbine generating systems. Many of the darkest areas on the map are mountainous and unsuitable for a wind turbine installation. Other areas have frequent wind gusts of such high velocity that a wind turbine would have to shut down to avoid damage to its rotor or generating mechanism.

Locations that are suitable for a wind turbine generating system are those that have a consistent average wind force strong enough to turn a wind turbine a majority of the time, but seldom, if ever, strong enough to force the turbine to shut down.

The wind resource of an area is a measure of the potential for generating electricity from wind power in that area expressed in kWh per year. It is calculated as the estimated annual electricity production per square meter of rotor-swept area for a small wind turbine. Areas that are favorable for wind turbine installation are those that have a wind resource potential of 510 kWh or more. Table 10-1 shows the wind resource classes depicted on the map in Figure 10-3 and the average wind speeds associated with each class.

TABLE 10-1 Small Wind Turbine Productivity Estimates

WIND POWER CLASS	PRODUCTIVITY PER M^2 OF SWEPT AREA IN KWH/YEAR	WIND SPEED AT 33-FT (10-M) HEIGHT—MPH	WIND SPEED AT 33-FT (10-M) HEIGHT—M/S
2	0–500	0–11.5	0–5.1
3	500–610	11.5–12.5	5.1–5.6
4	610–690	12.5–13.4	5.6–6.0
5	690–770	13.4–14.3	6.0–6.4
6	770–880	14.3–15.7	6.4–7.0
7	880–1170	15.7–21.1	7.0–9.4

To determine the wind resource in your area, consult the detailed wind resource maps at the National Renewable Energy Laboratory's (NREL) website, http://www.windpoweringamerica.gov/wind_maps.asp. Download the map for your state and check it to find the average wind speed in the area where your property is located. Average wind speeds on these maps are given in meters per second (m/s) rather than in miles per hour (mph). A wind speed of 5 m/s is equal to about 11.08 mph, and wind speed of 6 m/s is equal to about 13.30 mph. If the map shows an average wind speed for your area of 5 or more m/s, your property probably has sufficient wind to adequately power a small wind turbine.

If your home is near an airport, you can obtain further confirmation about wind in the local area by contacting the meteorological department there and asking them for wind data. For more detailed information about whether you have a good wind resource, check the consumer guide for your state published by the U.S. Department of Energy and available at the department's website Energy Efficiency and Renewable Energy at http://www.windpoweringamerica.gov/small_wind.asp.

Once you confirm that the area where you live has enough average wind speed to power a wind turbine, you should consider some additional factors as you decide whether to install a system. Among these are:

- Whether the property on which your home is located is large enough to accommodate a wind turbine. At least 1 acre of land is necessary to prevent a wind turbine from dominating the adjacent landscape.
- Whether the property is in an open and mostly rural area; urban home sites are nearly always unsuitable both because of limited space and for other reasons noted in this chapter.
- Whether the local zoning codes and property covenants allow wind turbines on residential property.
- Whether the home uses enough electricity so that the cost of installing a wind turbine will eventually be repaid in savings on electrical charges. If the home's average monthly electrical costs are not at least $150 or more, a wind turbine is not cost effective.
- Whether you are comfortable making an investment that will not return its cost of installation for 10 or more years and may never show any profit above its cost of operation.

How Large a Wind Turbine?

Small wind turbines are manufactured by many firms in the United States and other countries. The most common type of turbine is one with a horizontal axis and a two-, three-, or four-bladed rotor. The size of the wind turbine you need depends on your application. Small turbines range in size from 20 watts to 100 kilowatts (kW) of output. The smaller or "micro" (20- to 500-watt) turbines are often used for single-purpose applications such as charging batteries for electrically powered lights and recreational vehicles.

One- to 10-kW turbines can be used for pumping water. Although mechanical windmills still provide a low-cost method for pumping water in low-wind areas, wind-powered pumping using an electric motor-powered pump is less expensive and more versatile in areas with sufficient wind resource. An electric pump can lift water from deeper underground and the wind turbine that powers it does not have to be located over or even near the well.

Wind turbines for residential applications range in size from 500 watts to 100 kW. The largest size would only be appropriate for very heavy users of electricity. The size you select should be determined by the amount of electricity you want to generate, your available budget, and the wind resource in your location. Figure 10-4 shows a 10-kW wind turbine installed at a residence. The relative size of the turbine, which is mounted on a 98-foot monopole behind the home, can be clearly seen.

FIGURE 10-4

A 10-kW Hummer wind turbine installed behind a private residence on a 98-foot monopole.

A typical home uses about 830 kWh of electricity per month, which adds up to an annual total of around 10,000 kWh. Depending on the average wind speed in the area, a wind turbine capable of generating 5 to 15 kW will be able to provide 75% to 100% of the home's power requirements. A smaller 1500-W wind turbine will provide about 35% of the average home's electrical demand if placed in a location with a Class 5 wind resource, as defined in Table 10-1.

A 2-kW wind turbine, such as the one manufactured by Hummer Wind Power, can provide 1640 kWh per year in a location with a Class 2 wind resource and about 9500 kWh per year in a location with a Class 6 wind resource. A 2-kW wind turbine is shown in Figure 10-5 mounted on a guyed pole in a suburban back yard.

A wind turbine of this size may be the largest installation permitted in some residential areas. It is usually mounted on a 60-foot tower but can be placed on a lower pole if local regulations limit the height of wind power systems. The 2-kW wind turbine in Figure 10-5 is a compact and lightweight machine, as its specifications noted in Table 10-2 show.

FIGURE 10-5

A 2-kW wind turbine installed in a small yard using a guyed 28-foot pole.

TABLE 10-2 Two-kW Wind Turbine Specifications

Rated Output	2 kW
Maximum Output	3.5 kW
Charge Voltage	DC 120 V
Output Voltage	AC 110 V, 120 V, 220 V, 230 V, 240 V
Number of Blades	3
Blade Material	Glass fiber reinforced plastic
Rotor Diameter	3.8 m (12.5 ft.)
Rotor-Swept Area, m^2	11.34
Startup Wind Speed	3 m/s (6.7 mph)
Rated Wind Speed	9 m/s (20 mph)
Generator Weight	25 kg (55 lbs.)

Vertical-Axis Wind Turbines

Another type of small wind turbine design is the vertical-axis wind turbine (VAWT). A VAWT has a vertical shaft with vanes or a round "squirrel cage" attached to capture the wind's force. These turbines are generally more complex and therefore more expensive to construct than the blade type, but continued design research is improving VAWTs so that their efficiency and cost more closely approaches those of the horizontal-axis designs.

The main advantage of VAWTs for residential applications is their smaller size and more compact shape. These features make them better suited for installation where space is limited. They are also usually mounted on lower towers than blade-type turbines. A 2-kW VAWT typically has a wind wheel diameter of about 8 feet and is mounted on a 25-foot tower. A 10-kW model has a wind wheel diameter of 14 feet and a tower height of 40 feet. Some examples of VAWTS are illustrated in Figure 10-6.

FIGURE 10-6

Four vertical-axis wind turbines ranging in output capacity from 500 W to 10 kW.

The basic VAWT design for wind turbines is older than the designs for horizontal-axis wind turbines even though the latter are now the most common type. Some vertical-axis turbines are very effective at producing energy, but they lose much of their effectiveness, just as horizontal-axis models do, if they are mounted too close to ground level. VAWTs need to be placed on towers tall enough to give them adequate ground clearance. All wind turbines are more efficient when placed on tall towers because wind speeds are higher 10 meters or more above the surface than at ground level. There is also less air turbulence higher above the ground, and smooth-flowing wind together with higher velocity produces greater energy yield in all wind turbines.

Some manufacturers of VAWTs claim they are more efficient and operate in lower wind speeds than horizontal-axis wind turbines. Some designs may be able to start generating electricity at lower wind speeds, but horizontal-axis wind turbines are actually better at capturing wind energy at all but the lowest speeds than VAWTs given the same amount of swept area (the area of the circle swept in the air by the blades or wind wheel). This is because VAWT machines always have at least one wind wheel blade traveling against the wind. This creates additional drag on the wind wheel as it spins and reduces its generating energy. Horizontal-axis turbines, on the other hand, always have all their rotor blades facing into the wind to garner maximum generating energy.

Neither horizontal- nor vertical-axis machines can produce much electrical energy in low wind speeds because there is almost no kinetic energy to capture from low-speed wind. Wind energy grows geometrically as its speed increases; a 4-mph wind (the lowest wind speed at which any wind turbine can operate) has only one-sixteenth the energy of a 16-mph wind. A 32-mph wind has four times the energy of a 16-mph wind and 64 times the energy of a 4-mph wind.

VAWTs are available in sizes up to 100 kW or more, but the large diameter of the wind wheel on models with more than 15-kW output capacity makes them impractical for most residential applications. A new commercial design for a VAWT is now under construction by a British firm and will be placed in service in 2014. Its 262-meter (875-feet) diameter wheel will sweep an area of 17,226 square meters and give it a generating capacity of 10 megawatts, the highest output of any wind turbine, vertical or horizontal axis, developed to date (Figure 10-7).

FIGURE 10-7

This new design is being scaled up to a 262-meter-diameter, 10-megawatt-output VAWT to be deployed in shallow-water offshore locations.

VAWTs should not be mounted on rooftops, even though this is an easier and less expensive installation option than mounting on a tower. The building and other obstructions at ground level create turbulence in the wind and decrease its speed, reducing the wind turbine's ability to produce electricity. Small wind turbines, both horizontal- and vertical-axis types, should be mounted so that the lower tip of the rotor blades or wind wheel is an absolute minimum of 25 feet above any obstruction on the ground below. An increased mounting height, up to a maximum of about 100 feet, will improve the performance of the wind turbine. The ground around a wind turbine tower should be free of obstructions for as wide an area as possible, but a minimum of 100 feet in all directions.

Wind turbines have now been in use long enough that many used, refurbished, or remanufactured machines are available for purchase. Even though these machines may be less expensive than comparable new turbines, they are rarely a good economic investment. Their lower cost is often quickly offset by costly repairs for a machine that is nearing the end of its useful life. Like restoring an old car to like-new condition, bringing a used wind turbine up to the standard of a new one will probably cost more than the price of a new unit.

Manufacturers of small wind turbines can provide you with complete specifications for various types and models, including the annual electricity output of the turbine as a function of annual average wind speed, rotor or wind wheel diameter, rotation speed, tower height, and the maximum wind speed at which the turbine is designed to operate safely. Most turbines have automatic governing devices to keep the rotor or wind wheel from spinning out of control in very high winds. This information, along with your local wind speed and your energy budget, will help you decide which size turbine will best meet your electricity needs.

WIND TURBINE TOWERS

There are three basic types of tower options for home-sized wind turbines: monopole, guyed monopole, or lattice and free-standing lattice.

A monopole is the most visually attractive mounting, but also the most expensive. It consists of a single pole without supporting guy wires set in a concrete base that is strong enough to support the wind turbine in the highest winds that occur in the area where it is located. The pole must also support the additional weight of one or two service persons who may be required to climb the pole to service the wind turbine. The need for great strength to remain rigid and resist wind pressure means that a monopole must be made of high-quality steel alloy and set in a concrete foundation deeply anchored in the earth. The expense of such an installation, shown in progress in Figure 10-8, prompts most wind turbine owners to opt for one of the other less costly tower options.

FIGURE 10-8

Workmen prepare to pour the reinforced concrete foundation for a 90-foot wind turbine monopole.

A guyed monopole is generally the least expensive type of tower. Guyed monopoles usually have four guy wires set at right angles to one another around the pole for support. The height of a guyed monopole is limited by the weight of the wind turbine it must support against the force of the wind, and 10-kW machines are the largest that can be placed on this type of tower. Guyed monopoles can be permanently fixed, rigid installations or they can be a tilt-up-style tower that can be lowered to the ground for maintenance on the wind turbine. Rigid guyed monopoles must be climbed to service the wind turbine at the top, or the service person must be lifted to the turbine with a crane. Rigid monopoles and tilt-up towers require a wide area of ground space for anchoring the guy wires and for lowering the tower. If the tower is installed at a residence, the guy wires should be clearly marked so that they are not a hazard to anyone walking near them.

Guyed lattice towers are more expensive than monopole designs but can support larger turbines at higher heights. Some guyed lattice towers are designed to be able to tip down for turbine maintenance, but most are rigid installations that service personnel can climb more easily than a monopole tower. Like monopoles, both rigid and tilt-down lattice towers require considerable space for guy wires and for lowering the tower. A lattice tower can easily be constructed to carry the weight of the largest wind turbines for home use, and it can be as high as the owner's budget allows. Lattice towers are not as visually attractive as monopoles and when used in a residential area may arouse concern for looking like commercial drilling rigs or power transmission line towers. Painting the tower an earth color will help it fit better into a home setting, as will limiting its overall size and height.

Free-standing lattice towers are constructed with wide enough bases and deep enough foundations that they require no guy wires for lateral support. They can accommodate the largest sizes of wind turbines and are usually required for machines larger than 20 kW in size. The foundations of free-standing towers are similar to those of a monopole, and for this reason they are more expensive than guyed lattice towers. The installation of these towers requires the use of a crane to lift the tower and turbine into place. Maintenance is performed by climbing the tower. If major work is required on a large turbine, a crane may be required to remove the machine from the tower (Figure 10-9).

BALANCE-OF-SYSTEM EQUIPMENT

Wind turbine systems, like photovoltaic generating systems, produce DC electricity. AC current must be produced by a generator turning at a constant speed in order to maintain the correct frequency of the AC current. AC power generation is therefore not feasible for small wind turbines because their current output varies with the speed of rotation of the rotor and its connected generator shaft. The DC charge voltage produced by a turbine is constant, even when the rotor is turning very slowly or very fast. The output amperage varies, increasing with the speed of the rotor up to the maximum capacity of the generator. The wind turbine's DC power output must be converted to AC power for most home uses, although some direct use of the DC power may be practical.

The components needed for a wind turbine generating system, in addition to the turbine and the tower, are the balance-of-system parts and are similar to those needed

FIGURE 10-9

Raising a 10-kW cyclone wind turbine on a 100-foot-high free-standing lattice tower requires a heavy crane and a large working area around the tower.

for a photovoltaic generating system. The specific application of the system will determine variations in the balance-of-system parts. If a residential system is intended to generate DC current that will be used only for charging batteries or operating DC power equipment, no inverter for converting the current to AC is necessary. But a DC system cannot be connected to the electric grid and so can be used only as a independent supplement to the grid-supplied power connected to the residence.

For a residential grid-connected system that generates DC current, the balance-of-system parts will include at least a controller, an inverter with power conditioning capability, disconnects, and wiring to connect the system to the grid through a net meter or an independent meter. If power storage is wanted to preserve electric power capacity in the event of a grid failure, the wind generation system will also need to include a battery stack and additional wiring connections that carry current to and from the battery stack. Figure 10-10 shows a typical grid-connected wind turbine system with a battery stack.

Although a battery stack is not essential for a wind turbine system, it is a good investment for installations where there is substantial risk of losing grid power occasionally. When grid power is lost, a wind turbine system may not function well even though it is generating useable power. If the home's demand for power is not close to what is being generated by the wind turbine, the inverter and power conditioning components will not be able to match the AC power output to the DC input. If the AC demand is too high, the output voltage will drop because the DC current input from the turbine is not sufficient to maintain it. The voltage drop can damage some AC appliances and electronic equipment. Conversely, if the AC demand is too low, the inverter will be fed excess power from the turbine, which it has no place to use or store. This excess power can overload the inverter and cause it to burn out.

FIGURE 10-10

Components for a grid-connected wind turbine generating system.

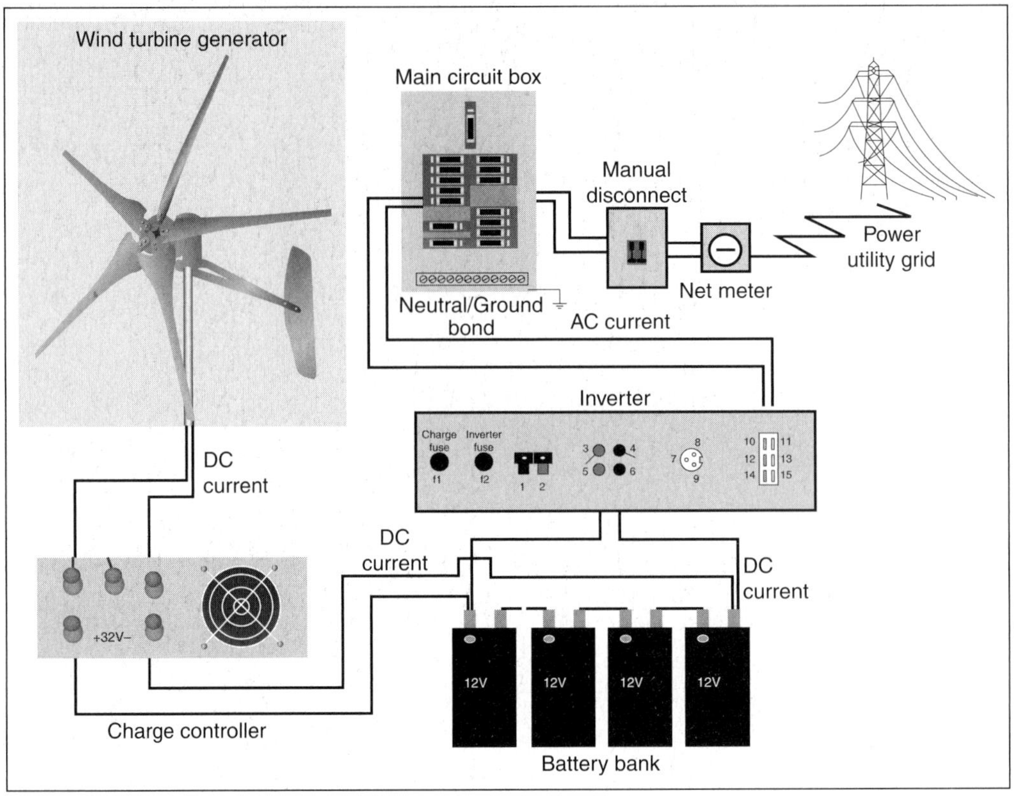

A battery stack solves both problems of too much or too little demand for AC power. If there is too much, the inverter draws more DC amperage from the charged battery stack to meet the requirement. If there is too little, the inverter draws only what power is needed from the battery stack and the excess DC current charges the batteries. If too little demand for AC current persists until the battery stack is fully charged, the system's charge controller reduces the power generated by slowing the turbine, if it has a braking system, or cuts out the power input entirely by stopping the rotor or by diverting the current produced to a DC diversion load. The diversion load can be a water-heating device or a high-resistance heating element connected to a heat sink.

Safety Equipment

Safety equipment for a wind turbine system is similar to what any outdoor electrical installation should have and is designed to protect the system from being damaged or harming anyone who comes in contact with it.

Automatic safety disconnects protect the wiring and components of a renewable energy system from power surges caused by lightning, short circuits, static electric discharges, and equipment malfunctions. Manual disconnects enable the system to be shut down safely for maintenance and repair. Safety disconnects ensure that a grid-connected systems can be isolated from the grid when necessary to prevent it from feeding power into the grid when people are working on the grid transmission and distribution systems.

Because they are tall structures and made of metal, wind turbine towers are subject to lightning strikes. The tower must have a lightning arrest that is solidly grounded to earth. This equipment provides a well-defined, low-resistance path from the wind turbine and tower to the ground to protect it against current surges from lightning strikes or equipment malfunctions.

Surge protectors installed in the wind turbine's charge circuit can also give some additional protection to the system in the event that it is struck by lightning. Nothing can prevent damage from a direct hit by lightning, but a surge protector can minimize the effect of high-voltage current running from the turbine to the battery stack and destroying the batteries with an overcharge.

Meters and Monitoring Equipment

Meters and other instruments allow you to monitor a small wind turbine system's output voltage and wattage, the wind velocity, the charge status of storage batteries, the amount of power being consumed in the home, and the amount of power being fed into the grid. These instruments are not required for the continuous monitoring of the system, but are useful to have for checking its performance periodically.

For a grid-connected system, the only required meter is a smart meter that will keep track of the electricity the home system produces and the electricity used from the grid. This can be a net meter that records the electricity the wind turbine system feeds into the grid and the amount the home draws from the grid. This type of meter enables the utility company to charge based on the net power usage. If the local power provider does not use net meters, the wind turbine system will need a separate meter, connected from the inverter to the grid, to measure the electricity the home energy system feeds into the grid.

Connecting to the Grid

Whether your wind turbine system is connected to the grid by a net meter or a separate meter, it will have to meet the other requirements that the local utility sets for any home energy to connect to the grid. These will include regulations regarding safety, power quality, metering standards, power rates, contracts for purchase of power, equipment maintenance agreements, and liability insurance. Before installing a wind turbine system, always contact the local power provider to find out about its requirements.

The IEEE 1547–2003 standard that provides technical requirements and tests for grid-connected operation covers all grid-connected distributed generation, including wind turbine generating systems. These systems should also use only balance-of-system equipment that meets the UL 1741 standard. Inverters, converters, charge controllers, and output controllers for power-producing stand-alone and grid-connected renewable energy systems should all be UL certified.

Most states and power providers have adopted the IEEE, UL, and NEC standards and codes, although frequently with local variations and additions. For this reason, it is best to get your installation information from the local power provider rather than relying on the national published documents.

To connect your wind turbine system to the grid, you will probably need to sign an interconnection agreement with your power provider. In addition to specifying the technical requirements that the home energy system must meet, the agreement will probably require you to carry liability insurance, and pay some connection fees and charges. Because wind generating systems are expensive, you may also want to consider carrying a homeowner's insurance rider that will cover physical damage to the wind turbine system as well as liability protection. Expect to pay an extra premium for this coverage.

Most homeowner insurance policies provide at least $100,000 of liability coverage, but you should verify that the policy covers events connected to your home energy system. Because wind turbines and their towers are not attached to the home, coverage may be limited to a lower amount for outbuildings or other structures on the property. Insurers may assign a higher risk to a wind generating system than to a solar generating system and charge more for liability coverage. Even if you decide not to have your system connected to the grid, you should have liability insurance covering any damage or injury that might result from your system's operation.

Identify any additional costs that an agreement requires you to pay and decide whether a home energy system is still a good investment with these charges included. Permit fees, engineering/inspection fees, charges for a second meter (or a met meter), maintenance charges, and insurance fees can add up to a heavy additional sum. They should all be figured into the overall cost of your wind energy system. When calculating whether the system is a good investment, remember that some of these charges (permits, meter cost, etc.) are one-time fees, but others, such as insurance premiums and maintenance costs, are recurring charges.

As with a solar generating system, a net meter connection will pay a better rate for the electricity your wind turbine system generates than a second meter connection. With a second meter connection, the power provider purchases the electricity your home energy system produces at its wholesale (cost-of-generation) rate, which will probably be considerably lower than the retail purchase rate you will get with a net meter. The net meter measures both the power the home uses and the amount the home energy system generates and subtracts the generated power from the total amount used. It credits all the excess electricity your home system generates to your electricity account at the retail rate, the same rate you pay for the electricity you use from the grid. If your system produces more than you've used, the power provider usually only pays for the extra electricity at wholesale cost, but you get credit at retail rate for all the power your system produces up to the full amount your home uses.

Legal Considerations

Each state and community has its own set of codes and regulations that you will need to follow in order to add a wind turbine generating system to your home. Because they are very visible structures and are sometimes regarded by local authorities as "attractive nuisances," wind turbines may be more heavily regulated than other renewable energy systems. Regulations may determine the size and particularly the height of the system you can install, who can install it, and whether it must be a grid-connected or a stand-alone system. Structural as well as electrical inspection may be required.

Electrical systems must always be installed according to local building code requirements and be inspected to ensure that they comply with standards. Your wind energy system will likely be required to pass one or more electrical inspections at the beginning and/or end of the installation process. To pass electrical inspection, a system should be composed of certified equipment that complies with local requirements and appropriate technical standards, and should be installed according to NEC requirements as adapted for the local area.

Structural inspections are to determine whether the tower will withstand high winds and other extreme weather conditions. The plans for the tower will be reviewed before installation to be sure that the design is sound, and the completed tower will be checked afterward to be certain that it has been properly constructed. Do not neglect to obtain all appropriate structural and electrical inspections for your wind turbine system. Failure to do so will probably void your insurance coverage in the event of a claim for loss, even if the loss was not caused by improper equipment or installation.

You may also have to obtain a conditional-use permit or a variance from the existing building code before the local zoning authority will issue a building permit for your wind turbine installation. Tower height will again be the most common point of contention, and a compromise may be required to get the project approved. High towers enable wind turbines to generate more electricity; lower towers are more visually appealing. Be prepared to defend the minimum tower height you need (25 feet to the bottom of the rotor blades), but expect to compromise on anything above that.

Before installing a wind turbine generating system, which must have wind to operate, you should check to be sure no structure can be built and no trees planted on adjacent property that will block or deflect the wind from your system. If this is a possibility, you might seek an easement from the property owner specifying that no such structures will be built. Such an agreement is binding regardless of changing land ownership.

You might also want to check your own property deed to determine if any language in it could prevent you from adding a renewable energy system to your own property. Some communities have covenants or other regulations specifying permitted or forbidden uses of residential property. Most regulations support renewable energy systems but some might prohibit them for aesthetic or noise-control reasons. In addition, it is good public relations to discuss your intentions with your neighbors to avoid any future objections.

HYBRID SOLAR PANEL AND WIND TURBINE GENERATING SYSTEMS

Hybrid generating systems consist of a combination of two or more types of generators that feed current through a common set of balance-of-system components to the end user. The generating units can be solar PV panels, a wind turbine, or a diesel generator. Combined solar and wind generating units are too expensive to be justified on a strictly economic basis, but the two technologies complement one another in providing a more constant generating capability than either can supply alone.

In northern latitudes, long summer days have the strongest sunlight for photovoltaic generation. During the summer months, winds are typically weak and can supply little energy to a wind turbine. In winter, when sunlight is weakest and photovoltaic generation consequently low, winds are stronger and a wind turbine can generate at maximum capacity.

Similarly, solar PV panels cannot generate current from sundown until sunrise, but wind force often remains constant through the nighttime hours and may even increase in some coastal locations due to temperature changes on land and over water. This day and night differential in generating potential again permits a hybrid system to generate more electricity, at a more constant rate, over a longer period of time than either a solar PV system or a wind turbine system could do alone.

The hybrid system may, in fact, be a good investment for homeowners seeking to minimize their dependence on the electric grid. With the addition of a third generating source, a fuel-powered generator to provide backup during the few times when solar and wind energy are insufficient, the home can be made entirely independent of drawing any power from the electric grid. The owner may, however, still want his hybrid system to remain connected to the grid in order to sell its excess power to the utility. Figure 10-11 shows a hybrid solar and wind generating system on a rooftop and monopole in a yard.

FIGURE 10-11

The rooftop solar PV array and monopole-mounted wind turbine of a hybrid generating system.

A typical hybrid generating system includes the following components:

- Solar array—A number of PV panels connected in series and/or in parallel to give the required DC charge current for input to the battery stack.
- Wind turbine—A wind generator with a capacity that can be more or less than the PV panel array, but that produces a charge current of equal DC voltage to be fed into the battery stack.
- Hybrid controller—This device controls charge current from both the solar array and the wind turbine into the battery bank to keep it within optimum operating and safety limits.
- Battery stack—A single battery or multiple batteries connected together either in series or in parallel to create essentially one large battery of the required voltage and amp-hour capacity.
- Inverter—A converter that changes the DC power drawn from the batteries into AC power suitable for the required loads.

Because both the solar PV array and the wind turbine will at times produce power simultaneously, the controller, battery stack, and inverter in a hybrid system must be larger and have a greater current capacity than would be required for either generator operating alone. This increased component capacity also increases the overall expense of a hybrid system. Connecting such a system to the grid is the best way to recover some of the investment expense and contribute renewable energy power to the system.

SUMMARY

Wind is another manifestation of solar radiation, which heats the earth's surface unevenly and causes air currents to form in the atmosphere. These air currents are the winds that power weather systems. Wind turbines convert the kinetic energy in wind into mechanical power that runs a generator to produce clean electricity. Large turbines standing up to 400 feet high are the most efficient. Residential-scale wind turbines are far smaller and slightly less efficient than the commercial behemoths, but they are otherwise similar. Locations that are suitable for a wind turbine generating system are those that have a consistent average wind force strong enough to turn a wind turbine a majority of the time. A location with an annual average wind speed of 12 miles per hour or greater has sufficient wind energy to drive a turbine; a location with an average wind speed of 10 miles per hour or less does not have enough wind energy for a wind turbine installation. The wind resource of an area is a measure in kWh per year of the potential for generating electricity from wind power in that area. It is calculated as the estimated annual electricity production per square meter of rotor-swept area for a small wind turbine. Wind turbines for residential applications range in size from 500 watts to 100 kW. The most common type of turbine is a horizontal-axis turbine with a two-, three-, or four-bladed rotor. Another type of small wind turbine design is the vertical-axis wind turbine (VAWT) that has a vertical shaft with vanes or a round "squirrel cage" attached to capture the wind's force. There are three basic types of tower options for home-sized wind turbines: monopole, guyed monopole, or lattice and free-standing lattice. The components needed for a wind turbine generating system, in addition to the turbine and the tower, are the balance-of-system parts and are similar to those needed for a photovoltaic generating system. Whether the wind turbine system is connected to the grid by a net meter or a separate meter, it will have to meet the other requirements that the local utility sets for any home energy system to connect to the grid. Hybrid generating systems consist of a combination of two or more generators that feed current through a common set of balance-of-system components to the end user. The generating units can be solar PV panels, a wind turbine, or a diesel generator. Combined solar and wind generating units are too expensive to be justified on a strictly economic basis, but the two technologies complement one another in providing a more constant generating capability than either can supply alone.

KEY TERMS

Annual average wind speed The average speed of the wind at a location over a 1-year period; calculated by taking a daily reading, adding all readings together, and dividing by 365.

Balance-of-system equipment Additional equipment beyond the basic solar generation components or wind turbine components that is required to make a renewable energy DC generating system compatible with the home's AC electric circuits and the electric grid.

Blades The long arms of a rotor that are pushed by the wind.

Charge voltage The DC voltage of the current generated by a wind turbine.

Expansion gearbox A mechanical transmission that increases the speed of rotation from its incoming shaft to its outgoing shaft.

Governing device A mechanical brake that limits the speed of rotation of a wind turbine or locks it in place so that it cannot rotate too fast in high wind.

Ground clearance The distance from the ground to the bottom tip of the rotor or wind wheel of a wind turbine.

Guyed monopole A single upright pole set in a base and supported by guy wires on which a wind turbine is mounted.

Guy wire A brace wire that runs from the top or midpoint of a pole or lattice to a ground anchor some distance away from the base to give lateral support to the structure.

Horizontal-axis wind turbine A wind turbine with a horizontal shaft on which a two-, three-, or four-bladed rotor is mounted.

Hybrid generating system A combination of two or more generators, solar PV array, and wind turbine or fueled generator that feeds current through a common set of balance-of-system components to the end user.

Kinetic energy Energy of motion possessed by any moving object; a fast-moving object pushing against a slower-moving object imparts energy to the slower object, causing it to speed up and the fast-moving object to slow down or stop.

Lattice tower A wind turbine tower constructed of straight metal pieces fastened into rigid triangle shapes to give maximum strength with minimum weight.

Lightning arrest A metal device that provides a well-defined, low-resistance path for electric current from a wind turbine and tower to the ground to protect against current surges from lightning strikes.

Monopole A single upright pole set in a foundation on which a wind turbine is mounted.

Output voltage The AC voltage of current from an inverter that can be used in AC residential circuits or downloaded to the grid.

Rated wind speed The wind velocity at which a wind turbine will generate its designed power output in watts.

Rotor The rotating bladed wheel of a horizontal-axis wind turbine.

Rotor diameter The length of a rotor blade multiplied by 2; the width of the circle created by the tips of the rotor blades as they turn.

Startup wind speed The wind speed at which a wind turbine starts to generate electricity.

Surge protector An electrical device that prevents momentary high current flows from passing into a circuit.

Swept area The area of the circle that the blades of a rotor pass through as they turn.

Vane The panel or blade on a wind wheel that the wind pushes against to turn the wheel; also the panel set at a right angle behind a horizontal-axis wind turbine to keep the rotor facing directly into the wind.

Vertical-axis wind turbine (VAWT) A wind turbine with a vertical shaft on which a wind wheel with vanes or blades rotates when pushed by the wind.

Wind power density The kinetic energy level in the wind.

Wind resource A measure in kWh per year of the potential for generating electricity from wind power in an area; calculated as the estimated annual electricity production per square meter of rotor-swept area for a small wind turbine.

Wind speed Speed of movement of air over the earth's surface; measured in miles per hour (mph) or meters per second (m/s).

Wind turbine A mechanical device for converting kinetic wind energy into electricity by means of a rotor or wind wheel attached to a generator.

Wind velocity Same as wind speed.

Wind wheel The vertical vanes or blades attached to a shaft that wind pushes against to rotate in a vertical-axis wind turbine.

REVIEW QUESTIONS

1. True or False: Wind is a form of solar energy produced by the uneven heating of the earth's surface by the sun.
2. How does a wind turbine turn wind energy into electricity?
3. Generally, how much wind is needed for a turbine to operate? What amount of wind is too little to produce much electricity?
4. What is the wind resource of an area?
5. A large two-, three-, or four-bladed wheel that turns on the end of a wind turbine's central shaft is called a ____________.
6. What is a VAWT?
7. Why are VAWTs not quite as efficient in using the wind's force to generate electricity?
8. Which of these towers does not use guy wires for additional support, a monopole or a free-standing lattice?
9. How can maintenance be performed on a wind turbine mounted on a free-standing lattice tower?
10. How is a wind turbine protected from damage by winds more powerful than the maximum speed it was designed to use for generating power?
11. True or False: A wind turbine can generate either AC or DC electricity depending on the home's needs.
12. In a wind turbine generating system, where in the system is the controller placed?
13. True or False: A wind turbine does not need a manual disconnect if it is connected to the electric grid.
14. What are the three additional components that a grid-connected wind turbine generating system requires in addition to the turbine?
15. What is the function of an inverter in a wind turbine generating system?
16. If an independent meter is connected to a wind turbine generation system, how much of the generated electricity does the system's owner pay for?
17. What is the most likely area of contention when requesting a permit to build a wind turbine generating system?
18. What is a hybrid generating system?
19. True or False: A hybrid system is more expensive than a single-source generating system, but also has the ability to generate more power more consistently than a single-source system.
20. How does a wind turbine complement the generating capability of a solar PV array?

GLOSSARY

10BaseT RJ-45 Ethernet connection A rectangular connector on either end of an eight-wire cable commonly used to connect components (nodes) of a local area network (LAN) to one another.

3-digit code A color code for lights that is a combination of the CRI and color temperature of the bulbs; the first digit is the CRI of the lamp divided by 10 and the second and third digits are the color temperature divided by 100.

Absorber plate The black-painted metal sheet placed behind the tubing at the bottom of a collector panel to absorb heat for transfer to the tubing and fluid inside it.

Absorption coefficient The measure of how readily a material absorbs light; a small coefficient means poor light absorption; a large coefficient indicates the material easily absorbs light.

Active system A type of solar water-heating system that uses a pump to circulate the heat transfer fluid (HTF).

Advanced Metering Infrastructure (AMI) The wireless communication network that allows two-way communication between smart meters, utility companies, and consumers.

Air conditioning (A/C) The process of cooling the interior air in a building to a comfortable temperature.

Air duct Any metal, plastic, or other type of tube, round or rectangular, used to move air.

Alternating current Electric current that reverses direction several times each second; the most common type of electric current for home and commercial use.

Always-live duplex outlet An AC electrical outlet that is not controlled by a switch and always has current available.

Always-on mode The state of an electric appliance in which it always has current running through its circuitry so that it can commence operation instantly without any startup delay.

Amperage The amount of electric current flowing in a circuit; measured in amps.

Analog meter A meter that measures cumulative electricity, gas, or water use mechanically or electromechanically.

Annual average wind speed The average speed of the wind at a location over a 1-year period; calculated by taking a daily reading, adding all readings together, and dividing by 365.

Article 690 of the NEC A section of the National Electric Code that provides guidance on equipment and wiring safety for small renewable energy system installations, including photovoltaic systems.

Automation The process of controlling a lighting or other electrical system remotely, either with manual commands from a controller or by a programmed set of instructions from a computer.

Auxiliary energy source In a solar heating system, a second source of heat energy to provide additional heat when solar energy alone is insufficient.

Average monthly cost The cost to operate an appliance for 30 days.

Average monthly kWh The power consumed in 30 days by an appliance operating at its present level.

Baffle A cardboard or other rigid block across the space between joists or studs used to prevent insulation from obstructing vents on the outside of the baffle.

Balance-of-system equipment Additional equipment beyond the basic solar generation components or wind turbine components that is required to make a renewable energy DC generating system compatible with the home's AC electric circuits and the electric grid.

Ballast A voltage transformer used to raise the voltage of current flowing to a fluorescent light.

Bandgap (symbolized by Eg) The energy difference between the conduction band and valence band in molecules of a semiconductor material.

Base load The electric load in a home or business that is always in use by devices that are never turned off.

Base-load power The portion of home or business electric power consumption that is used continuously at all hours of the day and night; base-load demand is the minimum power required at any given time.

Battery stack A connected bank of batteries used to store electricity generated by a renewable energy system for later use.

Batt insulation Fibrous insulation made of fiberglass or similar material formed into thick sheets (batts), sometimes with a facing adhered to one side.

Berm An artificial hill or rise in ground level intended to conceal a view or to deflect wind.

Blackout An electric power failure over a large geographic area; usually caused by the demand for power exceeding the immediately available supply.

Blades The long arms of a rotor that are pushed by the wind.

Blower A device that moves air like a fan but does so with a rotor rather than a bladed impeller.

Blown-in insulation Any insulation composed of small bits of material that are blown into place through a hose using an air blower.

Boiler An appliance for heating water to be circulated in a radiant heating system.

Brownout A drop-in voltage that can damage electric equipment; usually caused by the demand for power exceeding the available supply.

BTU (British thermal unit) A measure of heat equal to the amount of energy needed to heat 1 pound (0.454 kg) of water (0.1198 U.S. gallons) from 39° to 40°F (3.8° to 4.4°C).

Building code The laws or ordinances that govern the construction of residential and commercial buildings so that they meet standards of safety, quality, and energy efficiency.

Carbon dioxide (CO_2) A combustion product of burning fuels; toxic in high quantities.

Carbon monoxide (CO) a poisonous combustion product of burning fuels; produced in place of CO_2 when combustion is incomplete.

Caulking A viscous (thick) semi-liquid material that can be squeezed into place to plug leaks or seal cracks. Good-quality caulking forms a thin skin after placement, but does not harden for many years.

CCF 100 cubic feet of gas.

Cellulose insulation Insulation made of cellulose fibers, primarily composed of recycled newsprint and other paper.

Central heating system Any heating system designed to warm an entire building rather than a single room or small part.

Charge controller An electronic device that regulates the amount of current going to the batteries in a battery stack to prevent overcharging.

Charge voltage The DC voltage of the current generated by a wind turbine.

Circuit breaker An electrical device that allows only a set amount of current to pass through it. If the limit is exceeded the breaker cuts off the entire flow.

Clicker igniter An igniter for gas appliances that sparks to start the gas fuel burning; same as spark igniter.

Cluster LED A combination light in which numerous LEDs are combined into a single unit that produces light equal in magnitude to that of large incandescent bulbs.

Coal fired Any appliance using coal as its heat source.

Coal-fired power Electricity generated by steam turbines fueled by coal.

Color-enhanced phosphors Light-emitting materials used in CFLs to change the color of their emitted light to be closer to that of incandescent light.

Color rendering index (CRI) A measure of how closely colors viewed under a light source match the same colors viewed in natural daylight, which has a CRI of 100.

Color spectrum The band of frequencies of electromagnetic radiation that produce light of all colors visible to the human eye.

Combustion chamber The part of a fuel-burning appliance where fuel is burned and heat produced.

Combustion products The elements and compounds created when a fuel burns; CO_2 and water are two usual combustion products but many others are possible.

Combustion residue Solid combustion products such as ash produced by burning fuel that must be removed from the combustion chamber periodically.

Combustion vent A vent to the outside that allows a fuel-burning appliance to draw its combustion air from outside the building where it is located.

Command module An X-10 technology controller that sends commands (manually generated or programmed) to control modules that control lights.

Compact fluorescent light (CFL) A miniature fluorescent light made in several shapes that produces fluorescent light with a CRI close to that of daylight or incandescent light.

Compressor In A/C systems, a device for squeezing a gas into a smaller space at a higher pressure until it becomes a liquid; other types of compressor produce compressed air, liquefy natural gas, or compress other fluids.

Condenser A device that extracts heat from gas, causing it to become liquid; often used with a compressor in A/C systems.

Conduction band The higher energy level of atoms in the molecules of a semiconductor material created by sunlight energy being absorbed so that the electrons are free to move.

Constant-rate charge Charging for energy or water use at the same rate no matter how much is used or when it is used.

Continuous-flow water heater A water heater that can produce a continuous stream of hot water at a set temperature.

Continuous-load use A designation of the work that an electric motor can perform continuously without overheating or other failure; always less than peak-load use.

Continuously operating appliance An electric device that is always working, such as an electric clock or timer; not the same as a device in always-on mode, which may have power connected but not be actually operating.

Controllers Devices for managing the operation of another device remotely, either through a wired connection or wirelessly; devices that send commands to control modules in an automation system.

Control module In X-10 systems, any of the devices that receive lighting control instructions from the controller and implement them by adjusting the lights.

Convection The movement of heat through a liquid or gas by internal circulation within the material.

Convective heater A heater that warms the air around it; the warmed air then warms other nearby objects.

Cooling coil The component of an A/C system in which heat is extracted from the air passing through it.

Correlated color temperature (CCT) A measure of the color of light emitted by a light source; measured in Kelvins (K), a unit of absolute temperature; the higher a light source's CCT, the more blue or blue-white it appears to the viewer. Natural daylight has a CCT of about 5000 K; bright white and cool white bulbs range between 3500 and 4000 K.

Cubic foot (ft^3) A unit of measurement used for natural gas and water. Equal to a block of water or gas 1 foot square and 1 foot high at a pressure of 30 inches of mercury (approximately 14.7 psi) and a temperature of 60 degrees Fahrenheit.

Cumulative cost The cost to operate an appliance since it was plugged into a power meter.

Cumulative watt hours The watt hours of electricity used since an appliance was plugged into a power meter.

Current transformer (CT) A device that measures electrical use without being a part of the circuit it is measuring; a small measurement current is induced in the CT by the passage of the larger current being measured.

Current watts The amount of electric current in watts that an electrical device is using.

Cutoff switch A switch used to shut off electricity to continuously running devices or those in always-on mode.

Daisy-chain wiring Wiring light fixtures in parallel one to another, with only the first light being connected directly to a power supply and each of the others receiving power in succession down the chain.

Damper A movable door in a duct that can be operated manually or automatically to permit or block the passage of air.

DC-to-AC inverter A device that changes AC current flow to DC flow; used in CFL lights; also called a rectifier.

Deciduous trees and bushes Plants that shed their leaves in fall and grow new ones in spring; used to provide shade in summer while allowing sunlight to pass in winter.

Dedicated radio frequency (RF) network A wireless network for connecting smart meters to a utility company; can be a private network or part of a public network.

Demand pricing A method of charging for electricity in which power used during high-demand times is billed at a higher rate than that used during low-demand times.

Demand-rate charge Charging for energy or water use according to the volume used and the time during which it is used; also called demand pricing.

Diffused LED bulb A cluster of LEDs covered by a dimpled lens that spreads light from the individual LEDs over a wider arc.

Digital communication network A real-time communication system for providing data monitoring and control in an electrical grid.

Digital smart meter A meter that measures electricity, gas, or water usage electronically and sends a constant digital signal of its measurement to a utility.

Dimmable CFL A CFL that can be dimmed using a dimmer switch matched to the low-power consumption range of the CFL.

Diode A semiconductor device made of a material that has a varying ability to conduct electrical current; diodes conduct electricity in only one direction.

Direct current Electric current that flows in one direction only; batteries and photovoltaic cells supply direct current.

Disconnect A safety switch used to cut off power from a generating circuit or storage battery.

Distribution line An electric line carrying current from 480 down to 120 volts that connects a higher-voltage transformer to an end user.

Disturbance (electrical) Any variation in the voltage, amperage, or current characteristics in an electrical circuit.

Door jamb The part of a door casing against which the door fits and seals to prevent air passage.

Double-pane glass Window glass consisting of two panes parallel to one another with an air or gas-filled space between them.

DTH A unit of heat measure that is equal to one million BTUs.

Duct tape Cloth-based tape, usually about 2 inches wide, used to seal ducts and for other uses.

Ductwork The system of ducts that moves air to and from an HVAC system.

Economies of scale The axiom that larger and more complex buildings and structures are more energy efficient on a per-unit basis than smaller ones. This rule also applies in many other areas.

Edison-type base The standard screw-type base on most incandescent light bulbs; CFLs and LEDs have been adapted to fit the sockets designed for Edison bases so they can replace incandescent bulbs.

Electrical capacity The maximum amount of electricity a generator is capable of producing or a transmission line is capable of carrying without damage or failure.

Electric grid The system of electric generating and distribution lines that provides electricity to all areas of the country.

Electricity monitor Same as a power meter.

Electric meter A device for measuring the amount of electric current flowing in a circuit.

Electric service box The metal box that contains the electric service panel and into which all AC wiring in a home or business is connected.

Electric service panel The panel inside an electric service box that contains the main and branch circuit breakers and the neutral bus.

Electromechanical A system composed of mechanical parts (switches, relays, etc.) that are operated electrically.

Electronic ballast A type of ballast that uses a circuit board and electronic components to transform electric current voltage in an LED.

Electronic ignition A method of igniting a gas appliance flame by means of an electric spark.

Emissivity value A measure of how much radiation (of all wavelengths) a material absorbs and then emits again as radiant energy; emissivity values range between 0 (no absorption, 100% reflectivity) and 1 (100% absorption, no reflectivity).

Energy audit A survey that is designed to tell where your home is using energy and where it is losing energy.

Energy conservation The process of reducing the amount of electricity required to accomplish a given function.

Energy consumption The total energy in any form (electricity, fuel, gravity, wind, etc.) used in a home or building.

Energy efficiency The ability of a building or device to perform its intended function using the minimum amount of energy; the less energy consumed over time, the more energy efficient the building or device is.

Energy Star rating A designation for electronics products that meet the strict energy-efficiency and price guidelines set by the Environmental Protection Agency and Department of Energy of the federal government.

Energy used per household The total amount of energy in all forms used by a household over a given period of time.

Evacuated heat pipe tube (EHPT) A type of solar collector composed of double glass tubes, one within the other, with the space between the tubes evacuated to reduce heat transfer; the inner tube contains an absorber plate fused to a heat pipe.

Evaporative cooler An air conditioner that works by the cooling effect of evaporating water rather than by refrigeration cooling; not effective in humid climates.

Exhaust vent A vent leading from an HVAC or fueled appliance to the outside air.

Expanding foam insulation Semi-liquid insulation material that when squeezed out of a container expands to fill a space and seal against air passage.

Expansion gearbox A mechanical transmission that increases the speed of rotation from its incoming shaft to its outgoing shaft.

Fall (autumnal) equinox The day in September of each year on which the amount of daylight and darkness are equal.

Feed-in tariff (FIT) Also known as a standard offer contract or an advanced renewable tariff; an agreement designed to promote investment in renewable energy technologies by offering a long-term contract to purchase the power produced at a rate based on the cost of generation.

Filament The fine coiled wire in an incandescent light that emits light (incandesces) when a current passes through it.

Filter A screen that removes pollution particles from conditioned air before it is distributed.

Flame arrester screen A device that inhibits an open-flame heating appliance from starting a fire.

Floor sucker pump A type of sump pump that draws water directly from a floor or other flat surface rather than from a sump pit.

Flow sensor A device that senses when water or another liquid is flowing in a pipe.

Fluctuation A variation from the normal flow of electric voltage or amperage.

Fluorescent lamp Any type of tube, bulb, or CFL that emits light from phosphors excited by an electric current.

Fluorescent light The light produced by phosphorescent materials that glow when excited by an electric current.

Fluorescent lighting Lighting tubes that contain phosphorescent material and glow when a high-voltage current is passed through them. The tubes are made in various lengths up to 8 feet.

Foam-in-place insulation Foam material that can be injected into small areas to control air leaks; same as expanding foam insulation.

Forced-air heating system A heating system that uses a blower to distribute warmed air through ducts to all parts of a building.

Frosted bulb A light bulb coated on the inside with translucent material that diffuses its light output.

Full-on/full-off tankless water heater A water heater that cannot adjust its rate of flow or its water temperature, but produces a set amount of hot water continuously.

Gas fired Any appliance fueled by natural gas.

Gas-fired power Electricity generated by steam turbines fueled by natural gas.

Gas meter A device for measuring the flow of natural gas through a pipe in cubic feet.

Gateway An electronic device that receives data from a measuring device and relays it to a display or recording device.

Generation The process of producing electricity by converting other forms of energy.

Generator A mechanical device similar to a motor that produces an electrical current when its shaft is rotated.

Geothermal energy Energy produced in the heated interior core of the earth that can be used to provide heat and generate electricity.

Gigawatt A measure of electric power equal to 1000 megawatts (mW). A megawatt is equal to 1000 kilowatts (1000 kW).

Glazed panel A heat collector panel that has a glass faceplate over the interior absorber plate and tubing.

Google PowerMeter portal/website A free website where consumers can register to have their electric smart meter data or their monitored electric power use sent for display and analysis using a graphic display on a home computer.

Governing device A mechanical brake that limits the speed of rotation of a wind turbine or locks it in place so that it cannot rotate too fast in high wind.

Grid-connected system A PV system that is connected to the local electric grid.

Ground clearance The distance from the ground to the bottom tip of the rotor or wind wheel of a wind turbine.

Ground water Water that flows beneath the earth's surface; source of well water and spring water; may also flow into basements or other below-surface openings.

Guyed monopole A single upright pole set in a base and supported by guy wires on which a wind turbine is mounted.

Guy wire A brace wire that runs from the top or midpoint of a pole or lattice to a ground anchor some distance away from the base to give lateral support to the structure.

Halogen bulb A type of incandescent light that uses halogen gas to increase the efficiency of its light output by up to 30% compared to that of ordinary incandescent bulbs.

Hard-wired Permanently connected to a building's electrical system, as opposed to a plug-in appliance.

Heat absorption The process by which heat passes from a warmer substance to a colder one; the cold object absorbs heat from the warmer one.

Heat exchanger Any device that transfers heat from one object or substance to another; a stove burner and an A/C cooling coil are two examples.

Heating element An electric component that heats when electricity passes through it because it has high resistance to the current.

Heating register The opening at the end of an HVAC system duct through which warmed or cooled air passes into a room.

Heating, ventilating, and air-conditioning (HVAC) system The integrated electromechanical system that supplies heating and cooling to a home or business. HVAC systems often use gas or oil in addition to electric power.

Heat loss The loss of heat energy from a home or other building caused by radiation from the exterior walls and leakage of warm air from the interior.

Heat transfer The passage of heat through a material from its warmer surface to its colder surface or from one material to another; also the passage of heat through the walls, roof, and floors of a home or other building in either direction.

Heat transfer fluid (HTF) The fluid that flows through the absorber tubing in a solar collector. It may be hot water from the storage tank or another fluid that goes to a heat exchanger.

High voltage (user) Electric power at 120 to 240 volts; the voltage at which electric power is delivered to homes and commercial users.

High-voltage transmission The movement of electricity at voltages above 480 volts; transmission lines can carry as much as 700,000 volts.

Home receiver An electronic device for receiving output data from the smart meter that measures a consumer's electric power usage.

Home run connection A wired connection that links a light directly to its power supply without the current to it passing through any other fixture except a switch.

Horizontal-axis wind turbine A wind turbine with a horizontal shaft on which a two-, three-, or four-bladed rotor is mounted.

Horsepower (hp) A measure of work used to define the power of electric motors and other engines; 1 hp is the amount of work (power) required to lift 565 pounds 1 foot in 1 second.

Household energy consumption The total amount of energy in any form used by a home during a given time period.

HVAC schedule The programmed operation of an automatic thermostat designed to get maximum efficiency from the HVAC system.

Hybrid CFL A lamp that combines an instantly lighting halogen lamp with a CFL that has a warm-up period; when the CFL reaches full brightness, the halogen lamp goes out.

Hybrid generating system A combination of two or more generators, solar PV array, and wind turbine or fueled generator that feeds current through a common set of balance-of-system components to the end user.

Hydrocarbon fuel Any fuel composed of hydrogen and carbon compounds; coal, oil, and natural gas are examples.

Hydroelectric power Electricity produced from generators driven by water-powered turbines.

IEEE 1547–2003 An IEEE standard that provides technical requirements and tests for grid-connected equipment and operation.

Impeller The paddle-wheel-like component in a pump that rotates to drive water out of the pump at high pressure.

Incandescent light Light produced by passing current through a high-resistance filament that produces both heat and light.

Infrared light Light with a wavelength longer than that of visible red light and that cannot be seen by the human eye but only detected as radiant heat.

Infrared radiation Invisible solar heat energy with a wavelength just longer than that of red visible light; most solar heat energy is infrared.

Infrared remote control A device for controlling electronic equipment by means of signals sent as infrared light pulses.

Infrastructure The integrated components of the delivery system in a utility network. In a power grid it includes transmission lines, distribution lines, transformers, meters, etc.

In-home-display (IHD) An electronic device for displaying electric power use in real time and over designated time periods for analysis.

Inside leak An opening from the interior of a building to the cavities in the walls that permits air to flow in either direction.

Institute of Electrical and Electronics Engineers (IEEE) A professional electrical industry group that develops standards for electrical equipment, safety, and operations.

Insulation Any material intended to impede heat transfer; materials through which heat passes slowly or not at all.

Interruption (electrical) The complete loss of voltage and current flow in a circuit for any amount of time.

Inverter An electronic device that converts direct current (DC) into alternating current (AC) and adjusts current voltage for use by AC appliances and the electric grid.

IP-based public wireless network An RF network that is available for public use, such as Wi-Fi, mobile phone networks, LANs, etc.

Joist The cross members that support the floor of a building.

Kilowatt hour (kWh) A measure of electricity used; equal to 1000 watts of electricity used for 1 hour. It is the usual unit on which electric charges are based (for example, $0.08 per kWh).

Kinetic energy Energy of motion possessed by any moving object; a fast-moving object pushing against a slower-moving object imparts energy to the slower object, causing it to speed up and the fast-moving object to slow down or stop.

LAN Local area network.

Latitude The distance of a location north or south from the equator, measured in degrees, minutes, and seconds.

Lattice tower A wind turbine tower constructed of straight metal pieces fastened into rigid triangle shapes to give maximum strength with minimum weight.

Lift The vertical distance water must be raised by a pump from its intake to the point of discharge.

Light-emitting diode (LED) A light source consisting of a solid-state diode made of a material that emits light when a current is applied to one side of its circuit; more efficient than an incandescent light.

Light-emitting phosphor Any material that emits light when excited by an electric current.

Lightning arrest A metal device that provides a well-defined, low-resistance path for electric current from a wind turbine and tower to the ground to protect against current surges from lightning strikes.

Light scenes Illuminated rooms or areas within a room that not only provide light but beautify a setting or invoke a mood.

Light spectrum The band of electromagnetic radiation with wavelengths from ultraviolet light (shortest) to infrared light (longest) and including all colors of visible light in between these extremes.

Load In electrical systems, the amount of current (amps) flowing in the wiring at a given time.

Loose-fill insulation Insulation made up of small bits of material that can be dumped into place; this type of insulation can also be blown into place.

Low-emissivity (low-E) glass Glass that has been treated to reduce heat transfer; the treatment may be in the composition of the glass or in a film or coating applied to one or both surfaces.

Low voltage (user) Electric current below 50 volts, usually in the range of 5 to 24 volts; battery-powered devices, electronic equipment (internal circuits), telephones, and low-voltage lighting are examples of equipment that runs on low-voltage current.

Low-voltage transmission Movement of electricity in distribution lines at 480 volts or less; high-voltage power is stepped down by transformers to low-voltage levels for delivery to end users.

Lumen A measure of light output.

Macros Sequences of commands in a specified order that a programmable device stores in memory and executes at preset intervals.

Main breaker The circuit breaker that controls all incoming electricity to a home or business; if the breaker is tripped, all power is cut off.

Mastic A viscous substance similar to caulk used to seal air ducts and air leaks.

Maximum watts The power consumed when the load on an appliance is highest.

MCF 1000 cubic feet of gas.

Measuring transmitting unit (MTU) A device that receives measurement data from a meter and transmits it to a display or recording device.

Megawatt (mW) A measure of electric power equal to 1000 kilowatts (kW). A kilowatt is equal to 1000 watts (1000 W).

Mercury vapor The gaseous form of the heavy metallic element; it is extremely toxic but is a required component of fluorescent lights.

Mesh network A wireless network composed of many nodes spread over a geographic area, each of which can send and receive data from those nodes near it.

Meter reader A person who physically visits utility meters to record their readings for billing by the utility. Not needed in areas where digital meters are installed.

Minimum watts The minimum power consumed when an appliance is plugged in but not operating.

Modulated tankless water heater A heater that can vary the temperature of the hot water it produces, making precise adjustment for comfort easier.

Monopole A single upright pole set in a foundation on which a wind turbine is mounted.

Multiple-unit HVAC A system consisting of several small HVAC units, each of which constitutes a separate zone in a building and can be independently controlled.

National Electric Code (NEC) A detailed set of specifications for electric wiring and construction developed by the NFPA and used nationwide with some local variations.

National Fenestration Rating Council (NFRC) An organization that tests, certifies, and labels windows, doors, and skylights based on their energy performance ratings.

National Fire Protection Association (NFPA) The professional organization that publishes and updates the National Electric Code and other fire protection codes.

Net meter An electric meter that measures current fed back into the electric grid as well as current being used by the consumer; the utility then charges only for the "net" electricity used.

Net metering An arrangement where electricity generated by a grid-connected solar energy system is measured by the home's electric meter as it is fed into the grid.

Neutral bus The neutral (neither positive or negative voltage) line on an electric service panel to which the neutral (white) wires of each circuit are connected.

N-type material A semiconductor material with extra electrons; so called because it has extra negatively charged particles.

Nuclear power Electricity produced by turbine-generators driven by high-pressure steam from a nuclear reactor.

Off-peak hours Those hours during a day when electric power demand is not as high as during the 2-hour period of highest demand for the day.

Ohm's law $V = I \times R$, or voltage (V) is equal to amperes (I) times the resistance (R).

Oil fired Any device in which the fuel burned is heating oil.

On-demand water heater A water heater that produces hot water only when a hot water faucet is turned on; this type of heater does not store any hot water, but can produce it continuously.

One-way electric system An electrical grid that allows power and communication to flow in only one direction, from the producer to the end user; now being replaced by the smart grid that permits electric flow and communication in multiple directions.

Operating cycle The complete programmed set of functions performed by an electrical device in a timed sequence or in response to preset conditions.

Operating mode The full-functioning operating status of an electrical system or device.

Output voltage The AC voltage of current from an inverter that can be used in AC residential circuits or downloaded to the grid.

Outside leak An opening in the exterior surface of a building that allows air to pass from the outside into the wall cavities.

Overcapacity The amount of generating capacity that exceeds the projected peak demand of an electric grid; some overcapacity must be maintained to prevent failure of the system if demand exceeds supply.

Overload Current flow in an electrical circuit above its capacity to carry safely.

PAR (parabolic aluminized reflector) light A designation of enclosed light fixtures and light bulbs that refers to the size of the fixture and its bulb; PAR lights are measured across the lens in eighths of an inch; a PAR 16 light is 2 inches across, a PAR 30 light is 3.75 inches across.

Passive solar energy Solar energy striking the earth's surface that can be enhanced or blocked to increase its usefulness for heating or cooling a building.

Passive systems A solar water-heating system that relies on municipal water pressure or convection to circulate the HTF.

Peak AC voltage (170 volts) The highest level of electric voltage in an AC current sine wave; peak voltage for AC house current is + or − 170 volts.

Peak demand The period during a given day when the largest amount of electric power is required to serve a home or community.

Peak load The maximum amount of electricity a home or business can require; also the maximum amount of natural gas or water a consumer can require.

Peak-load capacity The maximum amount of electricity a system is capable of generating to meet demand.

Peak-load use The maximum work load that an electric motor can perform for a short period using the most current it can draw; an electric motor will fail if run at peak load continuously.

Pedestal pump A type of sump pump that has its motor above the pump casing that sits in the sump pit; the motor is intended to always remain dry.

Per capita electrical consumption A measure of the average amount of electricity consumed by each person in a given community; calculated by dividing total energy used by the number of people in the community.

Per-household energy consumption A measure of the average amount of energy consumed by each household in a given community; calculated by dividing total energy used by the number of homes in the community.

Phantom load The load placed on a home's electric service from the electrical appliances and devices in the home that remain in standby mode when not in use.

Phosphor A substance that emits light when excited by an electric current.

Photoelectric effect The generation of an electric current in a semiconductor material when it absorbs sunlight; the solar energy knocks electrons loose from the atoms of the material, which then flow out of the material as electricity.

Photon of light A unit of light that is emitted when an electron drops from a higher energy level to a lower one in a phosphorescent substance.

Photovoltaic (PV) module or panel A large group of solar cells mounted in a frame and interconnected in series to produce a larger current at higher voltage than a single cell.

Pilot light A method of lighting gas- or oil-fired appliances that uses a small continuously burning flame to ignite the gas or oil.

Point-of-use tankless water heater A water heater placed as close as possible to the faucet from which its hot water is dispensed so that the delay in delivery will be minimal.

Polarity The direction of current flow in an electrical circuit; current flows from the positive pole of a circuit toward the negative pole.

Polyurethane foam (PUF) insulation A type of high-R-value insulation now used in tank water heaters in place of the older fiberglass or similar type of insulation.

Power analyzer Same as a power meter.

Power conditioning equipment Electronic devices required to regulate and monitor electric current to maintain its voltage, amperage, and waveform to precise standards.

Power consumption The amount of electricity used by a device when operating normally.

Power Law W = I x V, or power (expressed in watts [W]) is equal to current (expressed in amperes [I]) times the voltage (expressed in volts [V]).

Power-line technology (PLT) A communication and control system in which data signals are transmitted on AC wiring using a different frequency and voltage than the regular AC current flowing in the circuits.

Power meter A device for measuring the amount of electricity used by an appliance and calculating other information from that measurement.

Power meter partnerships with Google PowerMeter A means of displaying electric power use using data from a home power meter that is transmitted to the Google PowerMeter site and then displayed on a home computer.

PowerPortal home display An in-home display made by Comverges Corporation and used in conjunction with smart meters also manufactured by Comverges.

Power rating The maximum amount of power an electrical appliance is designed to use safely.

Power strip A multiple-outlet power extension cord that is frequently used to connect components of electronic systems to AC power; can be used as a cutoff switch for the system, if equipped with a master switch for all outlets.

Power surge A sudden, momentary rise in electric current flow or voltage; electric motors create a current surge when starting.

Power-usage monitor A device for measuring the amount of electricity a circuit is using.

Pressure boundary The physical barrier of walls, windows, doors, and so forth that prevents air from passing into or out of a building.

Pressure tank A tank that stores water pumped from a well under pressure (produced by compressed air in the top of the tank) so that a home water system will have continuous stable water pressure.

Pressurization test A method of testing a building for air leaks by slightly raising or lowering the inside pressure to cause air flow through the leaks.

Programmable Capable of storing instruction sets for later execution in sequence.

Programmable thermostat A thermostat that automatically raises and lowers an HVAC system's operating temperature according to a preset schedule.

PSI Pounds per square inch; the method of measuring water pressure in a pipe or the lifting power of a pump; one PSI is equal to a lifting power of 2.31 feet.

P-type material A semiconductor with holes for electrons; so called because it has extra positively charged particles that can attract electrons.

Public Utility Regulatory Policy Act of 1978 (PURPA) A federal law that requires power providers to purchase excess power from grid-connected small renewable energy systems.

Radiant heater A heater that produces infrared-spectrum heat that passes through air without heating it much and heats solid objects as it strikes them.

Radiant heating system A heating system that pipes heated water from a boiler to standing radiators, baseboard radiators, or pipes set in a building's floor; the radiators or pipes warm the building structure or the air within it.

Radiator Any device from which infrared-spectrum heat is produced; also a type of heat exchanger that transfers heat from a fluid inside the radiator to or from the air passing by the radiator.

RadioRA A proprietary wireless lighting control technology developed by Lutron and used for home lighting control.

Rafters The cross members that support the ceiling of a building and form the angled framing of the roof above.

Rated wind speed The wind velocity at which a wind turbine will generate its designed power output in watts.

Real-time data transmission The transmission of data at or near the same time as the data are created or measured.

Real-time information Data about the condition and operation of a system that is sent while the system is working so that responses can be made quickly.

Refrigeration cooling Reducing air temperature by compressing a gas in a closed-loop system into a liquid and then allowing it to return to a gaseous state, extracting heat from the surrounding air as it does so.

Regularly operating appliances An electric or electronic device that operates regularly (at least once per day) but not continuously; its operation is usually controlled by a sensor or timer.

Renewable energy source A source of energy for producing electricity that replenishes itself within a reasonable period of time; solar energy, wind energy, and tidal flows are examples.

Residential electrical account The account a homeowner sets up with an electric utility to obtain electric power in return for a specified charge per unit used.

Resistance The opposition of any material to the passage of electricity through it; measured in ohms.

Rigid foam insulation Sheets of rigid Styrofoam or polyurethane plastic used to prevent heat transfer.

Rolls and batts Insulation made in flexible pads of material composed of fibers, such as fiberglass.

Root mean square (RMS) The value of AC current that is the equivalent of steady direct current (DC) at the same constant voltage.

Root mean square voltage (VRMS) The AC voltage equivalent of a given DC voltage; it is equal to 0.707 (about 71%) of the peak AC voltage (Vpeak).

Rotor The rotating bladed wheel of a horizontal-axis wind turbine.

Rotor diameter The length of a rotor blade multiplied by 2; the width of the circle created by the tips of the rotor blades as they turn.

R-value A measure of an insulation material's thermal (heat) resistance; that is, its ability to prevent the passage of heat energy through the material from one side to the other.

Semiconductor material A material that conducts electricity in only one direction and can be made to generate electricity or light by means of energy being input to it.

Silicon The most popular semiconductor material for solar cells; available in at least four different types.

Single-pane glass Window glass composed of only one pane; usually found in older homes and buildings; not as energy efficient as double- or triple-pane glass.

Single-pole switch An AC basic switch that opens or closes a circuit to control power to a fixture.

Sleep mode A state of an electric device in which its operation is suspended, but its circuits are still powered and ready for instant use on demand.

Smart electric grid A multidirectional electric grid that can accept multiple power inputs, has digital communication throughout, and is monitored and controlled by computer software.

Smart meter A meter that transmits a continuous digital data signal on a radio frequency that communicates how much current is flowing through it.

Soft white bulb A light bulb with a CRI much lower than that of daylight; it emits light with a warmer feeling than a bright or pure white bulb.

Solar array A grouping of solar panels connected in parallel to generate an increased current flow over that of a single panel.

Solar cell A small electronic device made of a semiconductor material that generates an electric current when struck by sunlight.

Solar collector A panel made up of an absorber plate and tightly coiled tubing that collects solar heat and transfers it to an HTF in the tubing.

Solar heat gain coefficient (SHGC) The amount of solar radiation admitted through a window, door, or skylight, and subsequently released as heat inside a home. The lower the SHGC, the less solar heat a glass panel transmits.

Solar power Energy from the sun transformed into electricity using photovoltaic technology or concentrated into a heat energy source to generate electricity and/or provide other heating functions.

Solar radiation The general term used to describe the electromagnetic radiation emitted by the sun.

Space heater Any heater designed to heat a room or similar small space rather than an entire building.

Spark igniter An igniter for gas appliances that sparks to start the gas fuel burning; same as clicker igniter.

Split duplex outlet An AC electrical outlet in which one receptacle is always live and the other is controlled by a switch.

Split solar water-heating system A system that heats HTF in a closed loop and sends it to a heat exchanger for transfer to the water in a storage tank.

Spring (vernal) equinox The day in March of each year when daylight and darkness are of equal length.

Stand-alone system A solar generating system that is not connected to the grid, but provides power independently to a home electric panel.

Standby mode Same as sleep mode.

Startup wind speed The wind speed at which a wind turbine starts to generate electricity.

Storm door A glass-filled door placed on the outside of an exterior door to improve its resistance to heat transfer.

Storm window A second pane of glass or plastic placed inside or outside of an existing window to increase its resistance to heat transfer; the storm window makes the single-pane window a double-pane window.

Stud The upright frame members in the walls of a building.

Submersible pump A pump designed to be operated while submerged in the water it pumps.

Sump pump A pump located in a small pit in the basement or crawl space of a home that keeps the lower level free of water by pumping any leakage as needed.

Surface water Water flowing on the earth's surface from rain or other sources that can flow into the basement or crawl space of a home.

Surge protector An electrical device that prevents momentary high current flows from passing into a circuit.

Swept area The area of the circle that the blades of a rotor pass through as they turn.

Switch A device for controlling current flow in an electric circuit; a switch can interrupt current or change its flow to another part of the circuit.

Switched duplex outlet An AC electrical outlet controlled by a switch that turns current to it on or off.

Tank water heater A water heater that heats and stores a quantity of water for later use; the hot water is constantly maintained at a set temperature, regardless of how little or how much is used at a time.

Tankless water heater A water heater that produces hot water only when a hot water faucet connected to it is turned on and water begins to flow; an on-demand water heater.

Tempering tank An auxiliary tank connected in series with a water heater to passively raise the temperature of water entering the water heater.

Therm (thm) A measure of heat equal to 100,000 BTUs; 10 therms is equal to 1 decatherm (Dth).

Thermal boundary The insulation barrier that reduces or prevents heat transfer by convection through the walls, floors, and ceilings of a building .

Thermal cutoff switch A switch on gas- and oil-fired appliances that cuts off the fuel if it detects no heat being produced; the gas type is also called a thermocouple.

Thermographic inspection A means of measuring heat loss through walls and ceilings by measuring surface temperatures using infrared video and still cameras.

Thermostat A device for controlling an HVAC system; it can be manually operated or programmed to operate on a schedule.

Thin-film solar cells Solar cells made of layers of semiconductor materials only a few micrometers (millionths of a meter) thick that can be integrated into building materials to make them solar panels.

Three-way switch A type of AC switch that is used in pairs to turn a circuit on or off from two locations.

Timer A device used to turn an electric appliance on or off at set times during the day or over a longer time period.

Traditional electric grid The electromechanical grid constructed mainly in the 1950s that provides electric power across North America; it is currently being upgraded into a digitally controlled smart grid.

Transformer A device that steps electric voltage up to a higher level or down to a lower level.

Transmission The process of moving electricity from the point where it is generated to the points where it is used.

Transmission line A high-voltage wire used to move electricity at voltages above 480 volts over long distances; distribution lines then carry the electricity from step-down transformers to end users.

Transmission loss The loss of power that occurs when electricity is moved through a wire; the amount of loss varies with the voltage, amperage, temperature, wire material, and weather.

Triple-pane glass Window glass consisting of three panes parallel to one another with an air- or gas-filled space between each pair of panes.

Turbine An engine with a bladed, rotating central shaft that is driven by high-pressure steam or moving water and connected to a generator to produce electricity.

Two-phase current 240-volt AC current that is distributed to consumers in the form of two 120-volt "legs," each delivered through a separate wire and useable individually (120-volt circuit) or as a combined pair (240-volt circuit).

U-factor The rate at which a window, door, or skylight conducts heat flow.

UL 1741 Standard developed by the Underwriters Laboratories for certifying the utility and safety of power-producing stand-alone and grid-connected renewable energy systems.

Ultraviolet (UV) light High-frequency light not visible to the human eye that causes some other materials to emit visible light (phosphoresce); UV is highly destructive to human tissue and over time will destroy most other solid materials.

Ultraviolet radiation The invisible wavelengths of solar radiation just shorter than the violet light in the visible spectrum.

Undercapacity The amount of potential peak demand in an electric grid that exceeds its peak generating capacity.

Underwriters Laboratories A safety-testing and certification organization that has developed safety requirements for inverters and other renewable energy equipment.

USB connection Universal Service Bus; a connection port and/or matching plug used to connect electronic devices so that data or low-voltage power can be transferred between them.

Utility partnerships Agreements made between electric, gas, and water utility companies and manufacturers of smart meters and other similar devices; the agreements allow the partners to work together to provide services to the customers of both.

Valence band The lower energy level of atoms in the molecules of a semiconductor material.

Vampire appliance or device Any electrical device that draws power to maintain its readiness for operation when not in use.

Vane The panel or blade on a wind wheel that the wind pushes against to turn the wheel; also the panel set at a right angle behind a horizontal-axis wind turbine to keep the rotor facing directly into the wind.

Vapor barrier A waterproof sheet of material placed under attic or wall insulation to impede the passage of moisture through the insulation.

Variable-load power The portion of electric power used by a home or commercial building that exceeds the base power load for that structure.

Vent A duct connecting a fuel-burning appliance to the outside; a vent may exhaust hot gases from combustion or warmed humid air, as from a dryer. An intake vent permits outside air to be drawn into a combustion chamber or an HVAC system.

Vertical-axis wind turbine (VAWT) A wind turbine with a vertical shaft on which a wind wheel with vanes or blades rotates when pushed by the wind.

Visible light-emitting diode (VLED) An LED made of materials that emit light photons at frequencies between infrared and ultraviolet wavelengths.

Visible light spectrum The band of light wavelengths that can be seen by the human eye, ranging from violet to red and including all visible colors.

Voltage The electric pressure in a circuit measured in volts; the higher the voltage, the more current will flow through a given resistance.

Wall cavity The space in a wall between two studs that can be filled with insulation to improve energy efficiency.

Water heater blanket An extra layer of insulation added to a tank water heater to increase the R-value of its insulation.

Water meter A device for measuring the flow of water in cubic feet through a pipe.

Water vapor (humidity) Water in a gaseous state always present in the atmosphere that creates humidity in the air; as air cools it can hold less water vapor, and this may produce liquid water condensation.

Watt meter Same as a power meter.

Watt rating The amount of current that will flow through a light bulb or other electric device when operating.

Weather stripping Any sealing strip or material placed around the edge of a door or window to prevent air flow; can be made of rubber, fabric, metal, plastic, or any material that will improve the seal between the moving window frame or door and its casing.

Well shaft The hole leading from the surface to the ground water that supplies a well.

Wi-Fi A wireless radio frequency technology that permits high-speed data and voice communication throughout a smart grid.

Windbreak Any object that blocks, slows, or deflects wind.

Wind-generated power Electricity produced from a generator driven by wind power turning a multi-bladed rotor.

Window casing The wood or metal structure that fits in a window opening and contains the movable window frame.

Window frame The wood or metal square or grid that contains the glass of a window and can usually be opened and closed.

Wind power density The kinetic energy level in the wind.

Wind resource A measure in kWh per year of the potential for generating electricity from wind power in an area; calculated as the estimated annual electricity production per square meter of rotor-swept area for a small wind turbine.

Wind speed Speed of movement of air over the earth's surface; measured in miles per hour (mph) or meters per second (m/s).

Wind turbine A mechanical device for converting kinetic wind energy into electricity by means of a rotor or wind wheel attached to a generator.

Wind velocity Same as wind speed.

Wind wheel The vertical vanes or blades attached to a shaft that wind pushes against to rotate in a vertical-axis wind turbine.

Wireless MBus (sub 1-GHz) A wireless radio frequency technology that permits high-speed data and voice communication throughout a smart grid.

Working fluid Water or an antifreeze fluid that is circulated through a solar collector to pick up heat for transfer directly to the water or to a heat exchanger for heating water.

ZigBee (802.15.4 2.4 GHz) A wireless language that is used to connect different control and monitoring devices to work together; based on the IEEE 802.15.4 physical radio standard that operates at 2.4 GHz (global), 915 Mhz (Americas), and 868 Mhz (Europe).

Zone The section or part of an HVAC system that is separately operated and controlled; also a group of lights and other electrical devices controlled together to provide a specific scene or accomplish a specific task such as security lighting.

Z-Wave A wireless technology developed by the Danish company Zensys that is popular in Western Europe and is now available in the United States.

INDEX

Information in figures and tables is indicated by *f* and *t*.

A

Absorber plate, 239
Absorption coefficient, 239
Active system, for solar water heating, 223–224, 239
Advanced Metering Infrastructure (AMI), 34, 50
Air conditioning (A/C)
  central, 131–132, 132*f*
  definition of, 157
  room, 142, 142*f*
Air duct, 157
Alternating current, 5, 21
Alternative energy
  home smart grid and, 18–20, 19*f*
Alternative energy, grid factors in, 13–14
Always-on mode, 73
Amperage, 8, 21
Appliance(s). *See also specific appliances*
  chart, in energy use audit, 77, 77*t*
  continuously operating, 67–68, 73
  energy efficient, 177–182
  measuring power consumption of, 78–84
  regularly operating, 68
  replacement of inefficient, 70–71
  turning off, 70
  vampire, 185
  in variable load, 68–69
Article 690, 239
Attic hatch, 109
Audit, energy
  air leaks in, 86–88
  appliance chart in, 77, 77*t*
  appliance power consumption in, 78–84
  coal use in, 85
  conducting, 76–94
  definition of, 100
  gas fuel in, 84–85
  heat loss in, 90–94
  heat transfer in, 88–90
  insulation in, 91–94, 92*f*, 93*t*
  lighting chart in, 77–78, 78*t*
  loss, 85–94
  oil use in, 85
  report, 94–98
  use, 76–85
Automation
  command modules in, 202–205
  control components, 201–202
  controlled outlets in, 203–204
  controllers in, 202–205
  daisy-chain wiring in, 193–195, 194*f*
  definition of, 212
  grounding and, 190
  home, 189–195
  HVAC, 139–140, 140*t*
  installation of system, 205–209
  lighting zones in, 192–193, 207
  load requirements and, 190
  plan, 190–192, 191*f*
  power-line control systems, 196–199, 197*t*
  programming of system, 208–209
  remote-access controllers in, 203
  schedule, 208*t*
  sensors in, 205
  setup, 205–209
  troubleshooting, 209–210
  types of systems, 195–201
  of window treatments, 204–205
  wireless control systems in, 199–201
Autumnal equinox, 239
Auxiliary energy source, 239
Average monthly cost, 100
Average monthly kilowatt hours, 100
Awnings, 119–120, 119*f*
Aztech In-House Display, 38

B

Baffle, 124
Balance-of-system equipment
  in solar power, 229, 239
  in wind power, 254–259, 263
Ballast, 184
Bandgap, 239
Base load, 50
Baseboards, 109
Base-load power, 66–68, 73
Battery, in solar electricity system, 236–237, 237*f*
Battery stack, 239
Berm, 124
Billing, meters in, 27
Black & Decker Power Monitor, 38
Blackout, 8, 21
Blades, of wind rotor, 263
Blower, 157
Boiler, 157
Brick corners, 105

Brownout, 8, 21
BTU, 100, 157
Building code, 56, 73
Bulb
- frosted, 184
- halogen, 184
- incandescent, 162–165, 184
- soft white, 185

C

Canada, hydrogeneration in, 9–10
Capacity, electrical, 21
Carbon dioxide, 12–13, 157
Carbon monoxide, 157
Caulking, 124
Central air conditioning, 131–132, 132*f*
Central heating system, 129–130, 129*f*, 130*f*, 157
Charge controller, 239
Charge voltage, 263
Chimney joints, 106
Circuit breaker, 50
Clicker lighter, 157
Climate, consumption and, 54–55, 55*f*
Clothes dryers, 149–150
Clothes washers, 149
Coal consumption, in home, 85
Coal fired, 157
Coal-burning plants, 2, 3*f*, 4, 5*f*
- carbon dioxide and, 12–13
- sulfur and, 13

Color rendering index (CRI), 184
Color spectrum, 184
Color-enhanced phosphors, 184
Combustion chamber, 157
Combustion products, 157
Combustion residue, 157
Combustion vent, 87, 100
Command module, 196–197, 197–198, 202–205, 212
Compressor, 157
Computer systems, 179–180
Condenser, 132*f*, 157
Conduction band, 239
Conservation, 21
Constant rate charge, 50
Construction tools, 182
Consumption
- climate and, 54–55, 55*f*
- definition of, 73, 100
- household, 73
- measuring, of appliances, 78–84
- per capita, 55–56, 73
- per household, 73
- population density and, 56–57
- regional differences in, 55–57, 56*f*

Continuous operation, 67–68, 73
Continuous-load use, 184
Control module, 206–207, 212
Controllers, 50, 197–198, 202–205, 212
Convection, 239
Convective heater, 140–141, 157
Cooking, 151–152, 151*t*
Cooling coil, 157
Correlated color temperature (CCT), 184
Cost
- average monthly, 100
- cumulative, 100
- of LED lighting, 173–174, 174*t*

Cubic foot, 50
Current transformer (CT), 50, 57
Cutoff switch, 73

D

Damper, 157
DC-to-AC inverter, 184
Deciduous plants, 120, 121*f*, 124
Dedicated radio frequency (RF) network, 50
Demand
- base load and, 66
- in developing countries, 11
- increases in, 3–4
- lighting and, 8–9
- meters and, 28
- peak, 73
- pressures on grid, 10–11
- variable load and, 68–69

Demand pricing, 73
Demand-rate charge, 50
Developing countries, demand in, 11
Digital communications network, 7, 21
Dimmers, 204
Diode, 172, 184. *See also* Light-emitting diode (LED)
Direct current, 21
Disconnect, 239
Dishwasher, 150
Distribution line, 5–6, 21
Disturbance, 50
Door
- energy-efficient, 117–118
- storm, 124

Door casing, 106
Door jamb, 124
DreamWatts, 38
Dryers, 149–150
DTH, 100
Duct sealing, 136–138, 138*f*
Duct tape, 137, 157
Ductwork, 157

E

EcoDog, 38
EcoFactor, 38
Economies of scale, 73
Edison-type base, 184

Efficiency
definition of, 21, 73
home energy, 17–18
of HVAC system, 136–140, 138*f*
increasing, 69–72
of LEDs, 173–176, 173*t*, 174*t*
outside home, 119–122
reduction of total energy used for, 69–70
replacing inefficient appliances for, 70–71
timing and, 71–72
turning off appliances for, 70
of water heater, 144
Electricity
demand increases for, 3–4
generation of, 4
sources of, 2–3, 3*f*, 12*f*
Electromechanical
definition of, 21
traditional grid as, 6–7
Emissivity value, 114, 115*t*, 124
em•power, 38
Energate, 39
Energy Engage, 39
Energy Joule, 39
Energy Owl, 39
Energy Star rating, 184
Energy used per household, 73
EnergyHub, 39
Enrico Fermi Nuclear Generating Station, 14*f*
Environmental factors, affecting electrical grid, 12–13
EnviroScape, 39
Evacuated heat pipe tube (EHPT), 222, 223, 223*f*, 239
Evaporative cooler, 157
Exercise equipment, 181–182
Exhaust vent, 157
Expansion gearbox, 263

F

Fall equinox, 239
Feed-in tariff (FIT), 239
Filament, 162–163, 184
Filter, 157
Fireplace, 88
Fireplace dampers, 109
Fitness equipment, 181–182
Fixtures, 204
Flame arrester screen, 157
Flow sensor, 157
Fluctuation, 21
Fluorescent lamp, 184
Fluorescent lighting
advantages of, 168–169
color of, 169
compact, 165–171, 184
compact dimmable, 184
compact hybrid, 184
definition of, 184, 212
disadvantages of, 169–171
mercury vapor in, 169–170
phosphors in, 168–169
Forced-air heating, 129, 129*f*, 157
Foundation, meeting point of, 106–107
Freezers, 180–181
Furnace fuels, 131

G

Game consoles, 179
Gas smart meters, 32–34
Gas-fired, 157
Gateway, 50
Generation
definition of, 21
of peak-load power, 31–32
solar, 227–237
sources, 2–3, 3*f*, 12*f*
Geothermal energy, 239
Gigawatt, 73
Glass
double-pane, 115, 124
film overlay for, 115–116, 116*f*
low-emissivity, 124
single-pane, 115, 124
triple-pane, 124
Glazed panel, 239
Google PowerMeter, 50, 64–66, 65*f*, 73
Governing device, 263
GreenQuest, 39
Grid. *See also* Smart grid
alternative energy and, 13–14
demand pressures on, 10–11
digital revolution and, 7
as electromechanical system, 6–7
environmental factors affecting, 12–13
meters in, 6
overcapacity of, 9*f*
problems with, 8–10
solar electricity generation connected to, 228–235
traditional, 5–14
undercapacity of, 9*f*
water resources and, 14
wind power connection to, 257–258
Grid-connected system, 239
Ground clearance, 263
GroundedPower, 39
Grounding, 190, 232
Guy wire, 263
Guyed monopole, 263

H

Hard-wired, 100
Heat absorption, 157
Heat exchanger, 157

Heat loss
  definition of, 73, 100
  in energy audit, 90–94
Heat transfer
  definition of, 73, 100
  doors and, 117–118
  in energy audit, 88–90
  at foundation, 106–107
  insulation and, 109–114
  landscaping and, 120–122, 121*f*
  outside home, 119–122
  reduction of, 104–118
  sealing leaks for, 104–109
  skylights and, 118
  windows and, 114–117
Heat transfer fluid (HTF), 222, 225*f*, 239
Heating, ventilating, and air-conditioning (HVAC) system
  automation of, 139–140, 140*t*
  definition of, 50, 157
  design and operation of, 129–135
  efficiency of, improving, 136–140, 138*f*
  fuels in, 131
  maintenance, 138–139
  multiple-unit, 133, 158
  programmable thermostats for, 139–140, 140*t*
  schedule, 157
  single-unit, 133
  solar water heating and, 224
  zoned, 133–135, 134*f*, 135*f*, 139
Heating element, 157
Heating register, 157
High voltage, 21
Home Automation, Inc. (HAI), 40
Home run connection, 212
Home run-connected lights, 193–195
Home smart grid, 16–20
Home theaters, 178–179
Hoover Dam, 11*f*
Horsepower (hp), 184
Hunter Power Plant, 17*f*
Hybrid generating system, 259–261, 263
Hydroelectric
  in Canada, 9–10
  mechanism of, 5*f*
  as share of energy sources, 3*f*
  water resources and, 14

I

IEEE 1547-2003, 233, 239, 257
Ignition, electronic, 157
Impeller, 158
Incandescent light, 162–165, 184
  inefficiency in, 163, 163*t*
  power consumption of, 163–165
Infrared light, 184
Infrared radiation, 239
In-home display (IHD)
  data from, 64–66
  definition of, 35
  installation of, 41–48, 57–63
  PowerPortal Home, 35, 36*f*
  setup, 57–63
  smart meter and, 34–41
  smart meter communication with, 35–36
  transformers and, 37–38, 37*f*
  ZigBee technology in, 35–36
Institute of Electrical and Electronics Engineers (IEEE), 239
Insulation
  blown-in, 124
  cellulose, 124
  costs of, 110
  definition of, 100, 124
  of ductwork, 136–138, 138*f*
  in energy audit, 91–92, 91–94, 92*f*, 93*t*
  expanding foam, 124
  foam-in-place, 100, 124
  heat transfer and, 109–114
  loose-fill, 100, 124
  polyurethane foam, 158
  rigid foam, 100
  rolls and batts, 100, 124
  R-value of, 92, 93*t*, 100, 124
  in wall, 94
  of water heater, 144
  Internet, meter reading via, 29
  Interruption, 50
  Inverter, 239

J

Joist, 124

K

Kill A Watt, 40
Kilowatt hour, 50, 73
Kinetic energy, 263

L

Lake Mead, 11*f*
Landscaping, 120–122, 121*f*
Latitude, 239
Lattice tower, 254, 263
Leaks, air
  definition of, 124
  in energy audit, 86–88
  inside, 108–109
  outside, 104–107
  sealing, 104–109

Lift, 158
Light scenes, 192–193, 212
Light spectrum, 184
Light-emitting diode (LED), 50
  advantages of, 176–177
  cluster, 184
  cost of, 173–174, 174*t*
  definition of, 184
  diffused bulb, 175, 184
  disadvantages of, 177
  efficiency of, 173–176, 173*t*, 174*t*
  mechanism of, 171–173, 172*f*
  track, 175
  visible, 185
  white, 172–173
Light-emitting phosphor, 184
Lighting
  chart, in energy use audit, 77–78, 78*t*
  control design, 189–195
  daisy-chain, 193–195, 194*f*, 212
  demand and, 8–9
  electric, 3
  energy-efficient, 165–177
  fluorescent
    advantages of, 168–169
    color of, 169
    compact, 165–171, 184
    compact dimmable, 184
    compact hybrid, 184
    definition of, 184, 212
    disadvantages of, 169–171
    mercury vapor in, 169–170
    phosphors in, 168–169
  incandescent, 162–165, 184
    inefficiency in, 163, 163*t*
    power consumption of, 163–165
  UPB control systems for, 198–199
  X-10 control system for, 196–198, 197*f*
  zones, 192–193, 207
Lightning arrest, 256–257, 263
Load, 212
Luminaries, 204

M

Macros, 212
Main breaker, 50
Mastic, 158
Maximum watts, 100
Measurement transmission unit (MTU), 57
Megawatt (mW), 73
Mercury vapor, 169–170, 184
Mesh network, 50
Meter Reader EM-2500, 41
Meters. *See also* Smart meters
  analog, 26–28
  in billing, 27
  definition of, 21, 50
  demand and, 28
  in traditional grid, 6
Minimum watts, 100
Monopole
  definition of, 263
  guyed, 254, 263

N

National Electric Code (NEC), 233, 239
National Fenestration Rating Council (NFRC), 115, 124
National Fire Protection Association (NFPA), 239
Natural gas
  determining consumption of, 84–85
  pollution and, 13
  as share of energy sources, 3*f*
  water heaters and, 143
Net meter, 50, 229
Net metering, 239
Neutral bus, 51
N-type material, 184
Nuclear
  mechanism of, 15*f*
  pollution and, 13
  as share of energy sources, 3*f*

O

Off-peak hours, 73
Ohm's law, 100
Oil-fired, 158
One-way electric system, 21
Operating cycle, 100
Operating mode, 100
Outlet
  always-live duplex, 212
  controlled, 203–204
  split duplex, 212
Output voltage, 263
Overcapacity
  definition of, 21
  projected, 9*f*
Overload, 212

P

Parabolic aluminized reflector light, 175–176, 184–185
Partnership, utility, 73
Passive systems, 239
Peak AC voltage, 239
Peak demand, 73
Peak load, definition of, 51
Peak-load capacity, 21
Peak-load power, costs of, 31–32
Peak-load use, 185
Per capita consumption, 55–56, 73

Petroleum, as share of energy sources, 3*f*
Phantom load, 185
Phosphor, 185
Photoelectric effect, 239
Photon, 185
Photovoltaic (PV) module, 227–228, 240
Pilot light, 158
Polarity, 51
Pollution
 coal and, 12
 natural gas and, 13
 nuclear and, 13
Power conditioning equipment, 240
Power Law, 82, 100
Power rating, 100
Power strip, 73
Power surge, 100
PowerCost, 40
PowerHouse eMonitor, 40
Power-line control systems, 196–199, 197*t*
Power-line technology (PLT), 21, 51, 212
PowerNab, 40
PowerPortal Home, 35, 36*f*, 51
Power-usage monitor, 17, 18*f*
Pressure boundary, 100
Pressure tank, 158
Pressurization test, 100
Programmability, 212
PSI, 158
P-type material, 185
Public Utility Regulatory Policy Act (PURPA), 234, 240
Pulseworx, 199
Pump
 floor sucker, 153, 157
 pedestal, 152*f*, 158
 submersible, 152*f*, 153, 158
 sump, 152–154, 152*f*, 158
 well, 154–155

R

Radiant heater, 141, 141*f*, 158
Radiant heating system, 130, 130*f*, 158
Radiator, 158
RadioRA, 199, 212
Rafters, 124
Rainforest Automation, Inc., 40
Real-time information, 21
Receiver, in home smart grid, 17, 21
Refrigeration cooling, 132, 158
Refrigerators, 180–181
Remote control, infrared, 184
Remote-access controllers, 203
Renewable energy, 21
Residential electrical account, 73
Resistance, 100
Room air conditioners, 142, 142*f*
Root mean square (RMS), 240
Root mean square voltage (VRMS), 240
Rotor, 263
Rotor diameter, 263
R-value, 92, 93*t*, 100, 124

S

Safety
 in automation setup, 205–206
 in solar electricity generation, 231–232
 in wind power, 256–257
Screens, 119–120
Security lighting zones, 207–208
Semiconductor material, 240
Sensors, 205
Service box, 50
Service panel, 50
Shade, 120, 121*f*
Shaspa Smart Home Kit, 40
Siding corners, 105
Silicon, 240
Silver Spring Networks, 40
Skylights, 118
Sleep mode, 73, 100, 185
Smart grid
 alternative energy and, 18–20, 19*f*
 control devices in, 16
 definition of, 21
 efficiency devices in, 17–18, 18*f*
 home, 16–20
 overview of, 14–16, 15*f*
Smart meters, 17, 21
 adoption of, 30
 advanced metering infrastructure and, 34
 definition of, 51
 digital, 29, 30*f*, 50
 electric, 31–32
 gas, 32–34
 in-home displays and, 34–41
 Internet monitoring of, 29
 peak-load power and, 31–32
 transmission by, 29
 utility company use of, 30–34
 water, 32–34
Solar array, 240
Solar cell, 240
Solar collector, 240
Solar electricity generation, 227–237
 batteries in, 236–237, 237*f*
 grid-connected, 228–235
 hybrid, with wind, 259–261
 legal considerations in, 234–235
 liability insurance in, 233–234
 meters in, 232
 monitoring equipment in, 232
 photovoltaic module types in, 227–228
 safety equipment in, 231–232
 stand-alone, 235–237
Solar heat gain coefficient (SHGC), 124

Solar panel, in solar water heating system, 225–226, 226*f*
Solar power
definition of, 22, 240
home panels for, 19*f*
passive, 220–222, 239
Solar radiation, 216–220, 240, 244
Solar water heating, 222–227
Sources, 2–3, 3*f*, 12*f*
Space heater, 140–142, 141*f*, 158
Spark igniter, 158
Standby mode, 185
Stereos, 178–179
Storm door, 124
Storm window, 116–117, 124
Stoves, 151–152, 151*t*
Stringer, 107*f*
Stud, 124
Sulfur, 13
Surge protector, 263
Swept area, 263
Swimming pools, 224–225
Switch
single-pole, 212
three-way, 212

T

TED 5000, 37, 37*f*
Telecommunications advances, 7
Tempering tank, 145–146, 145*f*, 158
10BaseT RJ-45 Ethernet connection, 50
Tendril Residential Energy Ecosystem (TREE), 41
Therm (thm), 158
Thermal boundary, 101
Thermal cutoff switch, 158
Thermographic inspection, 90, 91*f*, 101
Thermostat
definition of, 158
programmable, 139–140, 140*t*, 158
Thin-film solar cells, 240
3-digit code, 184
Time, 73
Timing, efficiency and, 71–72
Transformer
current, 50
definition of, 22
in-home displays using, 37–38, 37*f*
Transmission line, 22
Transmission loss, 22
TVs, 178–179
Two-phase current, 51

U

U-factor, 124
UL 1741, 233, 240
Ultraviolet (UV) light, 185
Ultraviolet (UV) radiation, 240
Undercapacity
definition of, 22
projected, 9*f*
Underwriters Laboratories (UL), 240
UPB lighting control systems, 198–199
USB connection, 73

V

Valence band, 240
Vampire appliance, 185
Vane, 263
Vapor barrier, 101
Variable-load power, 68–69, 73
Vent, 158
Video games, 179
Visible light spectrum, 185
Voltage, 22

W

Wall cavity, 124
Washers, 149
Water
ground, 157
surface, 158
Water heater
blanket, 158
efficiency of, increasing, 144
full-on/full-off tankless, 147, 157
insulation of, 144
modulated tankless, 158
natural gas and, 143
on-demand, 158
point-of-use tankless, 147, 158
size, 144–145
solar, 222–227
split solar, 240
tank, 143, 158
tankless, 146–149, 148*f*, 158
tempering tank for, 145–146, 145*f*
Water heater, continuous-flow, 157
Water resources, 14
Water smart meters, 32–34
Water vapor, 158, 240
Watt meter, 78, 78*f*, 101
Watt rating, 101
WattsUp?, 41, 78, 78*f*
Weather stripping, 124
Weatherization Assistance Program (WAP), 110
Well pump, 154–155
Well shaft, 158
Wi-Fi, 22, 51

Wind power
- amount of wind needed in, 246–248
- balance-of-system equipment in, 254–259
- definition of, 22
- density, 247*f*, 263
- grid connection, 257–258
- home turbine for, 19*f*
- hybrid, with solar, 259–261
- legal considerations with, 258–259
- meters in, 257
- monitoring equipment in, 257
- safety equipment in, 256–257
- solar radiation and, 244

Wind resource, 247–248, 247*f*, 247*t*, 263

Wind speed
- annual average, 263
- definition of, 263
- rated, 263
- startup, 263

Wind turbine, 245*f*, 246*f*
- definition of, 263
- efficiency, 245
- horizontal-axis, 263
- size of, 248–249
- towers, 253–254
- vertical axis, 250–253, 251*f*, 263
- workings of, 245–246, 245*f*

Wind velocity, 263
Wind wheel, 263
Windbreak, 121–122, 121*f*, 122*f*, 124
Window casing, 106, 106*f*, 124
Window frame, 109, 117, 124
Window treatments, automated, 204–205
Windows, energy-efficient, 114–117
Wireless control systems, 199–201
Wireless MBus, 51
Wiring, daisy-chain, 193–195, 194*f*, 212
Woodworking tools, 182
Working fluid, 222, 240
Workshop equipment, 182

X

X-10 lighting control systems, 196–198, 197*f*

Z

ZigBee, 35–36, 51
Zone, 133–135, 134*f*, 135*f*, 139, 158, 212
Z-wave, 212